Hollow Core Optical Fbers

Hollow Core Optical Fbers

Special Issue Editor

Walter Belardi

MDPI • Basel • Beijing • Wuhan • Barcelona • Belgrade

MDPI

Special Issue Editor
Walter Belardi
Université de Lille
France

Editorial Office
MDPI
St. Alban-Anlage 66
4052 Basel, Switzerland

This is a reprint of articles from the Special Issue published online in the open access journal *Fibers* (ISSN 2079-6439) from 2018 to 2019 (available at: https://www.mdpi.com/journal/fibers/special_issues/hollow_core_optical_fibers)

For citation purposes, cite each article independently as indicated on the article page online and as indicated below:

LastName, A.A.; LastName, B.B.; LastName, C.C. Article Title. *Journal Name* **Year**, *Article Number*, Page Range.

ISBN 978-3-03921-088-6 (Pbk)
ISBN 978-3-03921-089-3 (PDF)

Contents

About the Special Issue Editor

Walter Belardi holds a research excellence chair in photonics at the University of Lille, in France. He obtained his PhD, on microstructured optical fibers, at the University of Southampton, United Kingdom. He then worked, first in industry, as a scientific consultant, and, later, as a researcher at the University of Bath (UK) and University of Southampton. His main research contributions are in the design, fabrication and use of novel optical fiber technologies, with key achievements that include the modelling and fabrication of novel hollow core fiber structures. Walter is an editorial board member of *Fibers* and has contributed overall to more than 100 scientific works. He has been a project evaluator for several research funding organisations and he has contributed to diverse personal and group research grants.

fibers

MDPI

Editorial

Hollow-Core Optical Fibers

Walter Belardi

CNRS, UMR 8523–PhLAM–Physique des Lasers Atomes et Molécules, Université de Lille, F-59000 Lille, France; walter.belardi@univ-lille.fr

Received: 15 May 2019; Accepted: 22 May 2019; Published: 24 May 2019

The possibility of guiding light in air has fascinated optical scientists and engineers since the dawn of optical fiber technology [1]. However, a remarkable progress in this area has been achieved "only" twenty years ago, when the first fabrication of a hollow-core photonic crystal fiber capable of delivering light over a length of few centimeters [2] gave rise to an increased interest in the field. Then, first the 20 dB/km attenuation barrier was overcome [3] and, few years later, the lowest loss (1.2 dB/km) hollow-core optical fiber (HC) was realized [4].

Since the beginning of this century, HCs have attracted the attention of a large worldwide research community working on the design, fabrication and device implementation, entering almost any specific application field of optics (from medicine [5] to security [6], telecommunication [7], industrial processing [8], instrumentation [9], biology [10], and so on and so forth). In parallel with the increased number of applications, still major advances are being made on the optimization of the hollow-core fiber designs and on the study of its underlying guiding properties, as well as in the use of different materials or fabrication techniques, which, in turn, are providing even more ways of exploitation of this technology and new technical challenges.

This special issue of *Fibers* wanted to ride the wave of this renewed interest in the field of hollow-core optical fibers by providing an overview of the recent progress in this field as well as an updated and indicative sample of current research activities worldwide.

Thus, the issue includes three outstanding reviews by leading institutions in the field of hollow-core optical fibers. The review *Hollow-Core Fiber Technology: The Rising of "Gas Photonics"* [11] by the *University of Limoges* (France) and the *University of Modena and Reggio Emilia* (Italy) moves from their first discovery and development of the Inhibited-Coupling hollow-core optical fiber to its application to gas photonics. It is an extremely rich, deep and detailed trip offered by some of the most renowned scientists in the field that highlights their key achievements in both design and fabrication developments, and, in particular, shows how this gave rise to the exploitation of gas/light interaction in an unprecedented way.

The review *Revolver Hollow-Core Optical Fibers* [12] by the *Fiber Optics Research Center (FORC)*, in Moscow, focuses on their specific simplified designs (HCs with only a single ring of tubular tubes in the cladding area), first pioneered and developed in their institution. Most properties, applications and fabrication approaches of this specific fiber type are addressed and discussed in all spectral domains. The review is not limited to silica glass, but also covers their demonstration of chalcogenide hollow-core optical fibers for the longer wavelength ranges.

The material and fabrication aspect is the object of the third review, *3D-Printed Hollow-Core Terahertz Fibers* [13] by *Instituto Tecnologico de Aerenoautica, Instituto de estudos avançados and Universidade Estadual de Campinas (UNICAMP)*, in Brazil. The realization and characterization of polymer-based HCs, in combination with 3D-printing fabrication, approaches is widely discussed. The review shows how the field of HCs is expanding also to the terahertz spectral regime and how it is starting to profit of the opportunities offered by the 3D-printing techniques.

After this overview on the last generation of hollow-core optical fibers, this special issue includes seven original contributions by scientists addressing current relevant issues involved in the design and application aspects of HCs.

On the design aspect, the paper *Effect of Nested Elements on Avoided Crossing Between the Higher-Order Core Modes and the Air-Capillary Modes in Hollow-Core Antiresonant Optical Fibers* [14], by the *Research Technology Organization of Photonics Bretagne (PERFOS)*, in Lannion (France), deals with the extremely important problem of mono-modality in the most advanced forms of HCs. The accurate numerical analysis made by the author provides an important insight in order to understand which HC geometry to use and how to simplify the analysis of its properties.

In the same way, the original paper *Understanding Dispersion of Revolver-Type Anti-Resonant Hollow-Core Fiber* [15] by the *Leibnitz Institute of Photonic technology* and the *University of Jena*, in Jena (Germany), is about the full comprehension of the dispersion properties of anti-resonant HCs, which is essential for applications involving high optical power and short pulse duration. Aside from providing useful analytical approximations, the authors perform a series of numerical simulations showing how the group velocity dispersion changes with the HC geometry.

Structure optimization is also the target of the last original contribution on fiber designs in this special issue: *Geometry of Chalcogenide Negative Curvature Fibers for CO_2 Laser Transmission* [16] by the *Baylor University* and the *University of Maryland Baltimore County*, in Baltimore (USA). In this paper, a large number of geometrical parameters are used in numerical simulations on HCs in chalcogenide glasses, in order to achieve the best possible attenuation performances at the CO_2 laser wavelength of 10.6 μm. This numerical work is of high importance in the field since anti-resonant HCs could be a valid alternative to other types of specialty optical fibers for the mid-infrared spectral range.

For example, passing now to the original experimental contribution of this special issue, the paper *Fabrication of Shatter-Proof Metal Hollow-Core Optical Fibers for Endoscopic Mid-Infrared Laser Applications* [17] by the *Sendai College*, the *Miyagi Gakuin Women's University*, the *Tohoku University*, in Japan, and the *Fudan University*, in China, concerns the experimental demonstration of innovative HCs for the 10.6 μm wavelength. Targeting medical applications of HCs, this paper addresses relevant implementation issues of this technology by looking, in particular, not only at the fiber attenuation and bending loss, but also at the characteristics of the material embedded inside the HC and at the ability of the same HC in guiding both mid-infrared and visible light for its practical operation.

Practicability in the device implementation is also the object of the second original experimental contribution to this issue. The paper *Combining Hollow-core Photonic Crystal Fibers with Multimode, Solid Core Fiber Couplers through Arc Fusion Splicing for the Miniaturization of Nonlinear Spectroscopy Sensing Devices* [18], by the *Wroclaw University of Science and Technology*, in Wroclaw (Poland), deals with the important problem of combining HCs and standard optical technology, in an effective and viable way. The optimization of the splicing parameters, by simply using a conventional arc fusion splicer, allows them to demonstrate adequate performances and the validity of their approach in a two-photon fluorescence spectroscopy experiment.

On the other hand, a method to process HCs via a CO_2 laser is used in the third original experimental work of this issue. The paper *A Method to Process Hollow-Core Anti-Resonant Fibers into Fiber Filters* [19], by the *Nanyang Technological University*, discusses how to modify the internal geometrical characteristics of an anti-resonant HC in order to use it as a filter device. It shows how the implemented methodology could also be employed in the dispersion control, a very relevant factor in optical pulse propagation and manipulation.

The study of the characteristics of high-power pulses delivered through an HC is the thematic of the last original experimental contribution to this special issue. The paper *Hollow-core Optical Fibers for Industrial Ultra Short Pulse Laser Beam Delivery Applications* [20], by *Photonic Tools GmbH*, in Berlin, shows relevant details on the implementation of a high-power laser-beam delivery device, in both the picosecond and femtosecond pulse duration regime. The suitability of HCs for flexible and efficient optical-power delivery was proved by the results when cutting different materials.

Besides providing a good balance between reviews (3 contributions), theoretical analysis (3 contributions) and applications (4 contributions), this special issue of *Fibers* represents a reasonable mix of the research activities from different geographical areas, with contributions from the *European*

Union (5 research institutions and 2 companies), *Russia* (1 institution), *Brazil* (3 institutions), *United States of America* (2 institutions), *Japan* (2 institutions), *China* (1 institution) and *Singapore* (1 institution). This shows the worldwide interest for a technology that is coming to better maturity and may largely affect industrial, economical and societal changes in the future years.

Funding: This research received no external funding.

Conflicts of Interest: The authors declare no conflict of interest.

References

1. Marcatili, E.; Schmeltzer, R. Hollow Metallic and Dielectric Waveguides for Long Distance Optical Transmission and Lasers. *Bell Syst. Tech. J.* **1964**, *43*, 1783–1809. [CrossRef]
2. Cregan, R.F.; Mangan, B.J.; Knight, J.C.; Birks, T.A.; St. J. Russell, P. Single-Mode Photonic Band Gap Guidance of Light in Air. *Science* **1999**, *285*, 1537–1539. [CrossRef] [PubMed]
3. Venkataraman, N.; Gallagher, M.T.; Smith, C.M.; Muller, D.; West, J.A.; Koch, K.W.; Fajardo, J.C. Low Loss (13 dB/km) Air Core Photonic Band-Gap Fibre. In Proceedings of the ECOC 2002, Copenhagen, Denmark, 8–12 September 2002. PD1.1.
4. Roberts, P.J.; Couny, F.; Sabert, H.; Mangan, B.J.; Williams, D.P.; Farr, L.; Mason, M.W.; Tomlinson, A.; Birks, T.A.; Knight, J.C.; et al. Ultimate low loss of hollow-core photonic crystal fibres. *Opt. Express* **2005**, *13*, 236–244. [CrossRef] [PubMed]
5. Lombardini, A.; Mytskaniuk, V.; Sivankutty, S.; Ravn Andresen, E.; Chen, X.; Wenger, J.; Fabert, M.; Joly, N.; Louradour, F.; Kudlinski, A.; et al. High-resolution multimodal flexible coherent Raman endoscope. *Light Sci. Appl.* **2018**, *7*, 10. [CrossRef] [PubMed]
6. Cruz, A.; Serrão, V.A.; Barbosa, C.L.; Franco, M.A.R.; Cordeiro, C.M.B.; Argyros, A.; Xiaoli, T. 3D Printed Hollow Core Fiber with Negative Curvature for Terahertz Applications. *J. Microw. Optoel. Electromagn. Appl.* **2015**, *14*, 45–53.
7. Wang, X.; Ge, D.; Ding, W.; Wang, Y.Y.; Gao, S.; Zhang, X.; Sun, Y.; Li, J.; Chen, Z.; Wang, P. Hollow-core conjoined-tube fiber for penalty-free data transmission under offset launch conditions. *Opt. Lett.* **2019**, *44*, 2145–2148. [CrossRef]
8. Michieletto, M.; Lyngsø, J.K.; Jakobsen, C.; Lægsgaard, J.; Bang, O.; Alkeskjold, T.T. Hollow-core fibers for high power pulse delivery. *Opt. Express* **2016**, *24*, 7103–7119. [CrossRef] [PubMed]
9. Digonnet, M.J.F.; Chamoun, J.N. Recent developments in laser-driven and hollow-core fiber optic gyroscopes. *Proc. SPIE* **2016**, *9852*, 985204.
10. Giovanardi, F.; Cucinotta, A.; Rozzi, A.; Corradini, R.; Benabid, F.; Rosa, L.; Vincetti, L. Hollow Core Inhibited Coupling Fibers for Biological Optical Sensing. *J. Light. Technol.* **2019**, *37*, 2598–2604. [CrossRef]
11. Debord, B.; Amrani, F.; Vincetti, L.; Gérôme, F.; Benabid, F. Hollow-Core Fiber Technology: The Rising of "Gas Photonics". *Fibers* **2019**, *7*, 16. [CrossRef]
12. Bufetov, I.A.; Kosolapov, A.F.; Pryamikov, A.D.; Gladyshev, A.V.; Kolyadin, A.N.; Krylov, A.A.; Yatsenko, Y.P.; Biriukov, A.S. Revolver Hollow Core Optical Fibers. *Fibers* **2018**, *6*, 39. [CrossRef]
13. Cruz, A.L.S.; Cordeiro, C.M.B.; Franco, M.A.R. 3D Printed Hollow-Core Terahertz Fibers. *Fibers* **2018**, *6*, 43. [CrossRef]
14. Provino, L. Effect of Nested Elements on Avoided Crossing between the Higher-Order Core Modes and the Air-Capillary Modes in Hollow-Core Antiresonant Optical Fibers. *Fibers* **2018**, *6*, 42. [CrossRef]
15. Zeisberger, M.; Hartung, A.; Schmidt, M.A. Understanding Dispersion of Revolver-Type Anti-Resonant Hollow Core Fibers. *Fibers* **2018**, *6*, 68. [CrossRef]
16. Wei, C.; Menyuk, C.R.; Hu, J. Geometry of Chalcogenide Negative Curvature Fibers for CO_2 Laser Transmission. *Fibers* **2018**, *6*, 74. [CrossRef]
17. Iwai, K.; Takaku, H.; Miyagi, M.; Shi, Y.W.; Matsuura, Y. Fabrication of Shatter-Proof Metal Hollow-Core Optical Fibers for Endoscopic Mid-Infrared Laser Applications. *Fibers* **2018**, *6*, 24. [CrossRef]
18. Stawska, H.I.; Popenda, M.A.; Bereś-Pawlik, E. Combining Hollow Core Photonic Crystal Fibers with Multimode, Solid Core Fiber Couplers through Arc Fusion Splicing for the Miniaturization of Nonlinear Spectroscopy Sensing Devices. *Fibers* **2018**, *6*, 77. [CrossRef]

19. Huang, X.; Yong, K.T.; Yoo, S. A Method to Process Hollow-Core Anti-Resonant Fibers into Fiber Filters. *Fibers* **2018**, *6*, 89. [CrossRef]
20. Eilzer, S.; Wedel, B. Hollow Core Optical Fibers for Industrial Ultra Short Pulse Laser Beam Delivery Applications. *Fibers* **2018**, *6*, 80. [CrossRef]

fibers

MDPI

Review

Hollow-Core Fiber Technology: The Rising of "Gas Photonics"

Benoît Debord [1],*, Foued Amrani [1], Luca Vincetti [2], Frédéric Gérôme [1] and Fetah Benabid [1]

[1]   GPPMM Group, XLIM Research Institute, CNRS UMR 7252, University of Limoges, 87060 Limoges, France; foued.amrani@xlim.fr (F.A.); gerome@xlim.fr (F.G.); f.benabid@xlim.fr (F.B.)

[2]   Department of Engineering "Enzo Ferrari", University of Modena and Reggio Emilia, I-41125 Modena, Italy; luca.vincetti@unimore.it

*   Correspondence: benoit.debord@xlim.fr; Tel.: +33-555-457-283

Received: 19 November 2018; Accepted: 18 January 2019; Published: 18 February 2019

Abstract: Since their inception, about 20 years ago, hollow-core photonic crystal fiber and its gas-filled form are now establishing themselves both as a platform in advancing our knowledge on how light is confined and guided in microstructured dielectric optical waveguides, and a remarkable enabler in a large and diverse range of fields. The latter spans from nonlinear and coherent optics, atom optics and laser metrology, quantum information to high optical field physics and plasma physics. Here, we give a historical account of the major seminal works, we review the physics principles underlying the different optical guidance mechanisms that have emerged and how they have been used as design tools to set the current state-of-the-art in the transmission performance of such fibers. In a second part of this review, we give a nonexhaustive, yet representative, list of the different applications where gas-filled hollow-core photonic crystal fiber played a transformative role, and how the achieved results are leading to the emergence of a new field, which could be coined "Gas photonics". We particularly stress on the synergetic interplay between glass, gas, and light in founding this new fiber science and technology.

Keywords: hollow-core photonic crystal fiber; gas photonics

1. Introduction

In the last twenty years, photonics has witnessed the advent of a new type of optical fibers named hollow-core photonic crystal fibers (HCPCF) [1], and has led to a huge progress in understanding the underlying physics of the guidance mechanisms, in its technology and in their applications. Indeed, HCPCF has been a unique platform for the demonstration of photonic bandgap guidance, the development of new conceptual tools such as "photonic tight binding" model to explain how these photonic bandgaps are formed in microstructured optical fibers [2], or the inception of "Inhibited Coupling" guidance, which is the fiber–photonic analog of bound state in continuum [3,4]. Furthermore, the motivation of fabricating HCPCF with exquisite control of its nanometric glass features has led to new fabrication techniques [5]. Finally, the ability to functionalize these fibers by introducing a fluid in its hollow-core to form photonic microcells (PMC) [6] proved to be a transformative and differentiating force in various fields [7].

In the course of the HCPCF continuing development process, a new landscape of research and technology, whose scope lies at the frontier of several fields, emerged, and is continuing to develop. These fields stand out by their variety and large range as they span from photonics, nonlinear and ultrafast optics, plasma physics, high optical field physics, atom and molecular optics, cold atom, lasers, telecommunications, and frequency metrology to micromachining and surgery. Despite this diversity and complexity, the landscape can be broken down into two main poles, which underpin all the aforementioned fields. The first one entails the research activities on the science and technology of HCPCF. It comprises the design and the fabrication processes of HCPCF and their derivative components, and which has witnessed not only a huge improvement in the fiber fabrication technology, but the development of novel concepts in the optical guidance mechanisms that is reshaping the field of guided optics. The second pole entails the HCPCF-based applications. Here, it was shown in a number of demonstrations that the combination of a HCPCF, a filling gas phase medium, and a judiciously chosen electromagnetic excitation are sufficient to provide a versatile and powerful tool to make various photonic components. These range from frequency convertors [8–11], supercontinuum generators [12,13], frequency standard cells [6,14], pulse compressors [15–17], high-power and high energy laser beam delivery cables [17], lasers [18–20] to quantum sensors, sources and memories [21–23], and even Raman gas spectroscopy for chemistry [24].

Remarkably, despite the variety of the aforementioned demonstrations, this landscape is chiefly built upon only three elements, which are gas, glass, and light. In a unique synergetic relationship, each one of these three elements plays a central role in controlling and structuring one of the two other elements. Figure 1a illustrates this synergetic "interfeeding" cycle between gas, glass, and light. Three representative examples on how to structure either of them are as follows. **(i) Structuring glass with gas**—In the process of HCPCF fabrication, one can shape the cladding glass structure by simply revisiting the glass blowing technique [5]. Here, the fiber cladding and core holes are pressurized with an inert gas (see Figure 1a) to achieve the desired fiber geometry whose features include glass web with nanometer scaled thickness and shapes as complex as the hypocycloidal core-contour (also called negative curvature), which strongly impacted the transmission performance in inhibited coupling guiding HCPCF (IC-HCPCF) [25,26]. **(ii) Structuring light with glass**—In turn, the HCPCF cladding nanostructured glass results in structuring the modal spectrum of the cladding modes so as to exhibit in the effective index and frequency space (i.e., $\left(n_{eff} - \omega\right)$ space) specific regions with no propagating modes (i.e., photonic bandgap) or with a continuum of modes whose transverse profile and spatial localization render their coupling to some core guided modes close-to-forbidden (i.e., inhibited coupling) (see Figure 1b). This structured modal spectrum allows ultralow loss optical guidance in hollow-core defects, and where the spatial optical profile of the guided mode can reach in IC-HCPCF an extremely low overlap with the cladding that led to the demonstration of ultrashort pulse (USP) energy handling up to millijoule energy level, and with a potential to withstand up to a joule level USP. It is noteworthy that this level of energy handling by the HCPCF implies the ability of engraving glass, which is the constitutive material of the fiber (see Figure 1e). **(iii) Structuring gas with light**—Finally, demonstrations have shown that light can also be used to structure the gas inside HCPCF. Among these, we count the generation of ionized gas plasma column generation in a HCPCF with microwave nonintrusive excitation [27] (see Figure 1d), or the nanostructuring of Raman gas [28] (see Figure 1c) or ultracold atoms [29] with particular optical excitation.

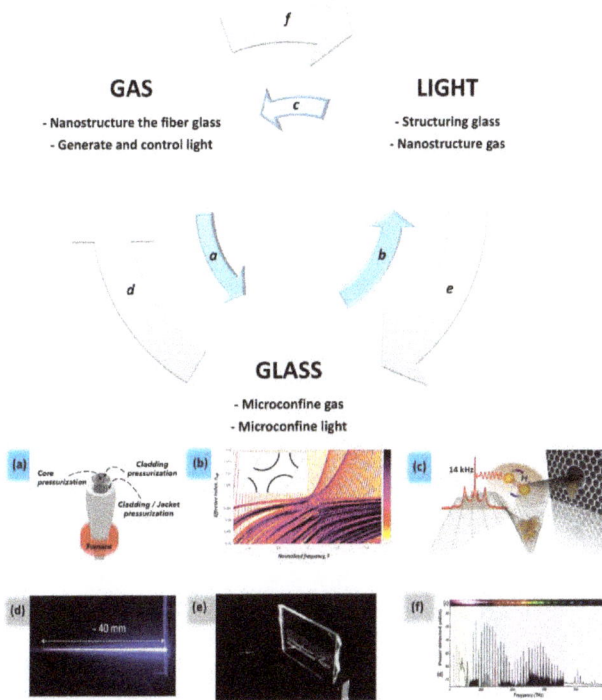

Figure 1. Synergetic cycle between gas, glass, and light in HCPCF science and technology and HCPCF-based applications. Representative examples of HCPCF related research activities. (**a**) Schematic of gas pressurization and evacuation during HCPCF drawing process. (**b**) Modal spectrum of an infinite cladding made with tubular lattice. Inset: unit cell of a tubular lattice (reprinted with permission from Reference [30], OSA, 2017). (**c**) Illustration of nanolayer of hydrogen molecules (Raman-active gas) formed by special stimulated Raman scattering configuration [28]. (**d**) Fluorescence from a plasma core photonic crystal fiber (reprinted with permission from Reference [27], OSA, 2013). (**e**) Engraving of glass sheet with HCPCF output laser (reprinted with permission from Reference [31], OSA, 2014). (**f**) Over five octave Raman comb generated and transmitted through hydrogen-filled-HCPCF (reprinted with permission from Reference [10], OSA, 2015).

What is noteworthy in some of these applications mentioned above is the ability of HCPCF to microconfine light and gases in extreme regimes. For example, laser intensity levels of PW/cm^2 and laser fluence that is several orders of magnitude larger than the silica laser damage threshold [17] are now generated and guided in HCPCF. The largest fiber transmission window is demonstrated via the generation of an optical Raman comb as wide as more than five octaves in hydrogen-filled HCPCF [10] (see Figure 1f), whilst the generation and guidance of high energy single-cycle compression was achieved thanks to HCPCF specific dispersion profile [16]. Conversely, HCPCF has proved to harbor gas media well beyond their common gas phase state such as the generation and microconfinement of ionized gas exhibiting high-power and electron densities combined with temperatures as high as 1000 K without damage to the structural integrity of the fiber [27], or the microconfinement of ultracold atoms with no collision with the micrometric core inner-wall. Finally, structuring molecular gas into an array of nanolayers has recently been demonstrated with hydrogen-filled HCPCF to create a new Lamb-Dicke-stimulated Raman scattering [28].

In this review, we present the major events that led to the development of HCPCF such as the key and seminal results and concepts. By highlighting the synergetic interplay between gas, glass, and light, we describe the contour of a research field landscape, which could be coined as "Gas Photonics", that is currently emerging thanks to the enabling power of HCPCF technology. We start by quickly reviewing the PCF fabrication process and the different microstructured fibers made in this way, and underlining the role of gas in successfully achieving intricate glass microstructures. Secondly, we show how the resulted cladding geometrical structure is exploited to engineer cladding modal spectrum, and thus to achieve the desired fiber guidance properties. In a subsequent section, we present the modal properties of the cladding defect (i.e., fiber core), by highlighting the salient features of the core fundamental mode such as its dispersion, its overlap with the silica, and how these properties differ between PBG-guiding HCPCF and IC-guiding HCPCF. The following sections of the review are dedicated to the applications, where we provide a nonexhaustive but illustrative list of the different applications that have been demonstrated in the last two decades.

2. Historical Overview of HCPCF

Photonic crystal fibers (PCF) [1]—optical fibers whose cladding is microstructured—were first reported in late 90s and are fabricated using an original process called "stack-and-draw" technique [32]. The versatility of this process and its ability to tailor the cladding modal spectrum by judiciously designing the cladding structure offered a platform to develop optical fibers with various core and claddings designs, and enabled novel optical guidance mechanisms and fibers with unprecedented linear and nonlinear properties. In turn, PCF has proved to be an excellent photonic component for multiple applications in varied fields such as supercontinuum generation in nonlinear optics, gas-based optics, and nonlinear optics [7].

Figure 2 illustrates, in a tree diagram, the PCF family and its diversity from the standpoint of the fiber structural designs, constitutive materials or the physics underlying their guidance mechanisms. If we had to classify these fibers by their structural architecture, we can identify two main families—solid-core and hollow-core fibers—each of them can be divided in several ways. For example, they can be classified by one of the three guidance mechanisms, which are (i) Modified Step Index (MSI), (ii) Photonic Bandgap (PBG), and (iii) Inhibited Coupling (IC). The fibers can also be categorized via their cladding geometry. The latter outstands with the impressive variety that can be found in each guidance mechanism, and the optical properties that can address. Among these, we can highlight the endlessly single-mode (ESM) fiber [33], which enables optical guidance in a single mode fashion regardless of the wavelength. This in turn led to the large mode area (LMA) single mode fibers [34], and subsequently to high-power fiber lasers [35]. The PCF tree diagram also shows other designs that were developed such as enhanced birefringence (Hi-Bi) fibers [36], dispersion compensation PCF (Disp-Comp) [37], all-solid PBG-guiding PCF [38], solid-core IC-guiding PCF [39], and hybrid guidance PCF [40] to mention a few. Finally, we can record PCF via their constitutive materials. Here, whilst silica remains the dominant material used, a lot of effort is currently undertaken to use alternative materials such as soft glass or chalcogenides [41,42] mainly driven by either further enhancing optical nonlinearities in PCF or extending their transmission well beyond the silica transparency window.

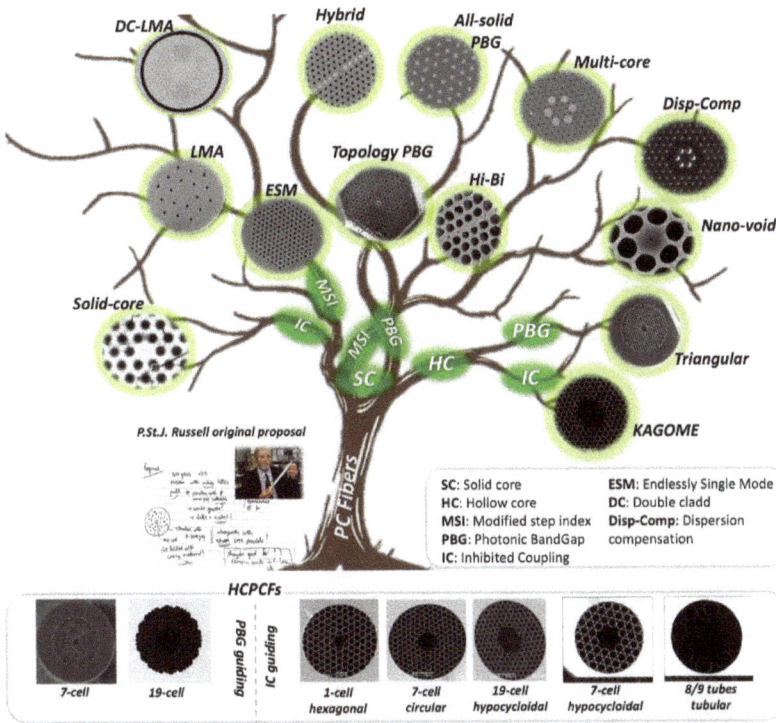

Figure 2. Photonic crystal fibers family tree diagram (top). Micrographs of HCPCF-based on PBG guidance and IC guidance.

Within this family, PCFs with a hollow-core defect [1] stand out from the rest of the PCFs because their optical guidance cannot rely on the conventional total internal reflection (TIR). As such, HCPCF was the fiber design of choice to explore novel guidance mechanisms such as PBG or IC, and whose main principles stem no longer from guided optics but from quantum mechanics or solid-state physics. The notion of PBG was first proposed by John [43] and Yablonovitch in 1987 [44,45]. This work represents a paradigm shift in optics, which led to a powerful conceptual transfer from quantum mechanics and solid-state physics to optics. Particularly, light propagation, confinement, and generation in dielectric microstructures, coined photonic crystals, is now casted as an eigenvector problem in a similar manner to solving Schrodinger equation and reconstructing the electronic energy diagram of a crystal.

In 1995, Philip St. J. Russell and coworkers extended this approach to optical fiber [32]. Here, the authors show for the first time the possibility for a fiber cladding structure made of silica and air holes to exhibit regions of the $\left(n_{eff} - \omega\right)$-space that are void of any propagating modes (i.e., PBG) and that extend below the air-line. This means that air guidance is possible within the PBG $(n_{eff} - \omega)$ region because of the absence of cladding modes to which a core-guided mode could couple to.

The proof of principle of fabricating a HCPCF was first reported by Cregan et al. in 1999 [1]. HCPCFs with sufficiently low loss were reported in 2002 [46,47]. The first one consisted of Kagome lattice HCPCF with ~1 dB/m, reported by Benabid et al. [46]. The second one, consisted with unambiguously PBG-guiding HCPCF by Corning, reported only few months later than reference [46] in a post-deadline paper in ECOC [47]. The fiber exhibited a transmission loss figure of 13 dB/km at 1500 nm and a cladding structure with then the largest air-filling fraction. This was a strong evidence of the concept of out-plane PBG proposed by P. St. J. Russell [48]. To date, the lowest transmission loss recorded for HCPCF is set at 1.2 dB/km at 1620 nm reported by Roberts et al. [49]. It is noteworthy that the Kagome lattice HCPCF, which outstands with a broadband guidance from Ultraviolet (UV) to Infrared (IR), does not guide via PBG despite exhibiting the lowest loss when it was first reported. Also, this loss figure was lower than predicted by Fresnel reflection in a capillary [50] or by antiresonant reflecting optical waveguide (ARROW) [51] to explain how light is guided in such a fiber. It was shown later that the fiber guides thanks to the strong coupling inhibition between core and cladding modes, leading to the term of IC guidance mechanism. Such a cohabitation between a core-guided mode (even though leaky) and cladding mode continuum, which has raised a lot of questions within the fiber optics community, stem from quantum mechanics. In 1929, Von Neumann and Wigner theoretically demonstrated that electronic bound states with positive energy can exist for a particular potential profile [3], thus leading to the notion of bound state in a continuum (BIC) [52]. Consequently, IC guidance mechanism, proposed by Benabid and coworkers in 2007 [4], is the fiber photonics analog of Von Neumann and Wigner BIC. Though it is important to stress that in Kagome HCPCF, the core-guided modes are not strictly "bound"; consequently, the guided modes of IC-HCPCF are referred as quasi-BIC (QBIC). In a following section below, we detail the nature of interaction between a core and cladding modes using the IC model. The latter proved to be a very powerful design tool, as it led to the advent of IC-HCPCF with hypocycloidal core-contour [25,26], also renamed negative curvature fiber [53,54]. This in turn, led to a renewed interest in HCPCF fabrication and design, which is illustrated by the proposal of cladding structures having hypocycloid core-contour, such as the tubular lattice cladding [53,55] and their modified versions [56–60]. This renewal in IC-HCPCF is also illustrated by the continuous and dramatically rapid decrease in their transmission loss. The progress is such that the loss reduction in IC-HCPCF has been decreasing at an average rate per year of 20 dB/km since 2011, and that today IC-guiding HCPCF, which previous typical loss figure was in the range of 0.5 to 1 dB/m, outperforms PBG-guiding HCPCF in wavelengths shorter than 1500 nm. Indeed, the loss figure has dropped from ~180 dB/km in the first negative curvature HCPCF reported in 2010 and 2011 [25,26], to 40 dB/km at 1550 nm in 2012 [61], 70 dB/km at ~780 nm [62], and 17 dB/km at ~1 μm [63] in 2013, and 70 dB/km in a 500 to 600 nm wavelength range [64] in 2014. Today's state of the art sets the loss figures in IC-HCPCF at below the 10 dB/km limit. For example, a reported hypocycloid core-contour Kagome HCPCF has been shown to have a loss as low as 8.5 dB/km at approximately 1 μm recently [65], and a tubular HCPCF to exhibit 7.7 dB/km at around 750 nm [30], and more recently, a modified tubular HCPCF is reported to show 2 dB/km transmission loss at the vicinity of 1500 nm [58]. Furthermore, the work in References [30,65] shows that the short wavelength (<1 μm) attenuation in these IC-HCPCF is limited by surface scattering loss (SSL) due to the capillary wave induced surface roughness, while for longer wavelength, improving the transmission will be determined by the cladding design. The details of this will be given in the next section.

In parallel with this continuous progress in the design and fabrication of HCPCF, this type of fiber has been the building block in a number of gas-laser related applications [7]. Among the salient features of these demonstrations is the generation of optical nonlinear effects with ultralow light level or the excitation with high signal-to-noise ratio of extremely weak spectroscopic signatures thanks to the fiber long interaction length and the small modal areas. Conversely, IC-HCPCF proved to handle unprecedentedly high level of laser pulse energy [31]. A relatively detailed account of these applications is given in a following section below.

3. HCPCF Fabrication Process: Using Gas to Nano- and Microstructure Glass

Fabricating microstructured optical fiber can be traced back to 1974 when Corning proposed an extrusion method to develop thin honeycomb structure thanks to extrudable material pushed through specific dies [66]. This extrusion technique was initially used during the very first attempts in making PCF. However, its impact on the PCF development was very weak because of the difficulty of the process, especially with hard materials such as silica and the surface roughness that it imprints on the extruded material. On the other hand, the explosive development of PCF was driven by then a new fabrication process coined "stack-and-draw" [67]. This technique has very quickly become widespread and most commonly used in the fabrication of microstructured optical fibers. It consists of a sequence of drawing rods or capillaries with typically a millimeter diameter and a meter in length and stacking them together by hand to form a "stack". The latter can be constructed into several forms depending on the final fiber design. Once the stack is built, it is drawn into preform canes, which are subsequently drawn into fibers. Figure 3a illustrates this sequence of stack and draw. One can readily notice the versatility and simplicity of this technique, which were the enabling factors in the development of the myriad of PCF designs that the scientific community gets to distinguish. Indeed, by simply judiciously stacking tubes or rods one can form different fiber microstructured architecture, and this hold for any material.

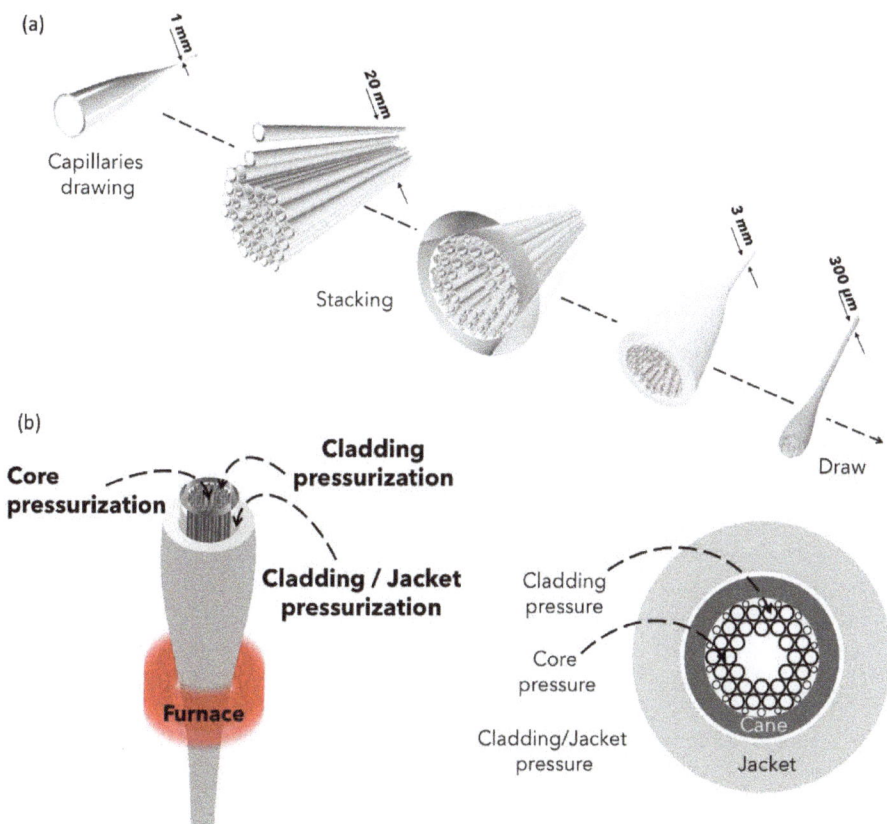

Figure 3. (a) Schematic of the HCPCF fabrication process highlighting; (b) the structuration of glass with gas.

Similarly, HCPCF, which is the topic of this review, are fabricated using stack-and-draw technique. However, because of the small thickness of the glass web that forms either the stack, the preform or the fiber, the stress on the material, which is induced via surface tension and the viscoelastic effect is too strong to keep the physical integrity of the microstructure during the draw. In order to prevent the fiber structure from collapsing via surface tension or to give some of its section a desired shape, gas pressurization in the different transversal segments of the cane is introduced. Typically, three gas control channels are used to independently pressurize the core, the cladding holes, and the cane–jacket gap, as shown in Figure 3b. This pressurization technique was first introduced during the original fabrication of Kagome HCPCF [5], and becomes since very common in HCPCF fabrication. With a careful pressure control, fiber-cladding lattice made with tens of nanometer glass struts are now readily fabricated. An example of the power of this technique is the successful draw of HCPCF with a hypocycloidal core-contour shape that led to the advent of ultralow loss IC-HCPCF. In conclusion, we can see how gas is used to nano- and microstructure the fiber glass, which in turn is crucial in how to confine and guide light as described below.

4. HCPCF Guidance Mechanisms: Micro-Structuring the Glass to Structure the Light

4.1. Introduction

4.1.1. Historical Account

HCPCF proved to be an excellent platform to investigate "exotic" guidance mechanisms and explore the predictive power of the "photonic crystal" approach that stemmed from the seminal works of John [43] and Yablonovitch [44]. This approach treats the problem of guiding, trapping, and generating light in dielectric microstructures—also called photonic bandgap materials—photonic bandgap structures, or photonic crystals, in exactly the same manner as that used in solid-state physics to derive the electronic energy band structure in a solid [68]. In solid-state physics, this is achieved by casting the time-independent Schrodinger equation as an eigenvalue equation. The resolved states of the equation map the energy–momentum space to give the range of energies that an electron within the solid may have (i.e., allowed bands) or may not have (i.e., band gaps). Similarly, in photonics, and following John and Yablonovitch, the frequency–wavevector space is mapped to identify the photonic states of a photonic crystal by casting Maxwell equation as an eigenvalue problem. Consequently, notions that were so far limited to quantum mechanics and condensed matter such as bandgap, Bloch states, density of state become critical conceptual components in designing and investigating dielectric microstructures.

Within this context, the manner on how to design PCF (especially HCPCF) departs from the conventional approach in fiber optics [69]. Akin to semiconductor and doped crystals, a PCF is treated under the framework of photonic crystal physics as a waveguiding 2D "crystal" whose order or symmetry is broken by introducing an optical guiding defect within its extended spatial structure. In other words, a PCF is a cladding photonic crystal structure to which a core defect with different geometrical shape or index than the unit cell of the cladding is introduced. The pertinence of the index and geometry profiles of the cladding and core is better assessed through the examination of the cladding modal spectrum. When the cladding is considered to be infinite and periodic, which is often done for calculation convenience so as to apply the Bloch theorem, the modal spectrum is simply the density of photonic states (DOPS) in the frequency–wavevector space $(\omega, \vec{k})$. Furthermore, in PCF, this modal spectrum involves only the propagating modes along the uniform direction of the 2D photonic structure (i.e., the fiber axis, which we refer as z-direction), which means the modes whose electric field amplitude is of the form $E \propto e^{i\beta z}$. Alternatively, the mapped frequency–wavevector space to derive the cladding modal spectrum is a subspace of $(\omega, \vec{k})$ Hilbert space, namely (ω, β), where β is the z-direction component of $\vec{k}$, termed propagation constant. The modal spectrum takes the form of a diagram showing the density of the cladding modes in (ω, β) or $\left(\omega, n_{eff}\right)$ space. Here, n_{eff}, called effective index of the mode, is given by $\beta = n_{eff}.k$. Similar to solid band-structure, the resulted DOPS of the cladding exhibits $\left(\omega, n_{eff}\right)$ regions that are populated with photonic states, or propagating modes, and other $\left(\omega, n_{eff}\right)$ regions that are void from any mode (i.e., PBG). The possibility of optical guidance in an introduced core defect within the cladding implies requirements on the core index and shape dictated by the type of guidance mechanism one is aiming for.

There are two strategies to guide light in the core. The first one relies on choosing a core index and geometry so at least some of its supported mode $\left(\omega, n_{eff}\right)$ lie in the cladding PBG range. Consequently, the core-mode cannot leak out because there is no cladding mode to couple to. This is the design strategy for PBG-guiding HCPCF. The second one relies on engineering a core and a cladding so the cladding modal spectrum is populated with modes which have very little spatial overlap and/or symmetry matching. Here, the core mode remains in the core because its coupling to the cladding modes is suppressed.

We conclude this section by adding the following comments on the terminology of "crystal" in the PCF field. First, it is noteworthy that the modal spectrum structure of a dielectric microstructure, meaning the existence of allowed and forbidden bands in the $\left(\omega, n_{eff}\right)$ space does not necessarily require a periodic material. In fact, the periodicity of a photonic structure is neither a necessary condition nor a defining feature for the existence of PBG [30,68]. The periodicity requirement is a mere mathematical convenience for applying the Bloch theorem [7]. This point draws its parallel from solid-state physics and crystallography, where amorphous materials can exhibit electronic bandgap, and a diffraction pattern can be produced from a solid with no crystallographic symmetry [70]. The latter surprising feature was discovered by D. Shechtmann and earned him the Nobel Prize in Chemistry in 2011 [70]. Consequently, and in addition to historical reasons, we apply the term of photonic crystal fiber (PCF) to any optical fiber whose cladding is microstructured and exhibits at least a short-range order.

4.1.2. Total Internal Reflection, Photonic Band Gap, and Inhibited Coupling

Before detailing, in the sections below, how PCF cladding modal spectra could be structured, we first give a simple and pictorial account on how the cladding modal structure affects optical guidance in fibers using the approach described above, and which is summarized in Figure 4. The figure shows schematically, and for a fixed frequency, the modal content of the cladding and of the core defect of the three types of optical fiber we can distinguish today. These fibers are (i) the well-known Total Internal Reflection (TIR), (ii) Photonic Band Gap (PBG), and (iii) Inhibited Coupling (IC). Using the solid-state physics approach, we consider the cladding to be infinite, or at least with a size much larger than any operating wavelength, whilst the core is considered to have micrometer scale size. The modal content is illustrated by colored regions on the n_{eff}-axis. The cladding modes are presented as the orange-color-filled rectangle on *lhs* of the n_{eff}-axis; the core modes are schematically shown by the intensity profiles on the *rhs* of the n_{eff}-axis. Finally, the material indices that are involved in the fiber structure are represented with dashed horizontal lines.

In the case of TIR (Figure 4a), where the cladding is a dielectric with uniform index n_g, the modal content is represented by a continuum of modes whose effective indices are necessarily $n_{eff} < n_g$. Hence, the cladding is void from any propagating modes for $n_{eff} > n_g$ (region labeled "gap" in the figure). In parallel, the introduction of a core defect with higher material index n_{dg}, shows discrete guided core-modes in the effective index range of $n_g < n_{eff} < n_{dg}$. These TIR guided modes are thus confined within the core because the cladding is void from any possible modes at their effective indices. Consequently, from this new standpoint, TIR guidance is simply one form of a PBG guidance, which is achieved by having a defect material with higher index than that of the cladding.

The requirement of higher index for the core material to have PBG can be lifted with microstructured cladding. Figure 4b illustrates this for the case of a PBG-HCPCF. Unlike with a uniform index cladding, the cladding modal content now shows a more structured pattern with bands corresponding to cladding mode bands (orange-color-filled rectangles) and gaps corresponding to effective index band with no possible cladding modes (white-color-filled rectangles). The latter can range even for $n_{eff} < n_{air} = 1$, allowing thus optical guidance in core defects that are hollow or filled with gases. The physical principles on how to engineer these "low index" gaps is detailed in a section below.

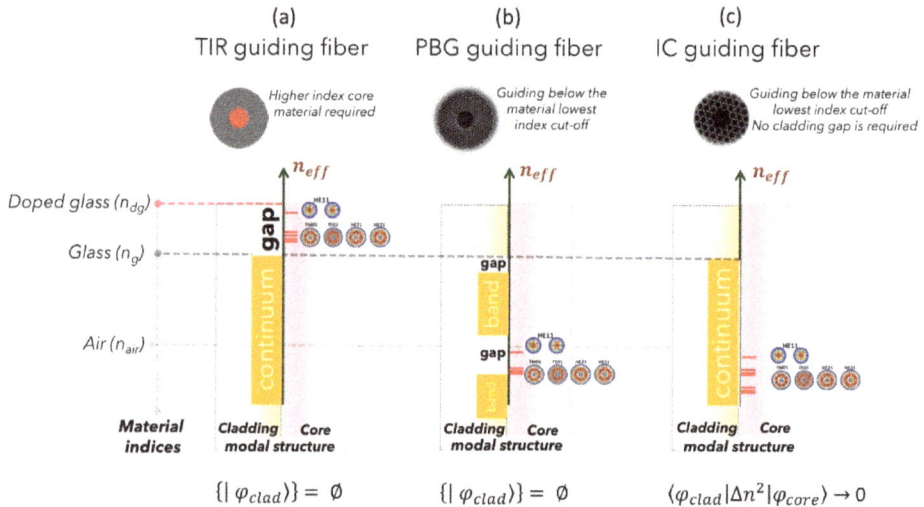

Figure 4. Modal content representation of the three different optical guiding fibers: (**a**) Total Internal Reflection (TIR); (**b**) Photonic Band Gap (PBG); and (**c**) Inhibited Coupling (IC) guiding fibers.

Finally, Figure 4c illustrates another type of guidance where neither a higher index core material nor cladding PBG is required. Indeed, core modes and cladding continuum can coexist with the same (ω, n_{eff}) without strongly hybridizing. Such a situation of having two modes with the same effective index does not violate the exclusion principle. In this case, indeed modes of heterogeneous structure with the same n_{eff} does not necessary mean having the same wavevector. The guidance mechanism, akin to BIC or QBIC, is called IC and was introduced in 2007 [4] to explain the Kagome HCPCF optical properties. According to this model the field of the core mode $|\varphi_{core}\rangle$ and the cladding mode $|\varphi_{clad}\rangle$ is strongly reduced (i.e., $\langle\varphi_{clad}|\Delta n^2|\varphi_{core}\rangle \longrightarrow 0$, with Δn being a transverse index profile function). This can be done by either having little spatial intersection between the fields of $|\varphi_{clad}\rangle$ and $|\varphi_{core}\rangle$ photonic states or by having a strong mismatch in their respective transverse spatial-phase. The details on how $\langle\varphi_{clad}|\Delta n^2|\varphi_{core}\rangle$ is reduced, is given below.

The above pictorial explanation highlights the crucial importance of the cladding modal spectrum in dictating the nature of a fiber guidance and its performance. Figure 5 shows three representative and most reported HCPCF cladding lattices and their associated modal spectra, or DOPS. The DOPS diagram is achieved by numerically solving Maxwell equations for infinite, periodic, and defect-free cladding, and displays the DOPS in the effective index and frequency space. The frequency is often represented by some normalized frequency such as $k\Lambda$, with k being the amplitude of the wave vector, and Λ is the pitch of the dielectric photonic structure. The structure is defined by its unit cell shown in the inset of each of the three DOPS diagrams. Also, the effective index range was mapped near the air-line (i.e., $n_{eff} \approx 1$) as we are interested in fibers that guide in air or diluted materials.

The first DOPS diagram (Figure 5a) is that of a triangular and packed arrangement of nearly hexagonally-shaped air holes with an air-filling fraction of ~93% [7]. This cladding lattice cladding is that of the most common PBG-guiding HCPCF. The PBG region corresponding to DOPS = 0, is shown in white. We note that for n_{eff} slightly below 1, the PBG spans from $k\Lambda \approx 14.5$ to $k\Lambda \approx 18$, which gives a transmission window of 330 nm centered at 1550 nm. In the section below, using the photonic analog of tight binding model [2], we review how PBG and cladding bands are formed in PBG-guiding HCPCF, and how they are related to the glass geometrical features. We particularly stress the role of the enlarged glass apices in the existence of PBG, and how their relative size can be optimized to increase the PBG bandwidth or to open-up higher order PBG. We then finish the section on PBG-guiding fibers by presenting properties of cladding lattice modes for high-normalized frequency range that relevant to IC-guiding HCPCF.

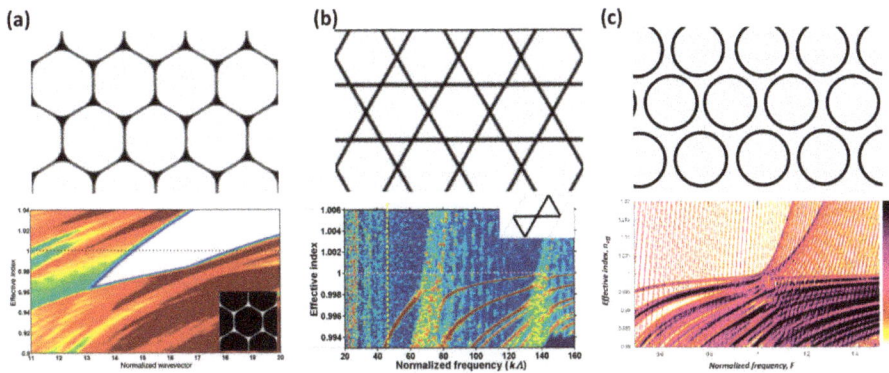

Figure 5. Representative DOPS of (**a**) PBG (reprinted with permission from Reference [7], Francis & Taylor, 2011), (**b**) IC Kagome (reprinted with permission from Reference [7], Francis & Taylor, 2011), and (**c**) IC Single Ring Tubular Lattice HCPCF (reprinted with permission from Reference [30], OSA, 2017).

Figure 5b,c shows the DOPS diagrams for Kagome and tubular cladding lattice, respectively. The Kagome lattice can be represented as an array of tessellated David's stars, and the tubular lattice as an array of isolated glass tubes. Several salient features can be drawn when compared to the previous one. First, in both modal spectra the DOPS does not reach the zero value. In other word, both photonic structures do not exhibit a PBG. Second, the normalized frequency range is much high than the case of the PBG HCPCF, that is why it is sometimes coined a large pitch regime [71]. In fact, we will see that in this regime, the pitch has secondary impact on the lattice modal spectrum. This is why the normalized frequency in Figure 5c, is represented by the "pitch-free" quantity $F = (2t/\lambda)\sqrt{n_g^2 - 1}$ rather than the most common $k\Lambda$. In the section below, and similarly with the PBG-guiding HCPCF, we will be using notions from solid-state physics to describe the IC guidance. This time, the driving concept in explaining IC optical guidance is BIC. We chiefly use it to draw physical rules to achieve a situation where modes with the same n_{eff} can "coexist without interaction" in a microstructured fiber. In turn, we use these rules to design cladding lattice geometry and defect core-contour in such a way the interaction between the cladding mode and the hollow-core mode is strongly suppressed despite having the same or comparable effective index. We particularly emphasize on the importance of the absence of enlarged glass nodes, which is in opposite requirement compared to the PBG HCPCF, and the benefit of having a cladding with a thin, "smooth" elongated glass membranes. Finally, we will see the advantage of working in the large pitch regime to an enhance the IC guidance.

4.2. Photonic Bandgap HCPCF: How to Engineer Photonic-Bandgaps below the Cladding Material Lowest Index

4.2.1. Photonic Tight Binding Model

The modal spectrum of the fiber cladding can be rigorously derived using solid-state physics concepts such as Bloch theorem and solving numerically the Maxwell equation. This, however, requires heavy numerical calculation and does not necessarily provide a direct physical insight on how this modal spectrum is formed or evolves. In parallel, the more intuitive and highly predicting model of the tight binding model (TBM) have been successfully applied to HCPCF by Benabid and coworkers in 2007 [2], and coined Photonic Tight Binding model (P-TBM), to explain how the cladding allowed bands and band gaps are formed within a PBG-guiding HCPCF. They found that the cladding bands in the modal spectrum of the PBG-HCPCF (shown in Figure 5a) are comprised with Bloch modes supported by the glass apices and struts that form the hexagonal shape of the unit cell, along with the modes which are supported by the air holes. Below, we review the basics of the tight binding model and how this is applied to microstructured fibers.

In solid-state physics, according to TBM the bands in the energy diagram of a crystal or solid result from the superposition of electronic state wavefunctions of the isolated atoms that form the solid. In this description the relationship between the solid physical structure in the real space and its energy structure in the Hilbert space is straightforward and is illustrated schematically in Figure 6a,b for the case of 1D crystal of identical atoms. We see that for a large number of atoms N, if the pitch Λ of the lattice is sufficiently large, the energy states of the crystal is reduced to discrete states of the isolated constitutive atom, $|j\rangle$, each state being N-degenerate. As Λ decreases, the atoms get closer and the wavefunctions of their energy states start to overlap. Consequently, the N-degenerate energy state splits by virtue of the exclusion principle, and creates a band of N-distinct extended states (i.e., Bloch states). The width of the band increases with decreasing Λ. Furthermore, the start of band formation and its width strongly depend on the state wavefunction of an individual atom of the set. The stronger the wavefunctions are confined (or bound), the closer the atoms must be to each other for sufficient overlap. The net result is an energy structure exhibiting allowed band separated by gap, i.e., bandgap. It is noteworthy that bandgaps result not only from the intrinsic separation between two allowed bands, but they can result from anticrossing between overlapping bands. This occur when the wavefunctions of the two bands exhibit strong symmetry matching. Hence they cannot occur in the presently considered monoatomic 1D crystal, because the atom states are orthogonal, and thus the bands here overlap without anticrossing (Figure 6b). However, for the case of a crystal molecule with different atoms, it is possible to observe anticrossing of overlapping band if the later results from two different atoms and exhibiting strong symmetry matching. This picture can be applied to guided photonics where the electronic state in an atomic site is replaced by the guided mode in a photonic site. The latter represented by a dielectric with higher refractive index, i.e., the waveguiding component of the structure. The electronic state energy is replaced by the effective index of the guided mode. Finally, the energy diagram to map the Hilbert space of the crystal is replaced by the dispersion diagram of the different modes supported by the photonic structure in the $\left(\omega, n_{eff} \right)$ space. Figure 6c,d reproduces the schematic picture shown in Figure 6a,b for the case of an array of N glass rods. The evolution of the dispersion with the pitch, plotted in function of normalized frequency $k\Lambda$, follows the same trend as for the energy state of a crystal, evolving from a N-degenerate dispersion line of a single rod into a band of propagating modes.

Figure 7 shows the above picture by considering an approximate and analytical model of the dispersion of a silica rod array suspended in air [7]. For the case of a single rod (see Figure 7a), we can retrieve the dispersion of the commonly known fiber modes (in the figure we limit to only LP_{01} and LP_{11} for demonstration purpose). When a much larger number of equally spaced rods is considered, the dispersion curve shows a band structure for the case of low $k\Lambda$ or sufficiently closed rods (small pitch regime). The width of these allowed bands narrow down with increasing $k\Lambda$ to the extent of forming a single dispersion line when $k\Lambda$ is above a certain critical value (large pitch regime). The formation of these bands in an extended physical photonic structure, such the considered rod array, implies the possibility of light PBG guidance in a defect within the structure if its index and geometry is judiciously chosen. Whilst this P-TBM toy model does not consider the modes below the lowest material index (i.e., $n = 1$), we can still draw several points from the formed modal spectrum of the 1D rod array, and whose importance becomes apparent in the following sections. First, for small $k\Lambda$ values, which we call the small pitch regime, the allowed bands strongly vary with the pitch. Hence, the resulted PBG transmission window edges of a guiding defect will depend on the pitch. This property is used to tune the transmission window spectral range of PBG-HCPCF by simply scaling accordingly the pitch. On the other hand, for sufficiently large $k\Lambda$ values, the bands are very narrow, and even reduced to a single dispersion lines. Under this large pitch regime, because the bands vary little with $k\Lambda$, a guiding defect will exhibit PBG transmission windows whose edge spectral location depends little with the pitch. Secondly, in the small pitch regime, the bands are formed by extended and spatially overlapping Bloch modes, whilst in the high pitch regime, the fields of theses Bloch modes exhibit very weak to no-overlap. Similarly with the high tight-binding regime in solid-state physics, these field wavefunctions are better presented by maximally localized Wannier functions than Bloch functions [72]. Below, we will be recalling these properties as we describe some PCF results or features.

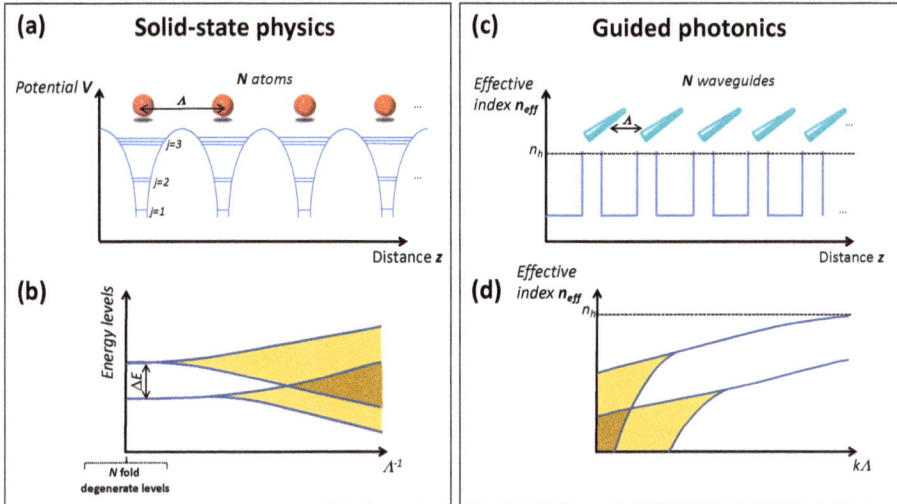

Figure 6. Analogy illustration between the Tight Binding model in an atomic structure and the Photonic Tight Binding model in an optical waveguide structure.

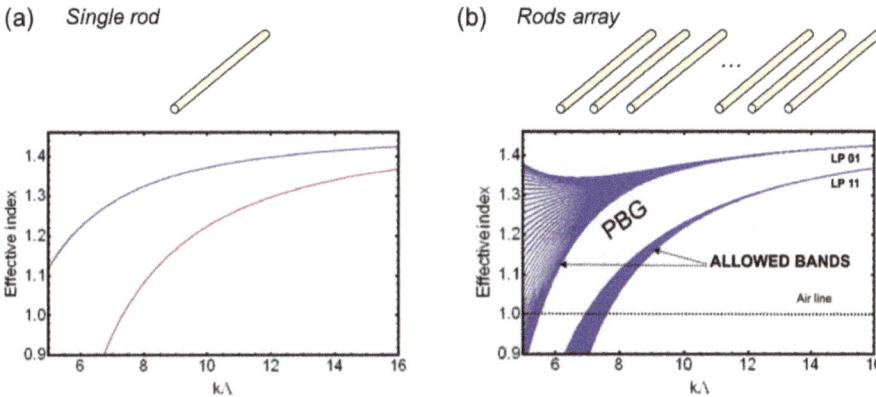

Figure 7. Curves of the dispersion of first modes for (**a**) an isolated rod and for (**b**) an array of similar rods (reprinted with permission from Reference [7], Francis &Taylor, 2011).

Figure 8 illustrates the aforementioned properties through a realistic PCF cladding structure. The latter consists of an array of high-index and isolated inclusions (typically doped silica) embedded in a silica matrix arranged in a triangular lattice. This type of all-solid PCF was first proposed by Birks and coworkers as a demonstration of PBG guidance with very low index contrast between the high-index and the low-index materials [73]. Figure 8a shows the cladding structure DOPS over a large normalized frequency range, from $k\Lambda = 0$ to $k\Lambda = 70$, spanning over the small pitch regime and large pitch regime. The index of the high inclusion is $n_{hi} = 1.5$ and that of the silica matrix is $n_{lo} = 1.45$. The band-structure of the modal spectrum is readily noticeable, especially for low $k\Lambda$. The bands (Figure 8a gray-colored areas), consist of propagating Bloch modes, which are spatially extending over the whole structure. In corroboration with the above toy model, the bands narrow down to a single line when $k\Lambda$ is sufficiently large. Figure 8b shows the intensity profile of the cladding unit cell modes for $k\Lambda = 70$. At this high-normalized frequency, all the bands corresponding to $n_{eff} > n_{lo}$ are reduced to a single line. This means that the Bloch modes of these photonic states exhibit a field that is highly localized at the high-index inclusions and weakly-to-no overlapping field between the high-index inclusions. Alternatively, and drawing the concept from solid-state physics, the photonic state modes at the large pitch regime are better represented by maximally localized Wannier functions [72]. This is shown by inspecting the mode intensity profile of the cladding modes at $k\Lambda = 70$ within the Wigner–Seitz unit cell. Each ultra-narrow band clearly shows the well-known profile of guided modes of conventional optical fiber, and labeled using the linear polarization (LP) approximation terminology.

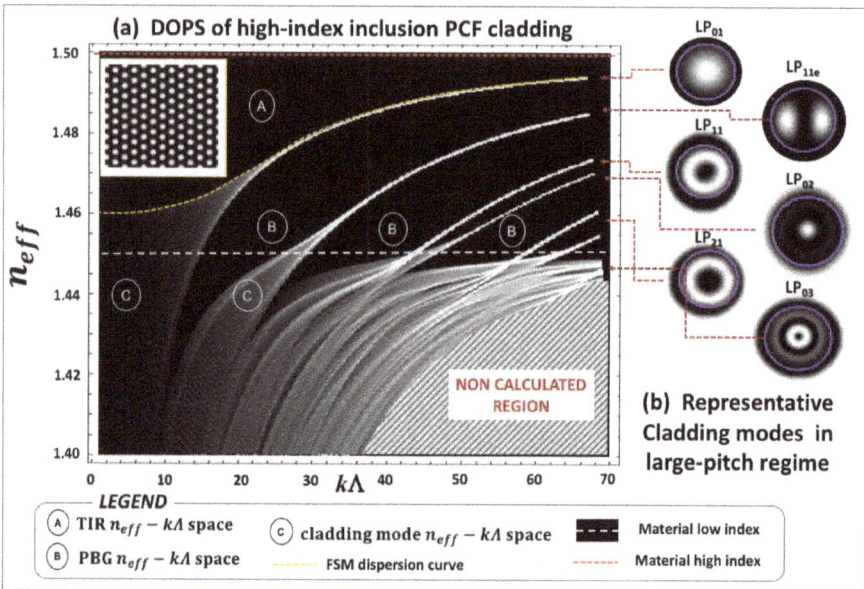

Figure 8. (a) Density of photonic state of high inclusion PCF. The PCF structure consists of a glass material with a uniform index $n_g = 1.45$ (dashed horizontal line) and high index cylindrical-shaped inclusions with diameter $d_{hi} = 0.46 \times \Lambda$ and index of 1.5; (b) The intensity profile of representative cladding modes shown within the Wigner-Seitz unit cell.

Using the lexicon of tight binding model, the highest effective index and the lowest effective index modes of each band correspond to the symmetrical mode ("bound photonic-state") and antisymmetrical mode ("antibound"), respectively. Conversely, using fiber optics formalism, we can recognize the highest effective index mode of the fundamental band, i.e., the band with the highest n_{eff}, to be the fundamental space-filling mode (FSM), introduced by Birks et al. in their seminal work on endlessly single mode PCF [33].

Moreover, when a core defect is introduced into this high inclusion PCF, an inspection of the PBG regions in the DOPS (Figure 8a black areas) reveals several points on its optical guidance properties. First, and after recalling that within the "photonic-crystal physics" approach TIR is only a particular regime of PBG guidance corresponding to a defect with higher index than that of its cladding, this fiber can guide via TIR if the index of the core defect fits inside region A. This means the core index is larger than the FSM n_{eff} (the superposed yellow curve on DOPS of Figure 8a), which can be considered as the cladding photonic structure "effective" material index. Second, the fiber can guide via PBG over a large $n_{eff} - k\Lambda$ range (black regions). Particularly, PBG guidance includes core materials with indices lower than the lowest cladding material index (white dashed horizontal line). Also, the expected fiber transmission spectrum presents multiple windows, and whose cut-offs are determined by the allowed band edges. Third, we can draw two important general properties from the inspection of this fiber DOPS in the large-pitch regime. Because the bands narrow down to the dispersion curves of the single high-index inclusion, the cladding modal spectrum depends little on the pitch. In turn, in the large-pitch regime a core defect exhibits a transmission spectrum whose cut-offs do not vary strongly when the pitch is changed. Instead, they only depend on the high-index inclusion index and size. This property led to refer to this particular regime of PBG guidance as antiresonant reflecting optical waveguide (ARROW) introduced in the 1980s [51]. In a section below, we detail the distinctions between ARROW, PBG, and IC. The second general property of note is that the large-pitch regime is an example demonstrating that photonic structure periodicity is not necessary to exhibit PBG. Indeed, because of the weak impact of the pitch on the DOPS, the initial triangular arrangement of the cladding lattice can be changed to an amorphous one without significantly change the modal spectrum for $k\Lambda > 50$.

4.2.2. Engineering PBG in HCPCF

Historically, the P-TBM was first demonstrated with PBG-HCPCF [2]; the results of which are summarized in Figure 9. Unlike the high-index inclusion PCF, the PBG-HCPCF (Figure 9(Ia)) cladding lattice has more complex cladding lattice. Consequently, it is difficult in identifying the constituent waveguiding components in the unit cell (Figure 9(Ib)). Before reviewing the results of Reference [2], we recall that the guided optics analog of an atomic site is a photonic site consisting of high index material surrounded by a lower index material and guides light via TIR. Furthermore, two waveguide components differ (i.e., have different modal spectrum) via the difference in their index or via the difference in their geometrical shape. Consequently, the first task in Reference [2] was to identify the unit cell waveguiding features whose modes form the bands in the DOPS diagram. Finally, a lower index material surrounded by higher index materials can support photonic states via reflection/interference at the interference between two different index materials. Consequently, the PBG-HCPCF unit cell considered here has the form of a set of glass sheet forming a hexagon with prominent apices surrounding an air hole.

In a similar fashion with a crystal made with a heteronuclear molecule, the results show that the PBG-HCPCF cladding unit cell is comprised with six enlarged glass nodes positioned at the apices of the hexagon and six thin glass struts forming the sides of the hexagon. This was achieved by visualizing numerically and experimentally the cladding Bloch modes of the bands that surround the PBG below the air-line. Figure 9(IIa) shows the DOPS at the effective index and frequency range close to this PBG. The latter is bordered at low $k\Lambda$ by the low-index edge of band 1 (red curve), which represents the low-frequency band edge of the hollow core transmission spectrum, and at the high $k\Lambda$ by the blue curve, which represents the high-frequency edge of hollow core transmission window. Furthermore, the PBG closes at the low effective index by the dashed green curve. The Bloch mode intensity profiles associated with these three dispersion lines have been calculated at the points of the DOPS diagram labeled by (b), (c), and (d) for the red, blue, and dashed-green curves, respectively (Figure 9(IIa)). The Bloch mode clearly shows an intensity profile dominantly confined in the apices at the point B (Figure 9(IIb)), and dominantly confined in the struts at the point C (Figure 9(IIc)). The Bloch mode at the point D shows an intensity profile dominantly in the air hole of the unit cell. However, a non-negligible power fraction resides in the struts. This is due to hybridization between the mode-band associated with the strut and the associated with air hole mode-band, which we detail below. These modes were experimentally observed using both scanning near-field optical microscope

(SNOM) to image their near field and by imaging their profile during their propagation in the Fresnel zone [2].

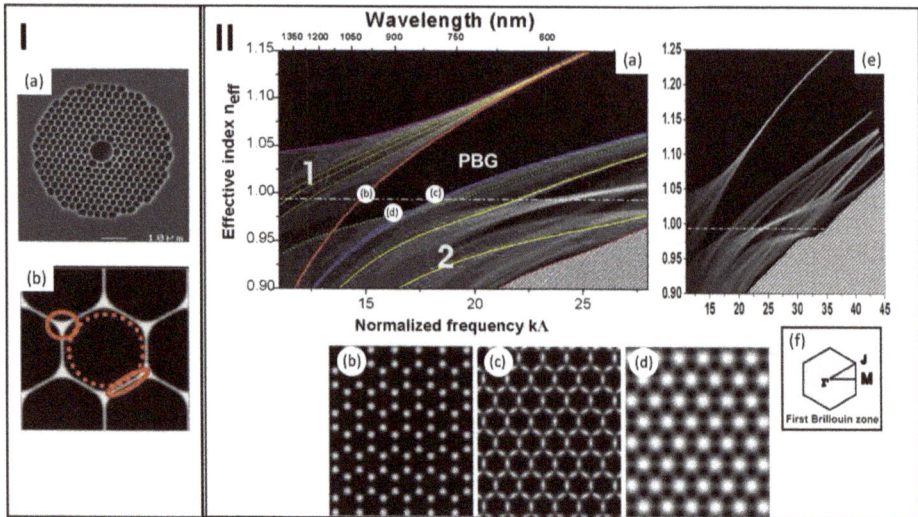

Figure 9. (I) HCPCF designed for 1064 nm guidance: (a) SEM of the fiber and (b) zoom in of its cladding structure. The red lines indicate the interstitial apex, the silica strut, and the air hole regions. (II) Numerical modeling: (a) Propagation diagram for the HCPCF cladding lattice; near field of (b) the interstitial apex mode, (c) the silica strut mode, and (d) the air hole mode, (e) Zoom out of (a), (f) Brillouin zone symmetry point nomenclature.

Figure 9(IIe) shows the DOPS over a large range of normalized frequency and effective index spanning $k\Lambda$ up to 45, and n_{eff} up to 1.25. Figure 10 reproduces this DOPS diagram supplemented with the Bloch mode intensity profile for several points of the DOPS (*lfh* of Figure 10a) so to (i) identify the modes of the different bands and (ii) show their dynamics and evolution with $k\Lambda$. In consistency with P-TBM and the features observed with the high-index inclusion PCF, it is easy to distinguish the structure of the bands and their narrowing to a single line when $k\Lambda$ increases.

Figure 10a (*lhs*) shows six Bloch mode profiles for different bands and at normalized frequencies higher than 35. The profile labeled 1 shows a spatially extended mode whose intensity is localized with the glass apices, as expected from the high pitch regime. In particular, we can identify from this the HE_{11}-like fundamental associated to an individual apex. Figure 10b shows the evolution of the intensity profile of this mode when the normalized frequency is increased from 12 to 30. The results show the enhancement in confinement of the light within the apex with increasing $k\Lambda$. At $k\Lambda = 12$, the intensity of the transverse field extends outside the apex with a relatively large space-filling (top of Figure 10b), whilst at $k\Lambda = 30$, the mode shows an intensity profile that is strongly confined with the apex with little spatial overlap with the silica strut or air. This is a fiber–photonics illustration on how photonic lattice modes evolve from a spatially delocalized state, which is suitably presented by Bloch functions, to a highly localized lattice mode, which is conveniently represented by Wannier functions. This feature of the cladding modes in the high pitch regime has been one of the driving principles in the development of IC-HCPCF (see below). A further substantiation of the P-TBM is shown in the mode profile labeled 2. This corresponds to a linear combination of the fundamental mode of an individual silica strut. Conversely, all the profiles of Figure 10a (*lfh*) shows mode profiles that are associated with either propagating modes of the individual apex or strut, or with supermodes from the coupling between the two glass features. Consequently, one can deduce that the high-index waveguiding components that make the PBG-HCPCF cladding lattice comprise a glass apex and glass strut. To this structural feature, we note the existence of photonic band related to modes localized in the air holes of the dielectric structure. The formation of these bands cannot be described by TBM approach, instead they are formed akin to modes formed between to potential wells via reflection off the low-high index interface.

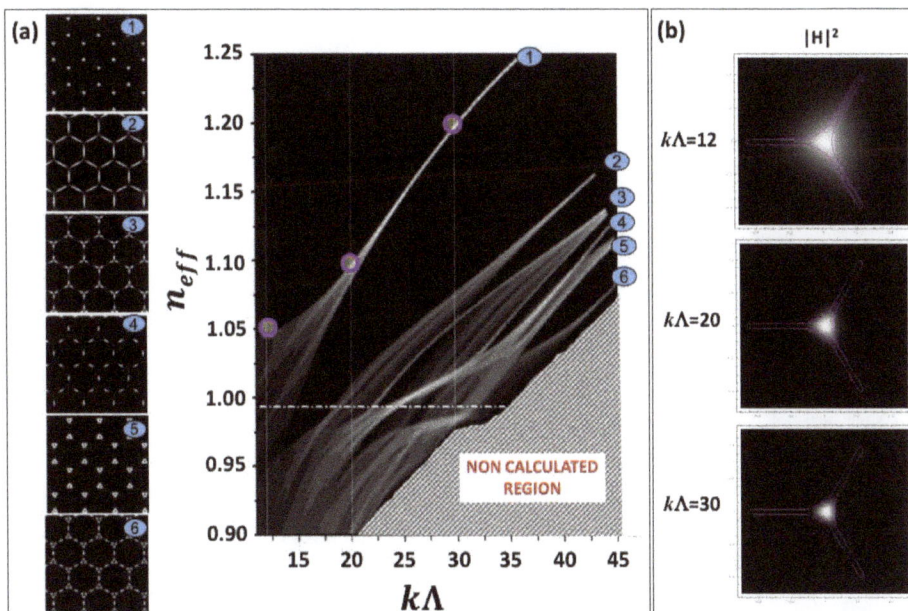

Figure 10. (**a**) Propagation diagram for the HCPCF cladding lattice with a gallery of the different guided modes. (**b**) Evolution of the interstitial apex mode for different $k\Lambda$.

A final remark on the structure of this DOPS relates to the dynamics at work when the photonic bands cross each other. For this purpose, we consider the high inclusion cladding lattice, which unit cell is formed by one single step-index waveguiding component (see Figure 8). Here, the DOPS shows bands that cross with no splitting or anticrossing. This is because the bands are mutually orthogonal because they stem from the same waveguide, and hence will not interact when they cross in the $\left(k\Lambda, n_{eff}\right)$ space. In the case of the PBG-HCPCF cladding lattice, we have seen that the unit cell is formed by two glass waveguiding components, and the bands stem from either apex, strut modes or air hole modes. This means that the modes of an "apex band" are not necessarily orthogonal with those from a "strut band". Consequently, when the mode crosses the same $\left(k\Lambda, n_{eff}\right)$ they can anticross. The DOPS of Figure 10 shows several of these anticrossings, such as the one near $k\Lambda \sim 23$ and $n_{eff} \sim 1$. Below, we show how these features are exploited by Light et al. [74] to design and fabricate a HCPCF with two bandgap transmission windows.

Figure 11 summarizes some of the results reported in Reference [74]. Figure 11a shows the near air-line DOPS of the above PBG-HCPCF cladding lattice but with different relative size between the apices and struts and with much higher air-filling fraction. This DOPS was achieved by optimizing the apex and strut size and shape, and using the P-TBM to control the position of the bands within the $\left(k\Lambda, n_{eff}\right)$-space. Another outcome from this work is the fact that the fundamental PBG results from a strong anticrossing between the "fundamental apex band" and the "fundamental strut band". This was illustrated in Figure 11b, which shows the DOPS of a lattice made with apices only (red band) and that of a lattice made with struts only (blue band). We clearly see that in the absence of apices, the strut fundamental band never fully crosses the air line around the $k\Lambda$-range of 20 to 27. Thus indicating that the shift of the band towards an effective index-range lower than one results from strong anticrossing with the fundamental apex band.

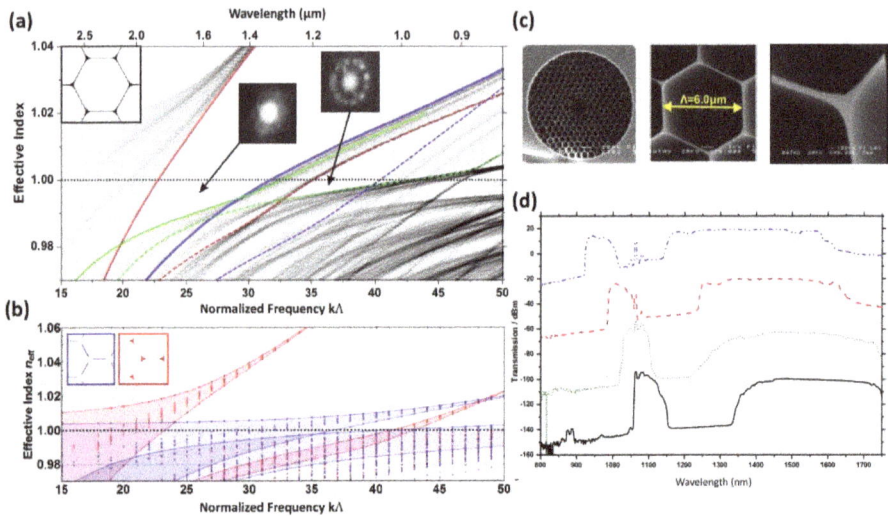

Figure 11. (**a**) Density of photonic states for the cladding structure illustrated inset top-left with a strut thickness of $t = 0.01\ \Lambda$ and apex meniscus curvature of $r = 0.15\ \Lambda$. The colored lines trace the cladding modes that form the edges of the two bandgaps. The wavelength in the upper x-axis is deduced for a pitch of 6.7 μm. (**b**) Evolution of the cladding modes of apexes alone (red) and struts alone (blue). (**c**) SEMs showing the fiber cross-section and the cladding structure. (**d**) Transmission spectra of 5 m lengths of the fabricated HCPCF with varying pitch, offset vertically for clarity (Reprinted with permission from Reference [74], OSA, 2009).

Figure 11c,d shows the physical characteristics of the fabricated double-PBG HCPCF tailored to guide at approximately 1.5 μm and 1 μm.

4.3. Inhibited Coupling HCPCF: How to Prevent Interaction between Longitudinally Phase-Matched Modes

4.3.1. Historical Account

This section deals with an optical guidance configuration where both the requirement of higher index core material and cladding photonic bandgap is no longer, as is schematically shown in Figure 4c. This configuration was first experimentally observed with the introduction of Kagome lattice HCPCF [46], and outstands with the peculiar situation in which a core mode is guided with relatively low loss at the same effective indices of cladding "continuum" of modes. The historical development of our understanding on how such a fiber guides followed the following sequence. The first fabricated Kagome HCPCF exhibited a very broad transmission spectrum, which ruled out the photonic bandgap scenario. This was then corroborated numerically by inspecting the DOPS of the Kagome lattice, which shows no PBG [75]. The fiber had a core diameter of only 15 μm, and showed attenuation loss of ~1 dB/m with relatively low bend sensitivity. This loss figure at such a core-diameter ruled out Fresnel reflection based optical guidance in dielectric capillary studied by Marcatili and Schmeltzer in the sixties [50]. Following this work, the predicted loss of the fundamental core-mode (i.e., HE_{11}) of a glass dielectric (refractive index: n_g) capillary of radius R_c, given by the expression

$$\alpha_{capillary} = (2.405/2\pi)^2 \left(\lambda^2/R_c^3 \right) \left(n_g^2 + 1/2\sqrt{n_g^2 - 1} \right), \tag{1}$$

results in a loss range of 360 to 5000 dB/m at 400–1500 nm spectral range, which is over two orders of magnitude higher than the experimentally observed with Kagome HCPCF. Also, the results in [46] ruled out the possibility of guiding via antiresonant reflecting waveguidance (ARROW), introduced by Duguay et al. [51]. ARROW guidance principle relies on the reflection enhancement off the interface between the air-core and the dielectric cladding when the dielectric-thickness is strongly reduced. This work aroused a number of effort in the eighties (see below on the difference between ARROW, PBG, and IC), among which is the work from Archambault et al. [76] who derived the confinement loss of a hollow-core fiber based on concentric antiresonant dielectric rings. The expression of the HE_{11} mode minimum loss of a single antiresonant hollow fiber for the case of $R_c \gg \lambda$, was found to be given by

$$\alpha_{ARROW,min} = \alpha_{capillary}(2.405\,\lambda/R_c)\left(2\pi\sqrt{n_g^2 - 1}\right)^{-1} \tag{2}$$

Similarly with the Fresnel reflection in the glass capillary, the ARROW model predicts a loss at 400–1500 nm in the range from 7 to 365 dB/m, which is higher than the measured loss of the Kagome HCPCF. Subsequently, these findings raised the question on how cladding modes and a core mode could have the same effective index without interacting. This interrogation was justified because the findings go against the conventional wisdom in fiber optics, which states that guided-modes with the same effective index strongly hybridize. However, this is not necessarily valid in heterostructures such as HCPCF. Firstly, two modes with the same n_{eff} do not necessarily imply that the two modes have the same wave vector, and subsequently, that modes with the same n_{eff} are not rigorously phase-matched, and thus can avoid interacting. One can envisage, for example two modes in a dielectric heterostructure, which are localized in different materials. Secondly, as is known in coupled-mode theory [77], the coupling-strength between two modes is ruled not only by the effective index matching, but also by the optical overlap between the modes. This means that if the optical overlap between two modes is nil or strongly suppressed, they will not strongly interact.

The explanation to the optical guidance in Kagome-like HCPCF was given by using once again concepts from quantum mechanics and solid-state physics. In 1929, Von Neumann and Wigner reported on the existence of a bound electronic state with positive energy in an artificial potential designed to extend to infinity in oscillatory fashion [3]. Such a counterintuitive situation describes localized waves coexisting with a continuous spectrum of radiating waves that can carry energy away. These localized waves were then later coined Bound States in the Continuum (BIC), also referred to as embedded eigenvalues or embedded trapped-modes. The occurrence of BIC happens because of symmetry incompatibility between localized waves and the radiation continuum, thus forbidding the former to couple to the latter. We can also picture BICs as resonances with infinite lifetimes, and quasi-BIC (QBIC) as resonances with high quality factor but finite. BICs and QBICs can be found not only in quantum mechanics but in electromagnetic, acoustic, and water waves (see the dedicated review to BIC by Hsu et al. [52]).

In 2007, Benabid and coworkers [4] introduced the BIC and QBIC concept to the field of fiber–photonics by showing that the guided core-modes in Kagome HCPCF are longitudinally phase-matched with a cladding mode-continuum. Analysis of the modal spectrum of the cladding structure (see Figure 12a) showed that the intuitively expected coupling between the core-mode and the cladding modes is inhibited by a strong transverse phase mismatch (i.e., symmetry incompatibility) between the modes and the highly localization of the cladding mode in the cladding glass web (i.e., spatial separation between the core air-mode and the cladding glass-modes). These results are summarized in Figure 12b,c. The top of Figure 12b shows the DOPS of an infinite cladding of a

Kagome lattice. The bottom of Figure 12b shows the effective index and confinement loss coefficient for a Kagome lattice with 1-cell core defect (i.e., 1-cell Kagome HCPCF). The DOPS clearly shows no photonic bandgap (i.e., DOPS = 0) in the mapped effective-index and frequency space. On the other hand, the 1-cell Kagome HCPCF shows a core-guidance with loss figures of 1 to 0.5 dB/km in a hexagon-shaped core with a size of 20 μm and with a large pitch of 12 μm [4]. The nature of interaction between the core mode and the cladding is illustrated in Figure 12c. The top of the figure shows representative mode profiles within the unit cell of infinite Kagome cladding near and below the air-line. We distinguish three kinds of modes. Proceeding by analogy with fibers with cylindrical symmetry, we associate the transverse phase of the silica-guided field with an effective azimuthal index number "m" which governs the azimuthal field oscillations (i.e., along the strut length axis), and a radial number "l", which governs the field variation along the strut thickness [78] (see below for a detailed account on m and l mode index numbers).

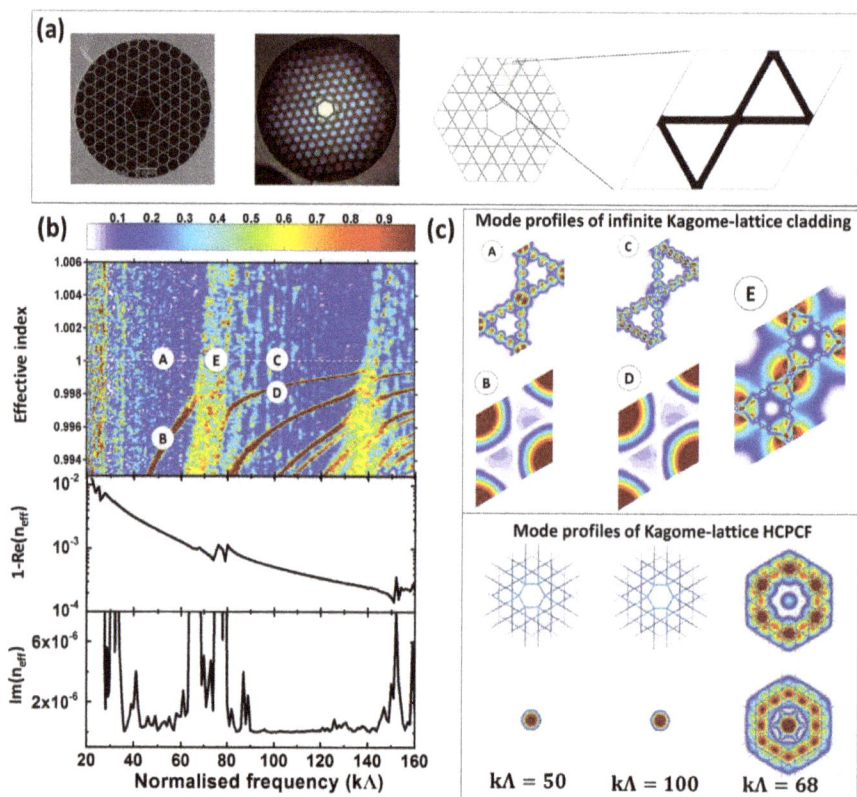

Figure 12. (a) Illustrations of Kagome lattice cladding HCPCF. (b) Calculated normalized DOPS diagram as a function of the real part of the effective index and normalized wavenumber. The two low DOPS intervals (blue colored regions) near the air light line n_{eff} = 1 (white dashed line) correspond to band (I-noted A) and band (III-noted C). Band (II-noted E) corresponds to a strong anticrossing between a lattice-hole mode (relatively flat red curve) and a particular type of mode associated with the network of connected silica struts. (c) Nature of the Kagome lattice cladding modes. The first two columns show the core mode and a cladding mode, respectively, for a frequency $k\Lambda$ = 50 in band I (top) and $k\Lambda$ = 100 in band III (bottom). The third column shows the mode of an infinite Kagomé lattice (Reprinted with permission from Reference [4], Science, 2007).

The first kind are highly oscillatory and highly localized photonic states (modes labeled A and C), presented in the DOPS by steep blue-colored dispersion curves. The field of these silica modes are associated with a very large m number. However, the mode labeled A shows no radial variation ($l = 1$, see below), whilst mode C exhibits one oscillation ($l = 2$). The second kind of modes is represented by lattice air modes, localized in the hexagon hole of the Kagome lattice (modes labeled B and D), and stand out with a weak interaction with the silica modes, illustrated by the relatively flat brown-colored dispersion. The third kind of modes are hybrid of silica and air modes (mode labeled E), and are localized within specific frequencies. The latter occur at normalized frequencies given by the identity $k\Lambda = j(\pi\Lambda/t)/\sqrt{n_g^2 - 1}$ ($j = 1, 2, 3, \dots$), and for the glass strut thickness considered in this work ($t = 0.044\,\Lambda$), the resonance condition for the hybridization corresponds to $k\Lambda = 68, 136$, etc. This resonance condition is often found in the literature in the following form $\lambda_j = (2t/j)\sqrt{n_g^2 - 1}$, where j is integer related to the radial number as we will see below [30]. At this wavelength, the air HE_{11}-like cladding lattice mode is transversely phase-matched with the silica lattice modes having the same propagation constant, leading to mode hybridization. When an air core is introduced within the photonic crystalline cladding, its HE_{11}-like mode experiences the same transverse matching with these cladding modes, thus leading to the enhanced leakage from the fiber core around the same wavelength values. This is clearly exposed in the bottom of Figure 12c, which shows cladding and core mode of 1-cell core Kagome HCPCF. At normalized frequencies away from the air–silica mode resonance (see the mode profiles at $k\Lambda = 50$ and $k\Lambda = 100$ in Figure 12c), the HE_{11} core mode coexists with silica cladding mode with no strong interaction. On the other hand, at $k\Lambda = 68$ the mode profile shows that of a hybrid mode between air and silica modes. Consequently, the resonant wavelength λ_j can be used to separate the Kagome DOPS near the air-line into multiple bands, which differ by the radial-number of their silica modes. A property that is also seen in the transmission of the Kagome HCPCF, and which further study when we consider below the tubular lattice.

It is worth mentioning that the Kagome lattice HCPCF core mode is leaky, and thus is a QBIC instead of proper BIC. To explore the possibility of a truly BIC-guided mode in HCPCF, Birks et al. [79] considered a cladding structure shown in Figure 13. This idealized Kagome-like structure consists of orthogonally superimposing two infinite 1D periodic arrays of thin glass membranes (index n_2) in air (index n_1) and having an intersection with the refractive index given by $n_3 = \sqrt{n_2^2 + 1}$ (see Figure 13A). The results show that the fiber supports a strictly localized core mode and has no leakage loss (Figure 13B) even though the mode has an effective index that lies within the continuum of radiation modes filling the infinite cladding (Figure 13C). Furthermore, IC optical guidance was also observed in solid-core PCF [39]. Finally, the results reported in Reference [4] provide the physical principles and the design tools for fabricating lower confinement loss fibers, coined as inhibited-coupling guiding fibers, which are detailed in the following section.

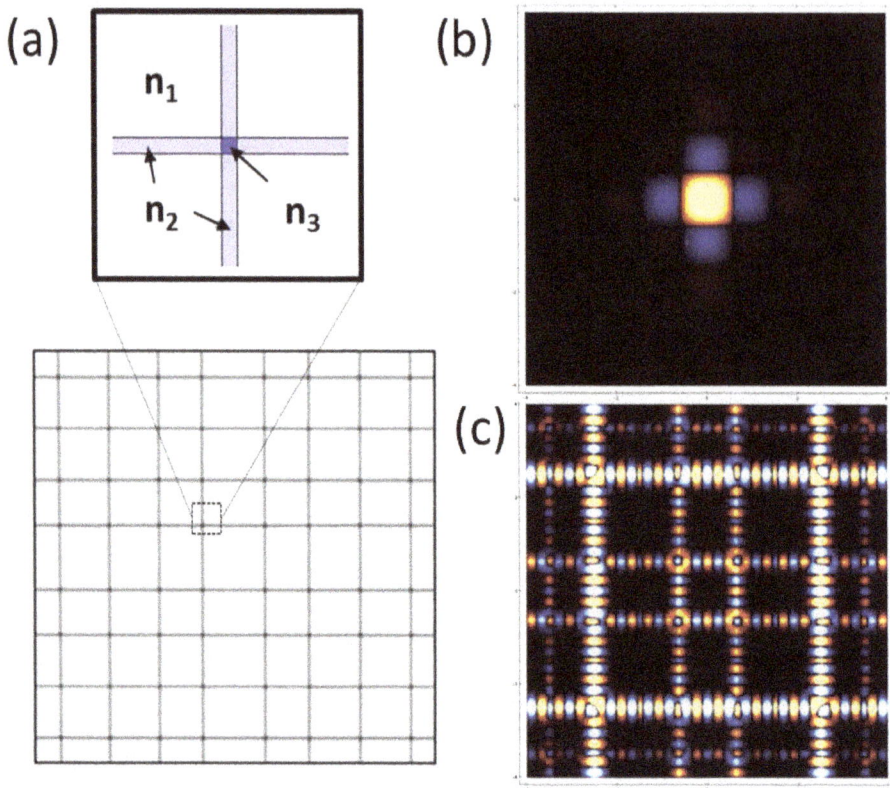

Figure 13. (**A**) Zoomed in crossing of glass webs separated by air of idealized Kagome structure. Index of the air, webs, and intersections are n_1, n_2, and n_3, respectively. (**B**) Density plots of a strictly-localized core mode (**C**). Continuum of radiation modes filling the infinite cladding (Reprinted with permission from Reference [79], IEEE, 2010).

4.3.2. Design Tools for Low-Loss IC-HCPCF

The work in Reference [4] shows that QBIC and BIC in optical fibers is an interesting and novel path in fiber photonics. The principles underlying the performance of IC-HCPCF are (1) a strong transverse phase mismatch and (2) a weak spatial overlap between the core mode and the cladding modes. This implies reduction in the overlap integral between the fields of the different modes that we want to suppress their mutual interaction. This overlap integral takes the form of the inner product $\langle \psi_{core} | \Delta n^2 | \psi_{clad} \rangle$ between the core and cladding mode transverse-fields. Denoting the latter by their scalar wave function, we can write the wave function of the two effective index matched modes as $\psi_{core}(r, \theta, z) = F_{core}(r, \theta) e^{i\phi_{core}(r,\theta)} e^{i\beta z}$ and $\psi_{clad}(r, \theta, z) = F_{clad}(r, \theta) e^{i\phi_{clad}(r,\theta)} e^{i\beta z}$. Here, the $F_{core}(r, \theta)$ and $F_{clad}(r, \theta)$ are real functions representing the field envelops of the core and cladding modes, and hence their spatial localization. The quantities $\phi_{core}(r, \theta)$ and $\phi_{clad}(r, \theta)$ represent their transverse phase, or how the field oscillates under the envelop formed by $F_{core}(r, \theta)$ or $F_{clad}(r, \theta)$. The term $e^{i\beta z}$ is associated with the common propagation constant β of the two modes.

Based on this premise, without the need to perform numerical calculation, even if the real function and the transverse phase terms are not necessarily separable we can state that inhibiting the interaction between the modes can be achieved by reducing the spatial overlap between the mode fields, i.e., by ensuring that $\iint F_{core}(r,\theta) \times F_{clad}(r,\theta) r d\theta dr \to 0$.

In the previous section, we have seen that operating in the large pitch regime (i.e., $\lambda \ll \Lambda$), the high-index material modes (i.e., glass mode in the HCPCF cladding) are highly localized with very little optical power overlapping with low index material (i.e., air is in the considered HCPCF cladding), and that are better presented as maximally localized Wannier functions than Bloch modes. Consequently, the "large pitch regime" is the first cladding design criterion [80].

The second means to reduce the overlap integral is to have the integral containing the transverse phase term, $\iint e^{i(\phi_{clad}(r,\theta) - \phi_{core}(r,\theta))} r d\theta dr$, vanishes. This infers having a strong symmetry incompatibility between the two modes. Expanding the transverse phase term in a similar manner as in Reference [30] (i.e., $e^{i(\phi(r,\theta))} = \sum_{m=0}^{\infty} A_m R(r) e^{im\theta}$, with m being the azimuthal index number, A_m a constant, and $R(r)$ a radial complex function), the integral can be written in the form $\int_{r_1}^{r_2} R^{clad}(r) R^{core*}(r) r \int_0^{2\pi} A_{m_{clad},m_{core}} e^{i(m_{clad} - m_{core})\theta} d\theta dr$. Consequently, the reduction of the overlap integral through "phase-mismatch engineering" implies a strong mismatch in the azimuthal-like index number between the cladding and core modes (i.e., $\Delta m = m_{clad} - m_{core} \gg 1$). Taking a core mode with no azimuthal number, such as HE_{11}, means that the coupling inhibition is achieved with a cladding mode with $m_{clad} \gg 1$.

This transverse phase mismatch-induced IC is summarized in a "toy model" shown in Figure 14. The figure considers the coupling between a glass rod HE_{11} mode and a semi-infinite glass membrane effective-index matched modes using a semivectorial and perturbative approach. Here we compare the coupling coefficients when the membrane mode transverse index number is increased. The transverse field profile calculations show that both overlap integrals C_{slab} and C_{rod} drop by four orders of magnitude when the slab mode transverse phase period (i.e., the equivalent to the azimuthal number) is increased from 4 to 19.

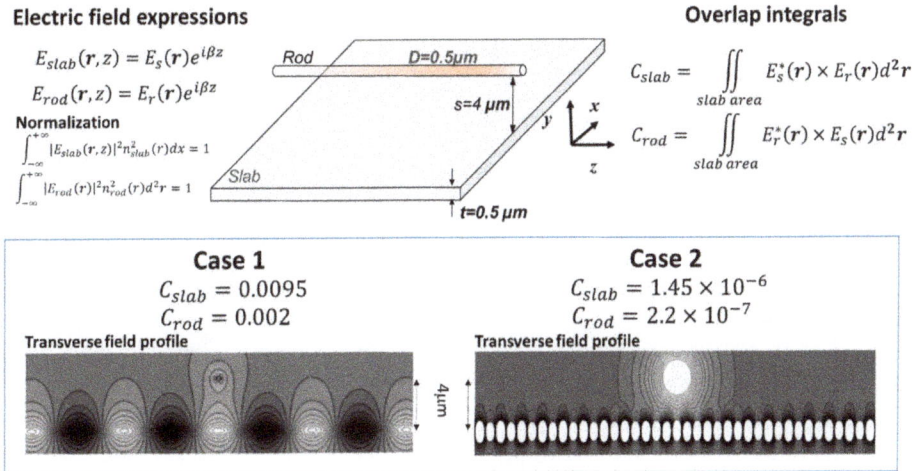

Figure 14. Top: Schematics of an optical coupling between a cylindrical dielectric rod and a semi-infinite dielectric slab. The rod has a diameter of 0.5 μm and a refractive index $n_{rod} = 1.45$, the slab thickness is set to 0.5 μm, and the distance between the rod and the slab to 4 μm. The operating wavelength $\lambda = 1.55$ μm for two cases. Case 1: the slab index is $n_{slab} = 1.1$. Case 2: $n_{slab} = 1.45$. Bottom: Calculated electric transverse field profile and the overlap integrals for the two cases.

In conclusion, IC optical guidance relies on a strong transverse phase mismatch between the core and cladding modes and on a very weak spatial overlap between them. These two criteria on the optical modes transform into the following geometrical criteria. IC fibers rely on a cladding with elongated and thin glass membranes with minimum corners and connecting nodes. It is noteworthy that the driving principles in designing IC fibers stem from structuring the modal spectrum of the cladding using chiefly quantum mechanics and condensed matter physics notions rather than those from guided optics. For example, tracing the path of the field and its reflection and/or interference is no longer required. Instead, we use the Hilbert space as the working space to identify the modal and spectral structure of the cladding lattice. We then use the P-TBM approach to pinpoint the waveguiding components of the lattice. Finally, we use the broad properties that underlie the dispersion, mode structure and symmetry of individual waveguiding lattice component to infer the most optimal cladding and core physical geometry.

These principles in designing low-loss IC-HCPCF were experimentally implemented by exploring large pitch Kagome lattice, but also square and honeycomb lattices HCPCF [71,81,82]. Here the confinement loss (CL) remained above 300 dB/km because of the unavoidable strut connections or nodes. A much more significant CL reduction in IC-HCPCF was achieved with the introduction of optimized core-contour. This contour took the form of a hypocycloid core-shape (also coined negative curvature core-contour), which consists of a set of alternating negative curvature cups with an inner radius R_{in} and an outer radius R_{out} and was presented by Benabid and coworkers in 2010 in a post-deadline paper of the CLEO conference [25], reported later in Reference [26]. This work was then quickly followed by several reports on negative curvature core-contour HCPCF [53,54]. The rationale behind the choice of such a core-contour profile is schematically illustrated in Figure 15, reproduced from [63]. Using the aforementioned design principles, the figure shows how the overlap integral between the core-mode and the highly oscillating (i.e., high azimuthal-like number m) silica core-surround mode (cladding-mode) is strongly reduced in an IC-HCPCF with a hypocycloid core-contour compared to a circular-like contour (see Figure 15a). The reduction in coupling between the HE_{11} core-mode and the cladding modes is reached via three avenues. Firstly, the spatial overlap of the core HE_{11} mode with the silica core-surround is reduced from one that spans over the whole circle perimeter in the case of the circular core shape to contour sections that are tangent with the six most inner cups in the case of hypocycloid core shape. Secondly, the HE_{11} mode-field diameter is related to the inner-core radius R_{in} of the hypocycloid contour. As such, because of the larger perimeter, L, of the hypocycloid compared to that of a circle of a radius R_{in}, the silica core-surround modes exhibit higher m, which is related to L by $m = n_{eff}\left(\frac{1}{\lambda}\right)\left[1 - \pi\left(\frac{t}{L}\right)\right]$ [26] (see Figure 15g). Hence, by virtue of a stronger transverse phase-mismatch when m is increased, we have stronger IC between the core mode and the cladding mode. Finally, the IC is enhanced by reducing the overlap between the core-mode with the connecting nodes, which support low azimuthal number modes. Here, the distance between connecting nodes and the circle associated with HE_{11} Mode Field Diameter (MFD) is increased when the contour is changed from a circular shape to a hypocycloid one.

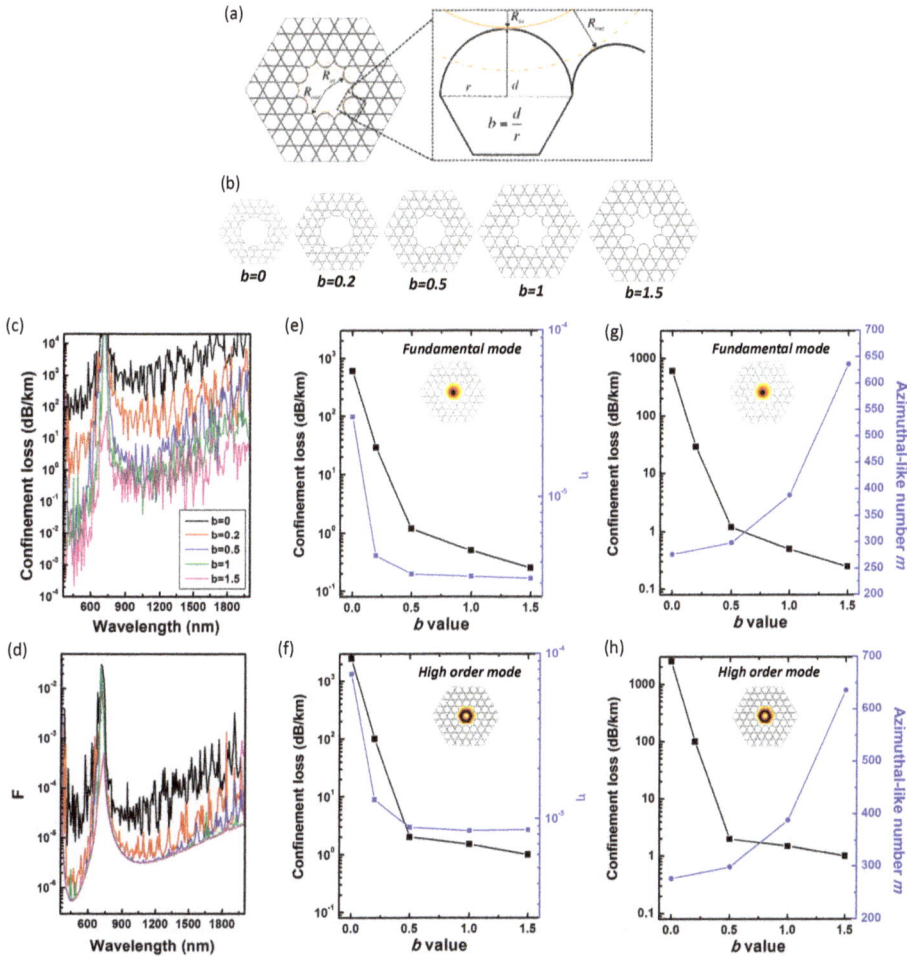

Figure 15. (**a,b**) Representation of the core contour curvature and Kagome IC HCPCF with different b. Spectra of calculated CL (**c**) and optical overlap with silica coefficient η (**d**) of Kagome HCPCF with the different b parameter (reprinted with permission from Reference [63], OSA, 2013). Evolution with b parameter of the CL and the optical overlap with silica coefficient η at 1030 nm for HE_{11} fundamental core mode (**e**) and for the lowest-loss higher order mode (**f**). Evolution with b parameter of the CL and m evolution at 1030 nm for HE_{11} fundamental core mode (**g**) and for the lowest-loss higher order mode (**h**).

The above principles were used in the recent development of low-loss IC-HCPCF, which we account below for both Kagome lattice and tubular-lattice cladding structure.

4.3.3. Hypocycloidal Core-Contour Kagome Lattice HCPCF

Following, the first demonstration of hypocycloidal Kagome IC-HCPCF, a comprehensive experimental and theoretical account was reported in References [63,83], showing the impact of the negative curvature and the cladding ring number on the CL. In particular, how strong CL reduction is achieved by optimizing the negative curvature. Figure 15 summarizes some of these findings by comparing Kagome IC-HCPCF with different core contours. Figure 15a defines the negative curvature by the parameter $b = d/r$, and Figure 15b shows the different fibers explored. Here, the silica strut thickness and the core inner-diameter were taken to be 350 nm and 60 μm, respectively.

Figure 15e shows respectively the evolution of the CL and the fundamental core mode HE_{11} optical overlap with the silica core contour, η, defined as $\eta = \iint_{S_{Si}} p_z \, dS / \iint_{S_\infty} p_z \, dS$, with p_z being the z-component of the mode field Poynting vector, and S_{si} and S_∞ are the cross section area for the silica and the full fiber, respectively. The confinement loss spectrum evolution with b clearly shows a drastic reduction when the core-contour evolves from a circular (i.e., b equal to 0) to elliptical cup-shape (i.e., $b > 1$). For a circular core-contour, which is representative of the initial Kagome HCPCF, the CL remains larger than 0.1 dB/m for the wavelength range of 400 to 2000 nm, in corroboration with the measured transmission loss of the previously fabricated Kagome fibers. For b larger than 0.5 the loss drops down to ~1 dB/km for the fundamental transmission band and down to ~0.01 dB/km for the 1st higher order band. These CL figures are comparable to those of PBG-HCPCF with finite cladding size. This is a seemingly remarkable situation when we recall the coexistence of the core mode with a cladding mode continuum. Conversely, the optical overlap η follows the same reduction trend with increasing b. The optical overlap drops by more than one order of magnitude from ~3 × 10^{-5} to down to ~10^{-6}. In addition of being correlated with the CL, the optical overlap with silica is a determinant factor in the application of IC-HCPCF in high-field optics (see the section on applications). Figure 15e,f shows the evolution of CL, η with b at a wavelength fixed at 1030 nm for HE_{11} and one of the lowest loss higher order mode. The curves clearly display the effect of increasing b on the reduction of the spatial overlap between the core mode and the cladding via the decrease of η. Similarly, Figure 15g,h shows the evolution of CL and m with b, which clearly shows the correlation between the increase of m (i.e., increase of the transverse phase mismatch) and the reduction in CL on one hand and between the increase of m and the increase in b, in consistency with the IC model predictions. We note that for $b > 0.5$, η saturates, whilst the CL continues to decrease with increasing b. Particularly, increasing b above 0.5 is associated with both CL further decrease, and further increase in m. Thus indicating the CL reduction is dominated by the transverse-mismatch when b increases from 0.5 to 2. This work in designing IC-HCPCF showed that CL scales with t, b, and λ as $\alpha_{CL} \propto t \, \lambda^{4.3} b^{-2}$, and the minimum loss for higher order bands to at wavelength $\lambda_{min} = \left(2t/(l - \frac{1}{2})\right)\sqrt{n_g^2 - 1}$, with l being the order of the transmission windows (i.e., the radial number of the silica lattice modes) [65].

Today, state-of-the-art Kagome IC-HCPCF is represented in Figure 16. The first fiber (Figure 16a) has a $b = 0.95$, and exhibits a minimum loss of ~8 dB/km at its 1st order transmission band, tailored to operate near 1 μm. The second fiber, designed to have a broadband fundamental band, exhibits a single window spanning down to 700 nm with a loss below 100 dB/km over one octave. Here, the CL is set by the limited achievable b at a thickness of 300 nm, which was found to be ~0.45. Such a state-of-the-art was one part of a series of transmission loss records at different wavelength ranges, which occurred since the first experimental demonstration of negative curvature HCPCF in 2010. Within eight years the loss figure dropped from ~100 dB/km to 8 dB/km, which represents a "drop rate" of more than 10 dB/km per year. This trend is bound to continue following two emerging trends. The first trend impacts the loss in longer wavelengths (>1000 nm), and is a continuation of last decade effort, which consists of reducing the CL by exploring alternative fiber cladding and core designs with lower confinement [56–60]. The second one consists of lowering the loss in the short wavelength range (<800–1000 nm), and which solution requires reducing the surface-scattering loss (SSL), which is now, like with the PBG HCPCF, the limiting factor in IC-HCPCF.

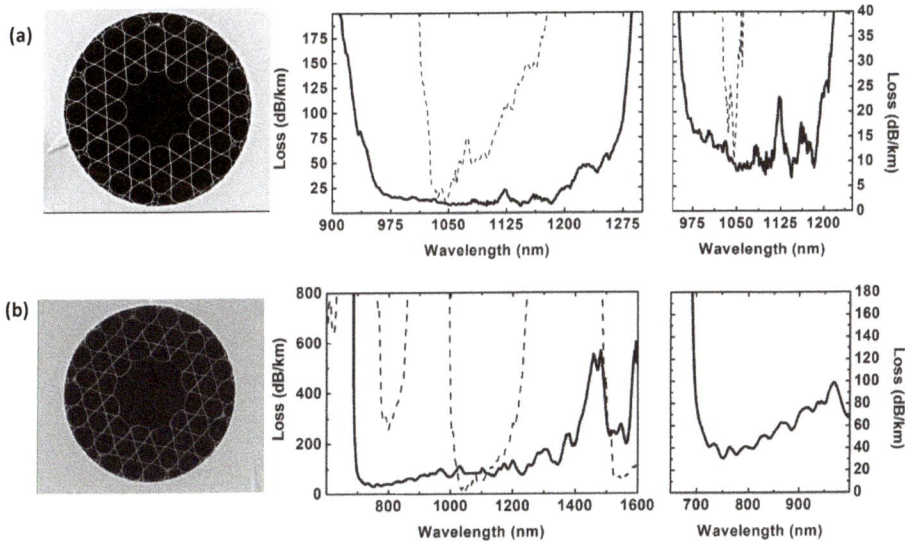

Figure 16. (**a**,**b**) Experimental loss spectrum of the loss record Kagome IC-HCPCF at an approximately 1030 nm spectral range and broadband-guiding Kagome IC-HCPCF (reprinted with permission from Reference [65], OSA, 2018).

Indeed, in addition to the CL, the modal propagation is attenuated by two additional sources: the bend loss and the scattering loss.

The bend loss is affected by both core-size and the cladding structure as illustrated in Figure 17. Figure 17a are results reported by Maurel et al. [65], which show experimentally measured and numerically calculated HE_{11} core-mode confinement loss evolution with bend radius at $\lambda = 1064$ nm for a hypocycloid-core Kagome with $b = 0.95$, core inner radius of 25.5 μm, and silica strut thickness of 800 nm. The numerical values give a bend loss of below 5 dB/km for $R_b > 20$ cm, of ~47 dB/km for $R_b = 10$ cm, and almost 400 dB/km for $R_b \leq 2.5$ cm. These results show excellent agreement with a R_b^{-2} fit, in consistency with the findings of Marcatili and Schmeltzer for the dielectric capillary tubes [50].

Figure 17b shows loss spectrum evolution of the fiber HE_{11} core mode 1st high-order transmission window for the different R_b. In addition to the confinement loss increase with the decreasing R_b, the spectrum shows bend induced resonant loss at several wavelengths, in agreement with the findings by Couny et al. (see Figure S3 and its associated text [4]). These loss resonances occur when the bend-induced effective index-variation induces transverse phase matching between the core mode and the cladding modes. As pointed out in Reference [4], because the Kagome silica lattice modes are associated with a very high effective azimuthal number m, a perturbation which may induce coupling between the core mode and such cladding modes necessarily requires a large Δm. On the other hand, a fiber bend is primarily associated with a change in m of just 1. Consequently, the observed bend-induced resonant coupling is caused by either cladding air modes or by cladding silica modes with low azimuthal number. The latter are spectrally localized near the red edge of the transmission window. Indeed, the red edge of each of the higher-order transmission windows in an IC-HCPCF corresponds to a unity-increment in the radial number of the cladding silica modes. In turn, over a frequency range close to the band red edge, these additional modes exhibit low azimuthal number modes. They manifest in Figure 17b as several narrow absorption peaks over the wavelength range of 1350 nm and 1500 nm.

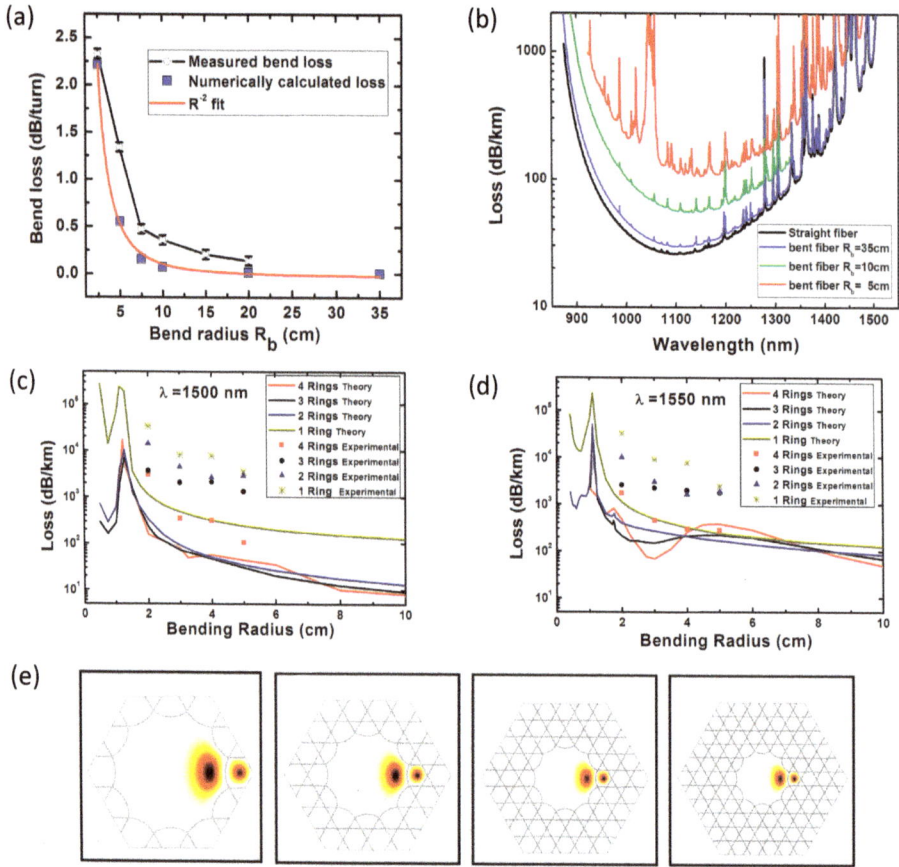

Figure 17. Calculated and measured bend loss evolution at $\lambda = 1064$ nm with bend radius (**a**), and (**b**) loss spectra for different bend radii of HE_{11} core mode of a hypocycloid-core Kagome HCPCF (reprinted with permission from Reference [83], OSA, 2013). Calculated and measured loss evolution with bend radius for a hypocycloid-core Kagome HCPCF with $b = 0.3$, $t \approx 440$ nm, and R_{in} between 23.5 and 29 μm, and having different cladding ring number at 1500 nm (**c**) and 1550 nm (**d**). (**e**) The calculated intensity profile of the mode when the fiber is under a bend of a radius of 1.1 cm for Kagome IC HCPCF with different cladding ring number (reprinted with permission from Reference [83], OSA, 2013).

The wider peak shown in the red curve of Figure 17b relates to a coupling with an air cladding-mode. This was previously shown by Alharbi et al. [83] in their experimental and numerical study of the cladding effect on confinement loss. Figure 17c,e shows, both experimentally and theoretically, the sensitivity to bend for a Kagome HCPCF with different cladding ring number. Figure 17c,d shows the evolution of loss with bends at 1500 and 1550 nm wavelengths, respectively. The latter are representative wavelengths of the fundamental transmission window that are further from those corresponding low azimuthal cladding modes. At both wavelengths, the loss evolution with the radius shows a peak at approximately $R_b = 1.1$ cm. Figure 17d shows the intensity profile of the fiber mode at this bend value and shows a coupling between the core mode and a mode residing in one the cladding holes. Outside this resonant bend radius range, the bend loss decreases with R_b following R_b^{-2}. Finally, Figure 17c,d shows the effect of cladding ring number on the CL for different bends. The results show strong dependence on the wavelength. For 1500 nm, a second ring drops the CL by more than one order of magnitude. This can be explained by the reduction in the leakage through tunneling to outside the cladding. However, this argument is not necessarily true for all IC-HCPCF and strongly depends on the cladding/core structure and wavelength. This is illustrated by the impact of adding 3rd or 4th ring on the CL. Here, we observe that the latter additional rings have little effect. Furthermore, at 1550 nm, the effect of adding additional rings is marginal and sometimes disadvantageous as the CL is increased for a given bend radii. This illustrates the key difference of IC guidance mechanism when compared to PBG guidance (including TIR guidance), where the reduction of the core more leakage cannot simply be achieved by increasing the cladding thickness. Instead, both the tunneling to the outside the fiber and coupling to cladding modes must be taken into account. Within this framework, we argue that adding an extra cladding ring IC-HCPCF increases the cladding mode number, and hence the residual coupling between the core and the cladding modes. In the case of Kagome lattice, adding more than two rings implies increasing the connecting struts to a level that weakens the IC compared to the 2-ring cladding design.

Finally, the third source of propagation loss is the surface roughness scattering (SSL). This is caused by the surface roughness of the glass web of the HCPCF, and has first been reported by Roberts et al. in the case of PBG fibers [49,84]. Indeed, SSL was identified as the limiting factor in PBG-HCPCF because of the large optical overlap between the core mode and the core silica surround. The surface roughness results from the frozen capillary waves that are present during drawing process. This type of propagation loss is expressed as

$$\alpha_{SSL}(\lambda) = \varsigma \cdot \eta(\lambda) \cdot \lambda^{-3} \tag{3}$$

where σ is a constant related to the surface roughness root-mean-square height, and $\eta(\lambda)$ is the core mode overlap with the core-contour [49,56]. The typical value of $\eta(\lambda)$ is ~1% for 7-cell PBG HCPCF and ~0.1% for a 19-cell PBG HCPCF.

For IC HCPCFs, the SSL was often ignored because of the very small values of $\eta(\lambda)$, which are typical in the range of 10^{-4} and 10^{-6} (see, e.g., Figure 15d), and because of the higher CL loss that characterized IC-HCPCF before the introduction of hypocycloid core-contour. However, recently measured propagation loss of less than 10 dB/km are common with IC-HCPCF. For example, one the fibers shown in Figure 16 shows a transmission loss of ~8.5 dB/km at a wavelength range near 1030 nm [65]. In this work, it was shown that the CL is no longer the limiting factor for IC-HCPCF at wavelength shorter than 1 μm, in consistency with the results reported in Reference [30] for tubular amorphous lattice IC-HCPCF (see following section). In fact, for wavelengths shorter than 1 μm, all the measured loss spectra of the different reported Kagome IC-HCPCF show increase with decreasing wavelength, which is contrary to the predicted CL. These results indicate that the surface roughness induced scattering loss is the dominant factor. Hence, any future improvement in IC-HCPCF transmission performance in these short wavelength ranges implies either a reduction in the surface roughness or further reduction in the optical overlap η.

4.3.4. Hypocycloid Core-Contour and Nodeless Tubular Lattice IC HCPCF

One of the cladding structures that followed the seminal introduction of hypocycloid core-contour in 2010 is a tubular cladding, which has shown a lot of promise and interest. The fiber form of this dielectric structure exhibits a hypocycloid core-contour, and consists of an amorphous lattice of isolated tubes arranged to form a circular layer around a hollow-core. An example of this single-ring tubular lattice HCPCF (SR-TL-HCPCF) is shown in Figure 18a, where the fiber cladding consists of a single ring made with eight nontouching tubes. What is particularly significant with such a fiber is the absence of connecting nodes at the cladding region surrounding the hollow-core, which are the primary source of coupling between the core mode and cladding modes in IC-HCPCF because their cladding low azimuthal number modes.

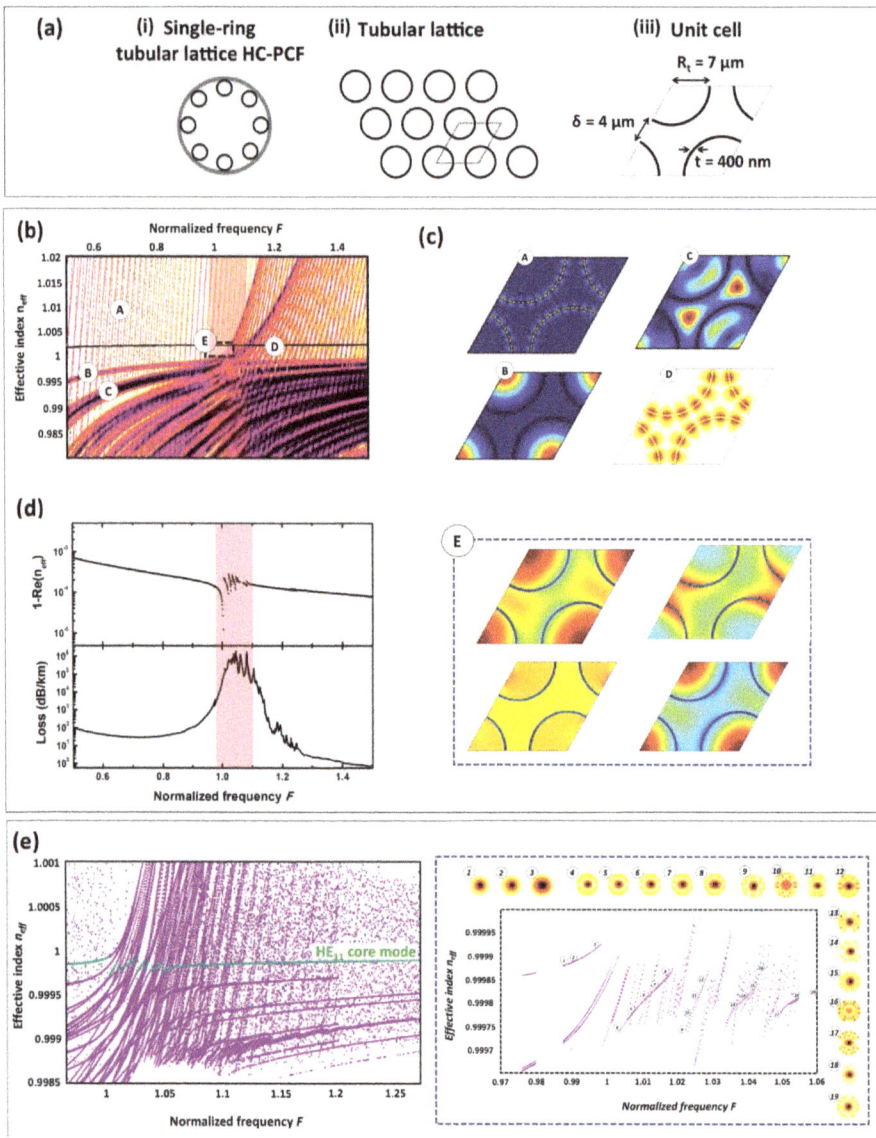

Figure 18. (**a**) Schematic of HCPCF with a single-ring tubular amorphous lattice (**i**), tubular lattice in a triangular arrangement (**ii**), and the details of its unit cell (**iii**). δ is the intertube gap distance, Rt is the tube radius, and t is the tube ring thickness (reprinted with permission from Reference [30], OSA, 2017). (**b**) DOPS of an infinite cladding of a triangular arrangement of tubes (reprinted with permission from Reference [30], OSA, 2017). (**c**) Modal spectrum near the air-line formed by silica ring modes ((A) and (D)), air modes (B), and silica–air hybrid modes (C–E). (**d**) Effective index and loss spectra of fundamental mode of a tubular lattice hollow-core defect. (**e**) Eight tubes SR-TL-HCPCF modal spectrum. Zoom in of the modal spectrum over the normalized frequency range of 0.97 to 1.06 (*rhs*).

As we have seen in a previous section, when we operate in the large pitch region, the modal spectrum is not very sensitive to the pitch. Consequently, as shown in Reference [30], the arrangement of the tubes has little impact on the modal spectrum of the resulting lattice. Figure 18b shows the resulting DOPS of an infinite cladding of a triangular arrangement of tubes (see middle of Figure 18a for the lattice and the *rhs* of Figure 18a for the lattice unit cell). Here, the normalized frequency is set to be $F = (2t/\lambda)\sqrt{n_g^2 - 1}$, and spanning from 0.5 to 1.5. The DOPS (see Figure 18b) and the lattice unit cell mode profiles (see Figure 18c) show the same essential features of the Kagome lattice. The tubular-lattice DOPS shows no bandgap and is populated with a continuum of photonic states. Near the air-line, the modal spectrum is formed by lattice modes consisting of silica ring modes, air modes, and silica–air hybrid modes (see top of Figure 18c). For the frequency range away from the anticrossing region near $F = 1$, we note the high azimuthal oscillations of the silica tube modes observed with the Kagome lattice (see Figure 18c mode profiles (A) and (D)). Figure 18c shows also the lattice air-modes (unit cell mode profiles (C) and (B)), and lattice hybrid-modes that populate the DOPS region near $F = 1$ and near $n_{eff} = 1$ (unit cell mode profiles in the box (E)).

Figure 18d shows the effective-index and loss spectra of the fundamental mode when a hollow-core defect is introduced into the tubular lattice. Similarly to Kagome HCPCF, the SR-TL-HCPCF shows low loss transmission bands separated by high loss bands. The high loss band corresponds to the spectral location of the hybrid modes, highlighted in Figure 18b DOPS by the red-colored shaded box, and occurs at the vicinity of $F = j$, $(j = 1, 2, 3, \ldots)$, which is simply the air–silica mode resonance condition $k\Lambda = j(\pi\Lambda/t)/\sqrt{n_g^2 - 1}$ $(j = 1, 2, 3, \ldots)$, mentioned above. A close-up view of the modal spectrum of the eight tubes SR-TL-HCPCF near $F = 1$, and $n_{eff} = 1$ is shown in Figure 18e. Here, the plots contain both the defect core air-mode (horizontal curves in the effective range between 0.9998 and 0.99985), and the cladding modes. The *rhs* is a further zoom-in of the modal spectrum over the normalized frequency range of 0.97 to 1.06. The plot shows a large number of anticrossings and dispersion branches corresponding to resonant interaction of the tube air HE_{11} mode with different silica modes. The intricate modal structure of this high loss region clearly indicates that the resonance, $F = j$, which is extensively used in ARROW model (see section below), does not fully capture the dynamics of such hybridization between the air modes and the silica modes.

Below, we explore the fact that unlike with the Kagome unit cell, the isolated silica tubes modes have identifiable profiles and even analytical expressions [85] to use it as an educational platform for IC guidance mechanism and to draw its salient properties, which were cumbersome to draw with the Kagome lattice. In particular, we can extract a transverse phase relationship between the silica mode and the air mode when they are longitudinally phase-matched (i.e., having the same n_{eff}), and thus identify the modes behind the high loss band.

The silica tube modes are classified as HE_{ml} (i.e., electric field direction is azimuthal) or EH_{ml} (i.e., electric field direction is radial) [85]. In the work of Debord et al. [30], it was shown that their effective index was given by approximating the tube to a slab. Within this approximation, the propagation constant β of the silica tube modes can be written as

$$\beta_{ml}{}^2 = (n_g k)^2 - \beta_\perp^2 = (n_g k)^2 - \left(\beta_\theta^2 + \beta_r^2\right), \tag{4}$$

with

$$\begin{cases} \beta_\theta = m\dfrac{\pi}{2\pi R_t} \\ \beta_r = (l-1)\dfrac{\pi}{t} \\ k = \dfrac{2\pi}{\lambda}. \end{cases} \tag{5}$$

Here, $\beta_\perp$ is the transverse component of the silica mode wavevector. In cylindrical coordinates, the transverse wavevector can be composed of an azimuthal component β_θ and a radial component β_r. Thus, m and l are the index number of the azimuthal and radial variation of the glass mode, respectively.

On the other hand, the air mode of the tube can be approximated to that of a capillary given by Marcatili and Schmeltzer [50]. Within this approximation, a HE/EH air mode has a propagation constant given by

$$\beta_{\mu\lambda}{}^2 = k^2 \left(1 - \left(\frac{u_{\mu\lambda}}{k\,R_t} \right)^2 \right) \tag{6}$$

Here, $u_{\mu\lambda}$ is the λth zero of the Bessel function $J_{\mu-1}(u_{\mu\lambda})$. Under the condition of longitudinal-phase matching (i.e., $\beta_{\mu\lambda} = \beta_{ml}$), the tubular-lattice glass mode azimuthal and radial indices (i.e., m and l), the air mode azimuthal and radial indices (i.e., μ and λ), R_t, t are related through the following identity, which provides the spectral location of the intersection between the dispersion curves of glass modes and the air modes:

$$\left(\frac{m}{2\,R_t} \right)^2 + \left((l-1)\frac{\pi}{t} \right)^2 = k^2 \left(n_g^2 - 1 \right) + \left(\frac{u_{\mu\lambda}}{R_t} \right)^2. \tag{7}$$

Furthermore, in this case, the condition of longitudinal phase-matching between the HE/EH air modes and the silica tube $\mathrm{HE}_{ml}/\mathrm{EH}_{ml}$ modes' full phase-matching condition (i.e., the modes have also the same $\beta_\perp$) is impossible because of the difference in index of the respective mode media. However, coupling between the air mode and the silica mode can occur through directional coupling whose strength is set by the couple waveguide parameter $X = \Delta\beta/2\kappa$ [86]. Where $\Delta\beta$ is the longitudinal mismatch and κ is the power coupling coefficient between the two modes.

According to IC, we have seen that such a coupling can be reduced or strengthen depending on the mismatch in the azimuthal number of the modes. If we limit the case of the air mode HE_{11} on one hand, and consider silica modes with no azimuthal variation or very low one (i.e., $\frac{m\lambda}{R_t} \ll 1$), Equation (6) can be written for the case of tube radius much larger than the wavelength as [30]

$$\beta_r = (l-1)\frac{\pi}{t} = \sqrt{k^2 \left(n_g^2 - 1 \right)} \tag{8}$$

The above expression can be written by setting $l - 1 \equiv n$ in a form that is commonly used in ARROW literature (see following section), such as

$$\lambda_n = (2t/(n))\sqrt{n_g^2 - 1} \tag{9}$$

According to the above, the IC approach shows that the ARROW formula from Duguay et al. [51] is an asymptotic limit when the azimuthal variation of the cladding mode is ignored, and the integer n is related to the radial number of the cladding silica modes. That being said, it is worthwhile that in ARROW approach, the cladding modal spectrum is not considered at all.

Below, we further examine the modal spectrum of the tubular lattice near $F = 1$, as it represents the spectral range where we can find both air and silica modes with inhibited coupling, and those with a strong hybridization.

Figure 19 summarizes the modal interaction dynamics near $F = 1$. Figure 19a shows the close-up of the DOPS and representative cladding lattice modes for $F < 1$ on the *lfh* panel and for $F \geq 1$ on the *rhs* panel. In similar fashion as with Figures 12 and 18, we retrieve the silica and air modes along with their hybrid forms. However, in this DOPS diagram, we color-coded the dispersion of the silica modes' radial number.

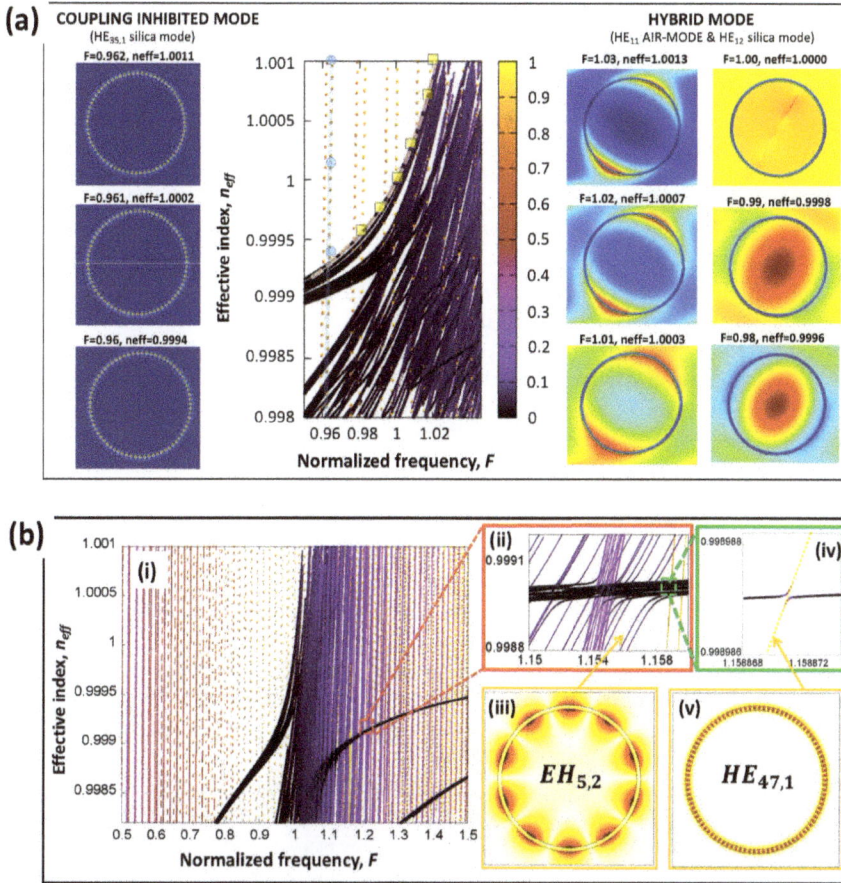

Figure 19. (a) Modal spectrum and lattice mode profiles near $F = 1$. Intensity profiles of a representative silica mode with high azimuthal number and radial number $l = 1$ ($EH_{35,1}$) and its evolution when the mode approaches HE_{11} air mode (*lhs*). DOPS diagram (middle). Intensity profiles of a representative silica–air hybrid mode (HE_{11} air mode and HE_{12} silica mode) and its evolution when the mode approaches HE_{11} air mode (*rhs*). The blue filled circles and the yellow filled squares in the DOPS diagram represent the $F - n_{eff}$ locations of the silica mode on the left and the hybrid mode on the right respectively. (b) Zoom out view of the DOPS diagram near $F = 1$ (i), Close-up near an anticrossing between the HE_{11} air mode and $l = 2$ silica mode ($EH_{5,2}$) (ii) and the intensity profile of the associated mode (iii). Close-up near an ultraweak anticrossing crossing between the HE_{11} air mode and $l = 1$ high azimuthal number silica mode ($HE_{47,1}$) (iv), and the intensity profile of the associated mode (v).

The lattice silica modes with radial number $l = 1$ (i.e., the $EH/HE_{m,1}$) are represented by orange dotted steep lines, and we distinguish them from the $EH/HE_{m,2}$ silica modes with radial number $l = 2$, which are represented by dark-blue colored thick steep lines. As it has been seen with the Kagome lattice [4], $EH/HE_{m,2}$ lattice silica modes populate transmission bands with $l \geq 2$, whilst $EH/HE_{m,1}$ lattice silica modes populate all the transmission bands. On the *lhs* of the DOPS diagram, we show the mode intensity profiles of the $EH_{35,1}$ lattice silica modes at three different points of its dispersion curve (blue line crossed by blue filled circles). We can see that this mode interacts little with the air modes, illustrated by both its highly silica-localized intensity, and its dispersion curve intersecting the HE_{11} air mode dispersion curve with no observable anticrossing (within the plot n_{eff} and F resolution). Based on the above and previous works [4,30], it is obvious to correlate this coupling inhibition between the air mode and the silica mode to the fast azimuthal oscillations.

On the other hand, the first lattice silica mode with radial number 2, $HE_{1,2}$, strongly interacts with the HE_{11} air mode. This is shown by the light-pink curve and the yellow square symbols; which are superimposed on the upper branch of the strong anticrossing between the two modes. The *rhs* of the DOPS diagram shows representative mode profiles along this curve. We can readily see the modal evolution from a $HE_{1,2}$, silica-dominated lattice mode for $n_{eff} > 1$ to HE_{11} air-dominated lattice mode when $n_{eff} < 1$. Additional anticrossings are also observed between low m number silica-$EH/HE_{m,2}$, and air-HE_{11} (see DOPS diagram).

Within this spectral range, the strength of these anticrossings decreases with increasing frequency. Figure 19b highlights the overall picture of the anticrossing between the lattice silica modes with the lattice air mode HE_{11} (Figure 19(bi)) and a close-up view of a DOPS region located on the lower n_{eff} branch of the anticrossing (Figure 19(bii)–(bv)). Within a small range of the normalized frequency $F \sim 1.15$–1.158, Figure 19(bii) displays the different anticrossing between the air-HE_{11} lattice mode and the low azimuthal number lattice silica modes $EH/HE_{m,2}$ (blue lines) and the high azimuthal number lattice silica modes $EH/HE_{m,2}$ (brown lines). An example of such low azimuthal variation the $EH/HE_{m,2}$ mode is shown in Figure 19(biii). This mode exhibits an azimuthal number of $m = 5$, and is associated with relatively large anticrossing at $F \sim 1.159$. On the other hand, at the vicinity of this frequency, the figure shows a much smaller anticrossing (indicating a strong coupling inhibition), which is zoomed in in Figure 19(biv), with its corresponding mode profile (Figure 19(bv)). The latter corresponds to the $HE_{47,1}$ silica lattice mode, and as expected it has a radial number $l = 1$, and a large azimuthal variation, with $m = 47$.

We conclude this analysis by noting that all the aforementioned features and the modal coupling dynamics are also valid when we consider a hollow core defect to form a tubular amorphous lattice HCPCF. In Equations (6) and (7) we only need to replace the lattice tube radius R_t with the inner radius R_c of the HCPCF core. This is illustrated in Figure 18e, which shows the modal spectrum for a SR-TL HCPCF.

Following the introduction of negative curvature in 2010 [25], the first fabricated fiber of the above described SR-TL HCPCF was reported in 2011 by Pryamikov et al. [53]. Here, the fiber was made with touching tubes whose thickness was larger than 1 μm. The fiber operated at 3.5 μm wavelength and longer, with transmission loss figure of 34 dB/km. Also, SR-TL HCPCF was analyzed theoretically for THz-wave guidance [55]. Since then, this type of fiber has been explored in various designs by many groups by changing the tube number and thickness, the core size, and even the jacket tube shape for fabrication convenience [87–89].

Figure 20 summarizes the current state-of-the-art (SOA) of SR-TL HCPCF in terms of loss and bandwidth of the fundamental transmission band. The left panel shows the electronic micrographs and the loss spectra of the two SR-TL HCPCF representing this SOA. Using the IC model described above as a design tool, the two fabricated fibers resulted from an optimization study in its architecture design (i.e., tube number, radius, and thickness and intertube gap) and its fabrication limit (i.e., minimum thickness and cladding structure shape) [30]. One particular finding of this work is the azimuthal distribution of the power leakage of SR-TL HCPCF and its evolution with the intertube gap δ (Figure 20e). The Figure 20e shows the evolution with δ of the Poynting vector radial component at minimum-loss wavelengths of the fundamental and 1st order transmission bands. The results showed that there is an optimum value of δ to achieve a low CL. Also, it was shown that when δ is smaller than a critical value (here ~2 μm), the power leakage "channel" is the connecting nodes between the tubes and the outer silica jacket. Conversely, when the gap size is larger than 6 μm, the loss channel changes to that of the intertubes gaps (see Figure 20e).

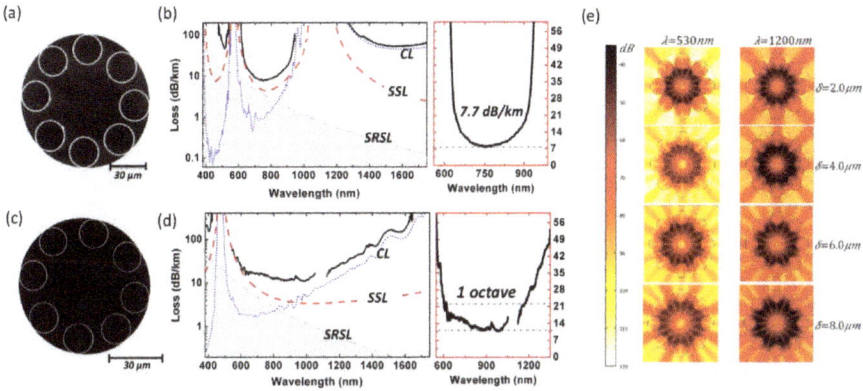

Figure 20. (**a,c**) SEM pictures of the fabricated 7.7 dB/km loss and broadband-guiding SR-TL HCPCFs, respectively. (**b,d**) Measured loss spectra of the two fibers (black curve) compared to theoretical SSL and CL evolutions. (**e**) Theoretical transverse distribution of radial component of Poynting vector for SR-TL fiber with different δ values from 2 to 8 μm (Reprinted with permission from Reference [30], OSA, 2017).

The fibers have eight tubes and an intertube gap of 2.7 μm. The first one demonstrated record transmission loss for an IC HCPCF with a figure of 7.7 dB/km at 750 nm (see Figure 20a for the SEM picture of the fabricated fiber and Figure 20b for the measured loss spectrum). Guidance in the UV spectral range down to 200 nm has also been demonstrated. The second fiber presents a thinner struts structure, and hence a broader transmission fundamental window with low loss guidance over one octave between 600 to 1200 nm. The losses in this band are in the range of 10 to 20 dB/km (see Figure 20c for the SEM picture of the fabricated fiber and Figure 20d for the measured loss spectrum). These fibers have shown close to single-mode propagation with an extinction ratio between the fundamental core mode and the first high order modes of ~20 dB. Finally, and similarly to Kagome HCPCF, the results show that the loss is limited by SSL for the short wavelength range and by CL for wavelength longer than 1 μm. Figure 20b,d shows the contribution to the measured fiber loss of the CL (dotted blue curves) and SSL (dashed red curves) based on a fitted peak-to-peak fluctuation height of 1.5 nm. For both fibers, we can observe that for wavelength >1 μm, the CL dominates the propagation loss, whilst for shorter wavelength, the SSL is the principal source of the transmission loss.

It is noteworthy that SR-TL HCPCF is still the subject of ongoing and active research, with progress being reported continuously. Among these recent results, we have the report on a 9-tube-based SR-TL HCPCF operating in the "green" spectral range [90]. The fiber demonstrated an impressive loss value of 13.8 dB/km at 539 nm [90], which is comparable to the Rayleigh scattering limited silica transmission loss of 10 dB/km at this wavelength. Also, this fiber is the first reported IC HCPCF to be drawn over a distance longer than 1 km with a variation of the outer diameter less than ±0.2%. Also, more recently, guidance in the UV with this fiber design has been targeted by scaling down the strut thickness of the tubes and the core size. Loss figure of 130 dB/km at 300 nm has been reported [91]. Finally, the modified tubular lattice IC-HCPCF, optimized for the telecom wavelength, range was reported to have a transmission loss as low as 2 dB/km at ~1500 nm [58].

4.4. Difference between ARROW, PBG and IC

The above sections showed how the optical fiber guidance mechanism is governed by the cladding structure and its modal spectrum. Here, we conclude this topic by touching on the guidance by antiresonance (i.e., ARROW). This is motivated by the recent surge of the ARROW terminology, which was used invariably to describe PBG-guiding [73,92] and IC-guiding fibers, thus creating confusion within the readership on how light is guided within microstructured optical fibers and on how to differentiate between them. Here, we give the reader some distinctive properties that could be useful in distinguishing between PBG, IC and ARROW.

The idea of antiresonant reflection was originally used in interferometry to distinguish the working conditions where a Fabry–Perot reflectivity is at its maximum [76]. In 1986, Duguay et al. extended this notion to optical waveguide by proposing a planar waveguide with a low-index core surrounded by higher index dielectric thin layers, and by studying light propagation in the core at glancing angles [51]. Here, the optical confinement is ensured by the reflectivity enhancement due to the high-index layer thickness reduction to the micrometer scale. Following Duguay et al., dielectric ring waveguide with micrometric thickness layers was studied [76,93,94]. The ARROW induced transmission improvement of these dielectric ring waveguides over a thick dielectric tubes [50] was found to be proportional to the ratio of the ring radius over the wavelength (see Equation (2)).

Figure 21 captures the main differences and similarities between a glass capillary, a glass ring and two negative curvature contour IC-HCPCFs by plotting the dispersion and CL spectra. On one hand, Figure 21a compares the glass ring to a glass capillary and, on the other hand, Figure 21b compares the glass ring to a Kagome and Tubular IC-HCPCF.

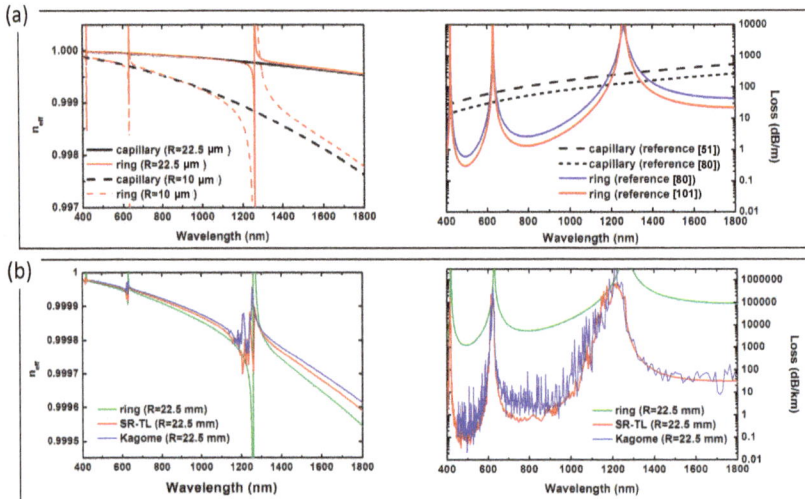

Figure 21. (a) *lhs*: Effective index spectrum glass capillary (black curves) and 600 nm thick ring (red curves) for the case of a radius of 22.5 μm (solid lines) and 10 μm (solid lines). *rhs*: Loss spectrum of 22.5 μm radius glass capillary calculated using the expression of Reference [50] (black dashed-dotted curve). Loss spectrum of 22.5 μm radius and 600 nm thick glass ring using the expression from Reference [76]. (b) Effective index (*lhs*) and loss spectra (*rhs*) of ring, SR-TL, and Kagome fiber with similar core diameter of 22.5 μm and struts thickness of 600 nm.

Figure 21a compares the dispersion and the loss between the glass capillary and the ring for different parameters and using different expressions. The left panel of Figure 21a shows the n_{eff} spectrum of the capillary and the ring for two diameters (black curves). The ring dispersion was taken from Zeisberger et al. [95] and reproduced below.

$$n_{eff}^{(ring)}(\lambda) \approx n_{eff}^{(cap)}(\lambda) - (u_{11}{}^2\lambda^3)((2\pi)^3 R_{in}{}^3)^{-1} ctg\left(\frac{2\pi}{\lambda}t\sqrt{n_g^2 - n_{eff}^2}\right)\left(\sqrt{n_g^2 - 1}\right)^{-1}\left(\left(n_g^2 + 1\right)/2\right) \quad (10)$$

Here, $n_{eff}^{(cap)}(\lambda)$ is the effective index of a dielectric capillary (see Equation (5)) and t and R_{in} are the silica ring thickness and radius, respectively. The results show that for wavelengths which are away from the resonant wavelength $\lambda_n = (2t/(n))\sqrt{n_g^2 - 1}$ (see above), the dispersion of the ring is very close to that of the capillary. Also, the approximation of the ring dispersion to that of the capillary improves with larger core diameters. On the other hand, at the vicinity of the resonant wavelengths, the ring dispersion curve shows an anticrossing which strength (i.e., spectral width) increases with smaller core diameter, as illustrated in the difference between waveguides of 22.5 μm radius ring (curve red) and 10 μm radius ring.

The right panel of Figure 21a shows the transmission loss of the HE_{11} core mode in 45 μm diameter glass capillary using the expression given by Marcatili et al. [50] (black dash-dotted curve) and by Archambault et al. [76] (black dashed curve) over a spectrum spanning from 400 to 2000 nm. The loss figures are in the range of 50 to 100 dB/m for wavelengths near 800 nm. The graph also shows the loss spectrum of a ring with the same inner diameter (i.e., 45 μm) and a thickness of 600 nm using the expression by Archambault et al. [76]. The results show a decrease in the loss figures by over a factor of 10, reaching ~2 dB/m near 800 nm. Very recently, and following the interest in IC-HCPCF, the dielectric ring optical waveguiding properties were reexamined by D. Bird [96] and Zeisberger et al. [95]. The red curve in Figure 21a shows the loss spectrum of the same dielectric ring using the expression from [95]. The spectrum corroborates the results reported in Reference [76], though a discrepancy of nearly a factor of 2 is observed. In any case, the found loss values are in the range of 1 to 2 dB/m at ~800 nm regardless of the expression used.

Comparing the CL results of the above antiresonant ring with those found with hypocycloid core-contour IC-HCPCF with the same core radius and the same silica thickness as the one used for the ring give values that are ~4 orders of magnitude larger. This is illustrated in the right graph of Figure 21b, which show the CL spectra for a ring, hypocycloid core-contour Kagome HCPCF and 8-tube tubular HCPCF, respectively. This discrepancy in CL clearly indicates that the ARROW is not appropriate as a guidance model for IC-HCPCF. In parallel, inspection of the effective index of the HE_{11} for the ring using ARROW and the two IC-HPCFs (see *lhs* of Figure 21b) shows that comparable dispersion trend for the three fibers. The main difference lies at the resonant wavelengths. Here, in contrast with the ARROW calculated HE_{11} effective index, the two IC-HPCF show broader and multiple resonant wavelengths on the blue-side of the ARROW resonant wavelength. This difference is also shown in the loss spectra (*rhs* of Figure 21b), which show broader high loss band and shifted to the blue. Using the discussion above, this can be explained by the fact that at the blue-side of the resonant wavelength, the cladding lattice supports silica lattice modes with low azimuthal number associated with the unity-increase in the radial number, and hence broader and blue-shifted high loss region compared to the ARROW based spectrum [4].

Conversely, ARROW model was also used to describe large-pitch PBG-guiding PCF because of their weak dependence on the cladding lattice pitch. As mentioned above, we have seen that PBG guidance can occur without the requirement of periodicity. This is exemplified with large pitch regime PCF, such as the one with the high-index isolated inclusion cladding, whereby the cladding-lattice Bloch modes are represented by Wannier modes. The latter are highly localized in the high index guiding constituents of the cladding dielectric microstructure. Consequently, and by virtue of the photonic analog of the tight binding model, the modal spectrum band-structure of the cladding-lattice is reduced to discrete dispersion curves associated with the individual waveguiding features making the cladding lattice.

We thus argue that the ARROW picture cannot be considered as a guidance model for microstructured fiber as it does not give an account of the cladding modal spectrum. The latter is necessary to identify the coupling dynamics between a fiber core mode and the cladding. We have seen that in the case of a PBG fiber, the cladding is engineered to support no mode at the frequency and effective-index of interest, and thus the core mode is bound to be guided by virtue of absence of any leakage channel. On the other hand, for IC fibers, the cladding is engineered in such that its modal spectrum is populated within the $n_{eff} - \omega$ region of interest by modes that are highly localized in the high index material and have a strong transverse symmetry incompatibility with the fiber core mode of interest. Such approach stems from the paradigm shift triggered by the work of John and Yablonovitch [43,44], whereby light dynamics in photonic devices, such as PCF, is explained no longer by tracing the electromagnetic field path, but by considering the reciprocal space (i.e., Hilbert space of the relevant photonic states) of its dielectric structure to solve the modal spectrum. Then, a simple use of well-established rules from condensed matter physics, such as tight binding model or spatial symmetry of the eigenmodes, confer the necessary conceptual tools for PCF design.

4.5. Core Modal Properties of HCPCF

In the above sections, emphasis was put on the modal spectrum structure and properties of the HCPCF cladding lattice. In this section, we give a short review on the main modal properties of the guided core modes in HCPCFs.

The fundamental properties of core modes in HCPCF don't strongly deviate from those obtained in fiber optics [69] or hollow dielectric waveguides [50]. As a matter of fact the HCPCF core modes show comparable propagation constants and transverses profile as those described by Marcatili et al. for hollow dielectric capillary [50]. Figure 22 summarizes the modes content and transverse profile for a Kagome IC-HCPCF. Using the linear polarization approximation, we can recognize the fundamental core mode LP_{01} (i.e., the linear polarization form of HE_{11}) with its largest effective index, and then the higher order modes following similar order in effective index as with a hollow capillary.

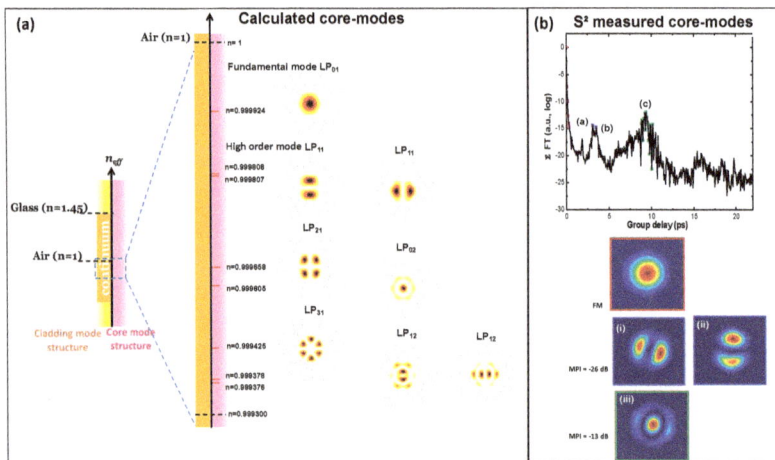

Figure 22. (a) *lhs*: Typical core mode profiles of a Kagome IC-HCPCF and their location in the n_{eff}-space. (b) A Kagome HCPCF S^2 measured trace and the intensity profiles of the retrieved modes around 1 µm.

The principal difference in the modal structure of HCPCF when compared to conventional optical fibers resides in the modal and polarization control. For example, in PBG-HCPCF, the mode number is set by the cladding PBG lowest effective index edge at a given frequency or wavelength. As seen above, this edge can be controlled to a certain level by adjusting the cladding structure to obtain single mode fiber. Another strategy to achieve single mode guidance in PBG-HCPCF is to introduce additional defects in the cladding that are phase-matched with the lowest loss higher order modes. When judiciously designed, the fiber can exhibit single-mode and polarization-maintaining operation [97].

For IC-HCPCF, the modal and polarization control requires different strategies than PBG fibers. Because of the intrinsic nature of the light guidance in IC-HCPCF, the notion of cut-off wavelength below which the fiber core does not support guided modes no longer holds because core modes co-propagate at the same n_{eff} with cladding modes (see *lhs* of Figure 22a). Consequently, the IC-guiding fiber can support an infinite number of core modes. However, an IC fiber can effectively operate in a single mode fashion at its lowest loss mode (typically the core fundamental mode) if the other modes are sufficiently leaky. Usually, in IC-HCPCF the lowest loss modes are those of LP_{01}, LP_{11}, and LP_{02}. The latter can exhibit lower loss than LP_{11} because its field has no azimuthal variation and thus lower coupling to cladding modes. Figure 22b shows experimental S^2 measurements [98] illustrating the above mentioned properties for a hypocycloid core-contour Kagome HCPCF. Here, the S^2 trace shows a fiber modal content dominated by LP_{01}, LP_{11}, and LP_{02} like modes. Furthermore, within this set of dominant modes, the light is chiefly in the fundamental core mode with an extinction ratio of ~13 dB for LP_{02} and ~26 dB for LP_{11}.

Another salient feature of IC HCPCF HE_{11} core mode is its extremely low overlap with the silica core-surround. The top of Figure 23 shows the profile of HE_{11} core mode and how its compares to that of a glass capillary [63]. Figure 23a shows the evolution of HE_{11} radial profile along two axes when the negative curvature b is increased whilst keeping the inner radius constant. The result shows that the mode MFD remains constant and its transverse profile is that of HE_{11}, regardless of the negative curvature value. Furthermore, the mode profile is related to the inner radius of the fiber, and the Kagome IC-HCPF mode profile and propagation constant can be approximated to that of a glass capillary with a radius equal to the inner radius of the hypocycloid core contour. Figure 23c shows the relative error of this approximation for different values of b. The results show that the relative discrepancy is less than 7% (which was found for a $b = 1.5$).

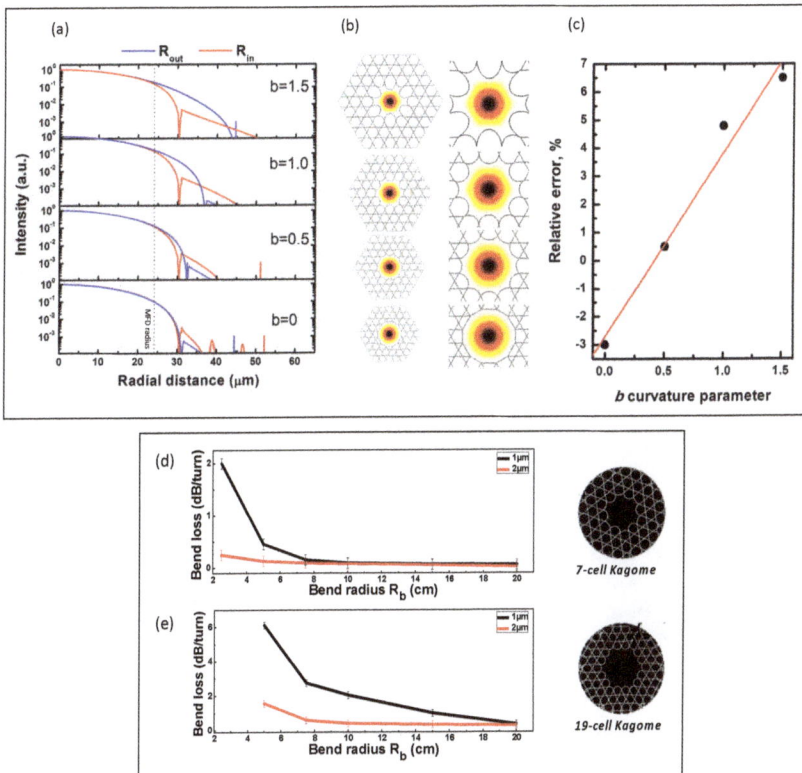

Figure 23. (**a**,**b**) Evolution of the mode field diameter of IC HCPCF HE_{11} core mode profile with negative curvature b. (**c**) Relative error on MFD when Kagome IC-HCPF is approximated to a glass capillary with a radius equal to the inner radius of the hypocycloid core contour. (Reprinted with permission from Reference [63], OSA, 2013.) Experimental bend loss evolution versus bend radius R_b for (**d**) 7-cell and (**e**) 19-cell Kagome HCPCF at 1 µm (black curves) and 2 µm (red curves) (reprinted with permission from Reference [65], OSA, 2018).

Figure 23d,e shows the bend loss for the case of 7-cell and 19-cell Kagome IC-HCPCF, respectively. The first fiber has an inner diameter of 57 µm, a silica strut thickness of 840 nm, and *b*-parameter of 0.75. The 19-cell fiber exhibits an inner core diameter of 119 µm, a silica thickness of 900 nm, and has $b = 0.88$. The bend loss was measured at 1 µm (black curve) and at 2 µm (red curve) at the fundamental and the 1st higher transmission bands, respectively.

4.6. HCPCF Prospects and Future Trends

Despite the field of HCPCF remains a nascent field and timely topic in research and technology, it is quickly developing in a mature subject both in guided photonics, in the design and fabrication of this type of fibers, and in its applications in various fields (see below). The performances achieved in the last 20 years with both PBG-HCPCF and IC-HCPCF were as exciting and insightful as counterintuitive and changing. Both HCPCFs morphed from an academic curiosity to a transformative and powerful technology platform in nonlinear optics, lasers, sensing, communications, and higher-power pulse delivery.

This being said, HCPCFs have not been adopted as extensively as expected. This is very much so in telecommunications, where HCPCF is yet to represent an alternative to the single mode fiber (SMF). Indeed, the lowest loss reported for HCPCF is ~1 dB/km, and was achieved with PBG-HCPCF. This is ~1 order magnitude higher than the SMF state-of-the-art loss figures, and seems to be a hard limit for PBG-HCPCF, especially when single-mode and polarization-maintaining are required. Indeed, achieving lower loss implies succeeding the difficult challenge of reducing the frozen thermal surface fluctuations induced surface roughness.

Within this context, a renewal in the hope of envisioning HCPCF as a future alternative to SMF is triggered with the progressed made in IC-HCPCF. Today, IC-HCPCF shows comparable loss to PBG-HCPCF, with losses below 10 dB/km being demonstrated over a large spectral range of the visible and NIR. Debord et al. [30] identified two possible future trends to further improve on the performance of IC HCPCF. These trends result from two limiting sources. Today, IC-HCPCF are surface roughness-limited for short wavelengths (typically shorter than 1 µm), and design-limited for longer wavelengths. Consequently, improving on the surface roughness-limited fibers implies lower SSL, and requires a reduction either in the surface roughness or in the optical overlap with the silica core-surround. Conversely, reducing the transmission loss of the design-limited HCPCF implies exploring alternative cladding and/or core-contour designs to those of Kagome HCPCF or SR-TL HCPCF. Among the proposed cladding designs we count nested tubular HCPCF designs [56,57,60] or double layer tubular cladding [58], which very recently demonstrated a loss figure of as low as 2 dB/km at 1500 nm.

5. HCPCF Applications

The advent of HCPCF has transformed laser–gas-based applications and is continuing to do so. The driving properties of such a transformative power are summarized in part in Figure 24. These are the laser–gas interaction ultra-enhancement, the laser power and/or energy handling (Figure 24), and specific and engineerable dispersion profiles.

The micrometer scale core and the long propagation length offered between the guided light and a gas phase medium that can be confined in its core permit to reach a very efficient gas–light interaction. This means that nonlinear phenomena and spectral contrast can be excited with extremely low light levels. This statement is quantified by the figure of merit $FOM = L_{int}\lambda / A_{eff}$, which is proportional to the ratio between the effective interaction length, L_{int}, and the modal area, A_{eff}. Figure 24a shows that the FOM to be enhanced larger in HCPCF by a factor between 1000 and 1 million relative to the usual capillaries. The colored regions in Figure 24a show the FOM magnitudes for the different and typical reported HCPCF. We note that for the same propagation loss, the PBG-HCPCF exhibit higher FOM because of a smaller fiber-core size than those of IC-HCPCF.

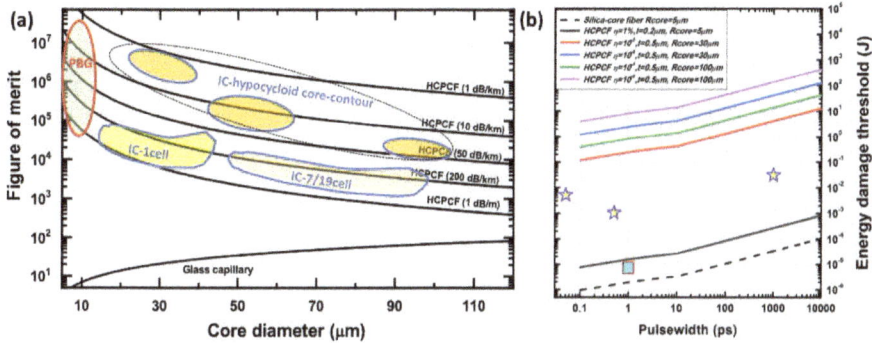

Figure 24. (a) Laser–gas interaction figure of merit evolution with the hollow core diameter for glass capillary and different HCPCF transmission loss figures. (b) HCPCF-guided laser energy damage threshold for different types of HCPCF. The star and square symbols are the experimental demonstrated energy with IC-HCPCF-guided lasers [31,99,100] and PBG-HCPCF [101], respectively.

Figure 24b shows the laser energy damage threshold (LIDT), with laser pulse duration for different silica-based fibers. The results show that for HCPCF with an optical overlap coefficient of $\eta = 10^{-6}$ and a core radius of 60 μm, which are fulfilled with current IC-HCPCF, the energy damage threshold is theoretically found to be 1 J for laser pulses of 100 fs and 100 J for pulses of 10 ns. The exceptionally low optical overlap in IC-HCPCF makes this type of fiber as an outstanding candidate for laser power/energy handling and for high field optics and photonics. The star symbols are the laser energies reported to be guided by IC-HCPCF. PBG-HCPCFs are represented by the configuration of $\eta = 10^{-2}$ and core radius of 5 μm. Here, the LIDT is around 10 μJ for sub-picosecond laser pulses. An example of an experimental demonstration using PBG-HCPCF is shown as a square symbol in the figure. This lower PBG-HCPCF's LIDT relative to IC-HCPCF is due first to the high optical overlap with silica in PBG-HCPCF and secondly to the smaller core size. Having said that both types of HCPCF show higher energy damage threshold than 5 μm core radius silica-core optical fiber with USP laser energy damage threshold of 1 μJ.

Below, we review a nonexhaustive list of experimental results in different fields, where the above properties, along with the dispersion, played driving roles in harnessing gases and light under surprisingly extreme regimes.

5.1. Non Linear Optics

In the field of nonlinear optics, gas-filled HCPCF proved to be an outstanding platform in transporting extremely energetic pulses, in pulse compression, and in generating extremely broad spectra with both very low light levels and with high laser pulse powers and energies.

5.1.1. High-Power Laser Beam Delivery

The HCPCF technology was a timely innovation for laser beam delivery, particularly for ultrashort pulse (USP) lasers. Indeed, the advent of the USP laser technology demonstrating high peak power and intensities of up to PW/cm^2, combined with pulse durations down to 200 fs calls for a fiber beam delivery that can handle this power and/or energy without distortion to the pulse spectral and temporal integrity. This need is becoming pressing with the growing and varied applications that are adopting USP lasers as a tool, and which span from laser surgery to surface marking and micromachining to mention a few.

Within this context, the PBG-HCPCF rapidly showed its inability to sustain such an intense pulse laser. The strong optical power overlap with the silica cladding (1–0.1%) implies a too low LIDT (see Figure 24b), and hence disqualifies this fiber family as an efficient mean for the intense USP delivery.

Among the reported work on PBG-HCPCF-based laser beam delivery, we count the report in 2007 of the delivery of 0.5 mJ energy and 65 ns duration pulses from of Nd:YAG laser through a 2m long 7cell fiber with 8 μm diameter [102]. Two years later, transportation of 1 mJ energy and 10 ns duration pulses has been demonstrated with a transmission coefficient of 82% in a 8 mm long 19-cell PBG-HCPCF with a core diameter of 15 μm [103].

Conversely, the very low optical power overlap with silica offered by IC-PCFs (10^{-4}–10^{-6}, see Figure 15d) and their larger core diameters allowed a breakthrough in USP beam delivery applications. Moreover, due to its broadband guidance, different spectral wavelengths have been targeted going from UV to mid-IR range. The conventional hexagonal core contour of Kagome HCPCF demonstrated guidance of fs laser beam of tens of μJ [11], and 1064 nm pulses up to 10 mJ for 9 ns pulse duration [100] leading to demonstration of gas ignition at the output of the HC fiber. An important milestone has been achieved in 2013 through the use of the hypocycloidal design that permits to drop further the optical overlap with the silica down to ppm level. The exceptional power of IC-HCPCF for ultra-energetic USP laser beam delivery was first shown with the demonstration of the transport of 600 femtosecond pulses at millijoule energy level of a 1030 nm emitting Yb-laser in robustly single-mode fashion using a 10 m-long piece of a IC Kagome HCPCF [31]. The energy record in term of launched energy for fs-lasers in a HCPCF is to date 2.6 mJ in a Kagome lattice HCPCF that has been designed to get a large core size of 100 μm [99]. One can notice that these fibers can be the host for transport of very high average powers of CW lasers. In fact, the delivery of 1 kW has been demonstrated with an impressive transmission coefficient of 90% with an excellent output beam quality defined by a M^2 of 1.1 [104]. Subsequently, much effort has been made to develop efficient fibers for covering most of the industrial lasers. For example, delivery of thulium-doped fiber laser technology at 2 μm (important wavelength range for spectroscopic applications, material processing of for High harmonic generation) has been addressed with transmission of 40–50 W of average power in a nanosecond [105] or femtosecond regime [106]. The green spectral range has been covered with a Kagome HCPCF, demonstrating delivery of 532 nm 10 W average power frequency-doubled Yb-fiber picosecond laser [64], and a with a tubular amorphous lattice HCPCF demonstrating delivery of 0.57 mJ and 55 ns pulses and 30 μJ and 6 ps pulses [107]. For shorter wavelengths, UV guidance has been reported with the launching of 15 mW of 280 nm CW laser with a hexagonal-core Kagome fiber with a transmission coefficient of 50% without appearance of UV induced damage over 14 h [108]. Recently, the SR-TL HCPCFs shows its capabilities with the delivering at 355 nm of 20 ps 160 μJ pulses thanks to a fiber exhibiting 130 dB/km loss level [91].

So far, laser beam delivery of micro- or millijoule through HCPCFs has been used for several demonstrations of glass micromachining [17], metal machining [109], laser ignition [100,110], or also tissue ablation [111,112] to mention a few. Figure 25 illustrates these different applications.

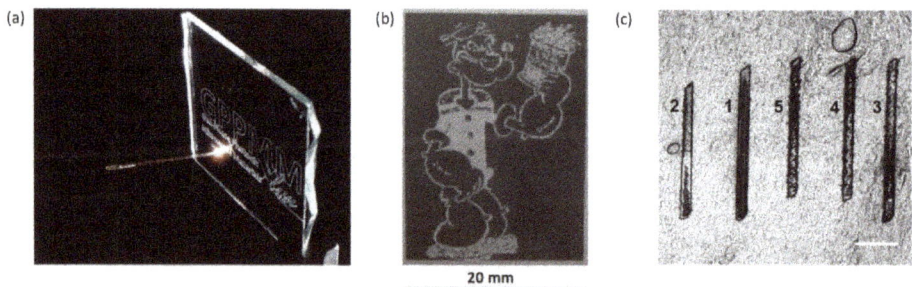

Figure 25. Illustrations of different applications of HCPCF fiber laser beam delivery, with (**a**) glass engraving ((reprinted with permission from Reference [17], OSA, 2013), (**b**) titanium micromachining (reprinted with permission from Reference [107], OSA, 2015), and (**c**) tissue surface ablation (reprinted with permission from Reference [111], OSA, 2016).

5.1.2. Pulse Compression

A large number of nonlinear optical applications require laser pulse widths that are much shorter than what is possible with today's high-power USP such as those of Yb-based materials [113]. An example of these applications is High Harmonic Generation (HHG) where few-cycle intense pulses are desired to reach high conversion efficiency and high photon energy cut-off. In the previous section, we have seen that the successful laser beam delivery of high energy ultrashort pulses in different spectral ranges using IC-HCPCF relies chiefly on the small optical overlap of the guided mode with silica. The pulse compression, on the other hand, relies on the gas-filled IC-HCPCF specific and controllable dispersion spectral profile. Thanks to a dispersion spectrum exhibiting both normal and anomalous regimes, dispersion values that are relatively low, and ultralow propagation loss, IC-HCPFS are particularly efficient in spectral broadening and pulse compression of such intense pulses. The many reported results demonstrated the stronger impact of this fiber-technology compared to the well-established technique gas filled capillaries to compress several millijoule USP. The latter technique presents the drawback of being limited to high energy pulses because of the high propagation loss of capillary, and of being physically cumbersome as the capillary is not bendable and must be stretched to avoid higher transmission loss. On the other hand, IC-HCPCF offers the best of two worlds: pulse compression over a large energy dynamic range and ease of use and integration.

One can distinguish mainly two types of compression schemes employed with the HCPCF technology. The first one exploiting the normal dispersion regime of the gas filled system is based on Self Phase Modulation (SPM) effect of the spectrum leading to spectral broadening. Pulse compression is then achieved by postcompression with dispersion compensation thanks to chirped optics. The second technique takes advantage of the IC-HCPCF low and anomalous dispersion to excite solitonic dynamics whereby pulse self-compression can take place.

Table 1 summarizes the most notable and representative intense pulse compression results obtained with both techniques, for various wavelengths and temporal domains. Among these compression results, one can notice the subcycle pulse generation by compression of 1.8 µm 80 fs pulses down to 4.5 fs in a Xenon filled 82 µm core HCPCF (20 cm long) with a self-compression dynamic [16]. To date, it represents the strongest compression ratio in self-compression regime with a value of 17 that has been achieved using HCPCF technology.

Table 1. Representative pulse compression performances in HCPCF technology.

Reference	Wavelength	Compression Scheme	Input/Output Energy	Input Pulse Duration	Output Pulse Duration	Compression Factor	Filling Gas
Mak et al. [114]	790 nm	Postcompression	10.3 µJ	103 fs	12.6 fs	~8	Krypton
Mak et al. [114]	790 nm	Self-compression	6.6 µJ	24 fs	6.8 fs	~3.5	Krypton
Hädrich et al. [106]	1030 nm	Postcompression	9 µJ	250 fs	30 fs	~8.3	Krypton
Guichard et al. [115]	1030 nm	Postcompression	70 µJ	330 fs	34 fs	~9.7	Ambient air
Debord et al. [31]	1030 nm	Self-compression	450 µJ	600 fs	49 fs	~12	Ambient air
Emaury et al. [116]	1030 nm	Postcompression	1.95 µJ	860 fs	48 fs	~17.9	Xenon
Balciunas et al. [16]	1080 nm	Self-compression	35 µJ	80 fs	4.5 fs	~17	Xenon
Wang et al. [61]	1500 nm	Self-compression	105 µJ	850 fs	300 fs	~2.8	Ambient air
Gebhardt et al. [117]	1820 nm	Self-compression	41 µJ/34.4 µJ	110 fs	14 fs	~7.8	Argon
Murari et al. [118]	2050 nm	Postcompression	227 µJ	1.8 ps	285 fs	~6.3	Argon

In the above, most of the reported pulse compressions were achieved using gas-filled HCPCF. This technique necessitates the cumbersome use of gas cells and gas pressure control. An ideal set-up scheme is the use the HCPCF in its simplest form, meaning the fiber core is filled with air at atmospheric pressure. In this context, compression of 330 fs/70 µJ pulses at 1030 nm down to 34 fs has then been demonstrated by Guichard et al. with a transmission efficiency of 70% in a 1.2-m-long piece of air-filled Kagome fiber [115]. This compression has been achieved by postcompression of a spectrum that has been broadened over 90 nm (at −10 dB). Self-compression of more intense 600 fs has been reported with a minimum pulse duration of 49 fs at the output of a 3 m-long piece of a 19 cell Kagome HCPCF exposed to ambient air for an input energy of 450 µJ [31].

5.1.3. Raman Comb Generation

The first demonstration of nonlinear optics in HCPCF was achieved in 2002 by Benabid et al. [46], via the generation of Stimulated Raman Scattering (SRS) in hydrogen-filled Kagome HCPCF pumped by 532 nm nanosecond pulses with conversion efficiency of ~30% on vibrational Stokes and anti-Stokes. Two years later, the same phenomenon was demonstrated with pure rotational resonance using hydrogen-filled PBG-HCPF excited at 1064 nm with an impressive higher conversion efficiency of 92% from the pump to the first rotational Stokes [119]. These results have been seminal to several demonstration based on SRS in HCPCF. For example, the first multi-octave Raman comb demonstrated in HCPCF has been reported in 2007 where nanosecond pulse-pumped hydrogen-generated Raman lines from 325 to 2300 nm by virtue of the broadband guidance offered by IC-HCPCF [4]. Also, in the same year, the same group reported on the observation of continuous-wave rotational SRS with a conversion efficiency of 99.99% to the first rotational Stokes [9]. Finally, in Wang et al. [120], the same team experimentally and theoretically demonstrated that within the single pump pulse and the spectral components of the comb reported in Reference [4] are phase coherent. This coherence results from the transient nature of the SRS, and is stronger when the pump pulse is narrower and the fiber is closer to single mode.

Multi-octave coherent comb generation are of great interest for optical waveform synthesis via Fourier synthesis and attosecond pulse generation for attoscience applications. Also, the generation of this large comb in the CW regime represents a very promising milestone for the advent of the optical analog of the electronic function generator.

Among the follow up results in the generation of several octave Raman comb, we can list the generation via transient SRS of over five octaves wide Raman combs in hydrogen-filled IC HCPCF pumped with a 27 picosecond laser at 22.7 W [10]. The Raman lines span from 321 nm to 12.5 µm (24–933 THz) with an output average power of 10 W (Figure 26). The use of a narrower pump pulse than the nanosecond pulses previously used enables stronger temporal filtering of the excited spatio-temporal modes duration the Raman generation, and thus ensuring a single wavepacket to be Raman-amplified from the quantum noise, and thus to generate more intrapulse coherent comb. Figure 26 shows the two output generated Raman combs optimized for vibrational emission (through linear polarization) and for rotational lines emission (with circular excitation polarization).

In parallel, the generation of dense Raman comb has been demonstrated by exciting with a 532 nm nanosecond laser a HCPCF filled with a gas mixture of H_2, D_2, and Xe [121]. With an input beam of 5 µJ of energy, 135 rovibrational Raman lines spaced by 2.2 THz have been measured from 280 nm to 1 µm. Finally, the generation of Raman comb in hydrogen-filled HCPCF has been for the first time used experimentally for a periodic train pulse waveform synthesis of 26 fs pulses duration with a frequency of 17.6 THz thanks to phase-locked Raman sidebands [122].

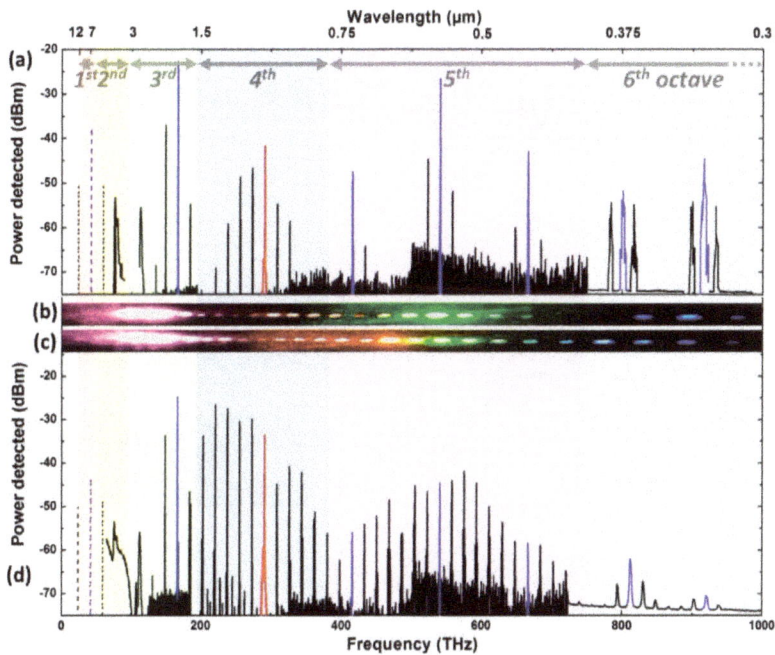

Figure 26. (**a**,**b**) Optical spectrum (**a**) and its associated diffracted output beam picture (**b**) generated with pump linear polarization. (**c**,**d**) same as (**a**,**b**) for pump circular polarization of a large optical Raman comb in H_2 IC HCPCF (reprinted with permission from Reference [10], OSA, 2015).

5.1.4. Supercontinuum Generation

Similarly to solid-core PCFs, many results have been carried out to generate supercontinuum in gas-filled HCPCF [123]. Thanks to the combination of high peak power and energy levels offered by the new generation of USP lasers and to the large FOM, one can even generate, with a low nonlinear medium such as ambient air with a nonlinear coefficient of ~5×10^{-19} cm^2/W, strong optical nonlinearities. This has been exemplified by the demonstration of nonlinear optical effects in air-filled IC HCPCF [13]. In fact, Debord et al. reported on the experimental demonstration of a very high energetic Raman comb based on excitation of N_2 of the ambient air in the fiber [13]. This has been obtained in a 3 m-long piece of Kagome HCPCF excited by 300 ps pulses with energy up to 1.3 mJ and an output efficiency of 75%. The generated N_2 Raman lines covers a large spectrum range of 300 THz between 600 and 1375 nm with five vibrational resonances spaced by 70 THz. By switching to shorter pulses of 600 fs, the dynamics have been completely modified to end up with the generation of a broad supercontinuum that spreads from 375 to 1500 nm with a transmission coefficient of 30%. It corresponds to a record energy spectral density of 150 nJ/nm.

Figure 27 summarizes the evolution of the input energy of the two spectra recorded at the output for the different temporal excitation regimes. The theory shows that this generation involves four-wave mixing, third harmonic generation, and soliton formation.

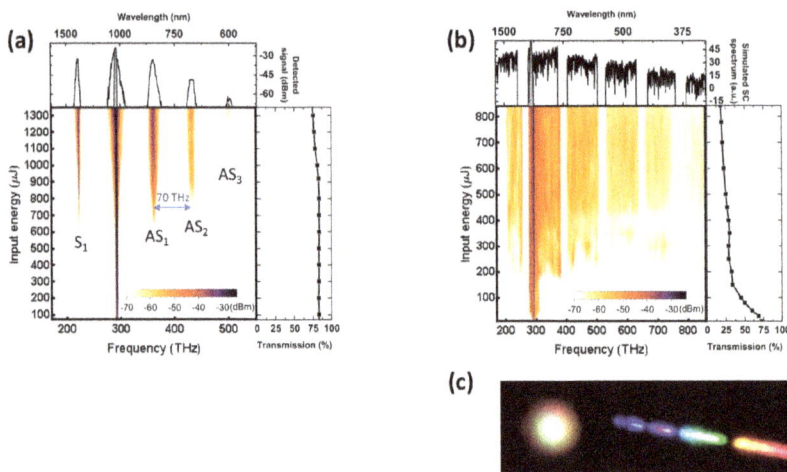

Figure 27. (**a**) Raman comb and (**b**) supercontinuum spectrum measured at the output of IC ambient air-filled IC HCPCF excited with millijoule 300 ps and 600 fs, respectively. (**c**) Dispersed-fiber output spectrum (reprinted with permission from Reference [13], OSA, 2015).

5.2. Plasma Photonics

5.2.1. Wave-Induced Plasma

In 2013, plasma-core PCF was reported [27], extending the gas phase materials that have been successfully loaded into HCPCF to ionized gases. The interest of confining the fourth state of matter is strong and represents a unique material for photonics because it is the only state of matter that can emit directly in the UV spectral domain. The confinement of such a phase in the HCPCF technology can be useful in building new generation of UV-DUV fiber gas laser.

Different attempts have been carried out to ignite gas discharges in hollow-core waveguides. The primary technique used was based on high-voltage DC discharge using two electrodes at both ends of the gas discharge tube [124]. Several limits are inherent to this ignition technique if one wants to target excitation in small core tubes with a limit around 150 µm of core diameter. In fact, important high voltage values are required to reach the ignition threshold due to important charge accumulations close to the core walls and on the electrodes implying instability of the plasma discharge. In fact, no successful discharges based on the CW excitation technique have been reported in a proper waveguide such as HCPCF. In order to mitigate this charge accumulation, the RF field is combined with DC excitation. In 1975, a new plasma ignition technique emerged based on the resonant microwave cavity using a specific coupler, named "Surfatron" [125]. It consists of a cylindrical microwave resonant cavity exhibiting a gap, where the field is strong enough so the coupled energy to the gas filling the hollow-core creates a discharge. This technique exhibits the advantage of being nonintrusive as the microwave excitation is applied on side of the fiber.

The Surfatron excitation scheme was successfully applied to the HCPCF in 2013 (see Figure 28a). For the first time, a stable 6 mbar argon plasma column sustained in a 100 µm core diameter IC HCPCF-based on Kagome lattice has been reported, with no damage to the fiber structural integrity. The plasma column length reached 4 cm for an incident microwave power of 30 W at 2.45 GHz. Figure 28b shows a transverse picture of the plasma column. The originality of this excitation relies on the fact that the plasma is created and maintained by a surface wave that propagates at the interface between the plasma core and the glass core contour in a self-consistent configuration where the surface wave creates the plasma, and in turn the formed plasma acts as a guiding medium for the surface wave itself. This surface wave transfers its power to neutral particles during propagation, and at a certain distance when the power becomes (or electronic density) too low the plasma column is stopped. The relatively low gas pressure used allows a stronger emission of the Ar atoms in the UV domain thanks to Ar+ ions excitation.

An efficient guidance of the generated Ar lines has been demonstrated at the output of a 20 cm long piece of the HCPCF, and the guided spectrum shows strong emission in the 350–550 nm spectral range (see Figure 28c). The experimental study reveals that the gas temperature was around 1300 K at the coupling zone. The achieved ionization rate reached the impressive levels of ~10^{-2}, corresponding to an electron density level of ~5×10^{14} cm^{-3} at ~1 mbar pressures and microwave power density of ~0.1 MW cm^{-3}. A theoretical study has been undertaken to give a clear explanation on the plasma processes involved that can explain this unusual plasma column excitation and confinement [126]. Particularly, the theoretical model enabled us to understand the counterintuitive situation of high-power densities and electron densities combined with the relatively moderate gas temperature (but strong enough to potentially starts softening the silica). The HCPCF confined plasma dynamic shows a strong ion acceleration, and thus an ionic mean free path which is larger than the core size of the fiber explaining the combination of the moderate increase of the gas temperature, and the large ionization rate.

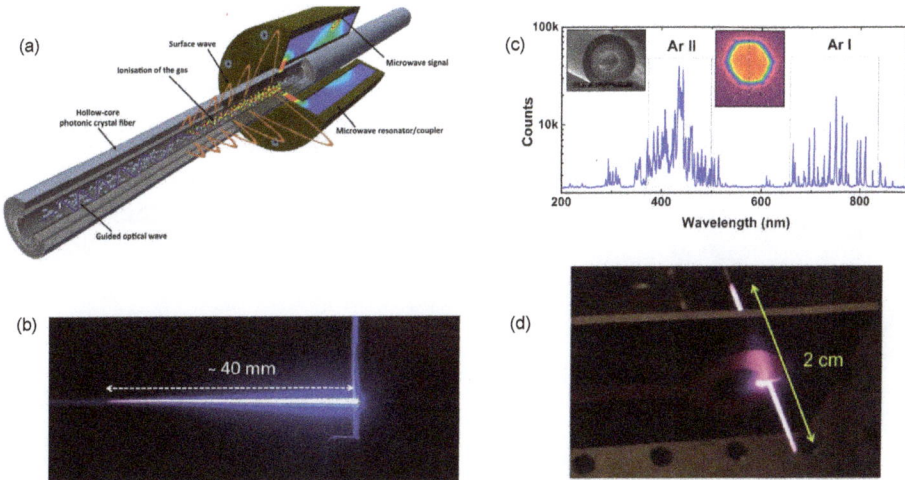

Figure 28. (**a**) Schematic representation of plasma column generation in HCPCF with a surfatron. (**b**) Side view picture of the generated 4-cm-long plasma column. (**c**) Measured spectrum lines at fiber output (reprinted with permission from Reference [27], OSA, 2013). (**d**) New microstrip split-ring resonator (MSRR) plasma column excitatory (reprinted with permission from Reference [127], OSA, 2016).

Later, a more compact microwave excitation scheme was reported [127]. It is based on the similar surface wave generation but a microstrip split-ring resonator (MSRR) was used as an alternative to the much large surfatron. Figure 28d shows the excitation of a plasma column of 2 cm with the use of a fabricated MSRR.

In order to enhance the emission of the plasma in the UV domain and to reach the DUV, gas mixtures have been more recently employed by using particularly the ternary mixture Ar-N_2-O_2 [128]. This has led to the demonstration of a tunable DUV source with the emission of fluorescence and several guided lines in the 200–450 nm wavelength range.

5.2.2. Photo-Induced Plasma

It is common today to have USP lasers emitting peak power of several GW. This corresponds to intensities in the range of several PW/cm^2 when the laser beam is coupled into HCPCF. Such an intensity level well surpasses the ionization threshold of most gases. Therefore, in HCPCF this phenomenon appears for lower laser average power values than the one required in simple capillaries in high field domain. This ionization process is intrinsically linked to a loss decrease [129,130].

The effect of this ionization regime has been both theoretically and experimentally studied in gas-filled HCPCF. One can distinguish two different induced effects. The first one is an induced blue-shift effect in the spectrum. The second is an impact on the emission of a resonant dispersive wave [129] in the deep-UV or mid-IR spectral ranges [131,132].

The free electrons directly impact the refractive index and induce a decrease of its value in the time domain and thus a self-steepening of the pulse. Spectrally, one can observe blue-shifting contrary to the Raman self-shifting in conventional fibers. Due to a longer time in the recombination of the electrons, the change in refractive index is still maintained for ultrafast pulses. The loss drop due to the photo-ionization implies a decrease of the pulse intensity and thus limiting progress of the blue-shift. Hölzer et al. illustrated this phenomenon experimentally with the propagation in a Kagome fiber of 800 nm pulses of 9 μJ and 65 fs [133]. Good agreement has been obtained between experiments and numerical simulations when an ionization term is taken into account in the model in term of spectral evolution and also in the energy losses. Further, the theoretical explanations of this blue-shift effect have been deeply developed in References [130,134].

Phase matching conditions can be obtained between a soliton pulse with a wavelength in the anomalous dispersion regime close to the zero-dispersion wavelength and a resonant dispersive wave in the normal dispersion regime. Various experiments results reported on the generation of UV radiation with the use of this resonance effect [129,135] and on the build of a tunable UV ultrashort pulse source in Ar-filled HCPCF for example [136,137]. The ionization phenomenon both experimentally and theoretically has revealed a role in the enhancement of the generation of the UV resonant emission.

5.3. Atom and Molecular Optics

5.3.1. Atom Optics

Atom optics in confined nano- and microstructured devices are becoming a timely topic. The pursuit of miniaturizing atom based functionalities such as frequency references, atomic clocks and quantum sensors is motivated by transferring the outstanding performances in frequency control and coherent optics achieved in laboratory environments to a broader community of users through the development of compact, friendly-user, and stand-alone atom devices. Among the atom devices that have been or are being developed, we cite HCPCF and its gas-filled form the photonic microcell (PMC). The latter outstands with its long interaction length and small modal area, making it thus an efficient platform to enhance gas–laser nonlinear interaction and/or absorption contrast by several orders of magnitude relative to free-space configurations. In this context, in 2006, vapor confined HCPCFs were first reported using Rubidium (Rb) [20,138]. In these seminal works, electromagnetically induced transparency (EIT) generation was demonstrated with a light-level as low as nanowatt power, which represents more than a 1000-fold reduction from the power required in bulk cell. The postprocessing of HCPCF by coating the inner core with specific materials to assess the atom surface dynamics and to reduce the physiochemical adsorption effects is one of the current challenges to improve the alkali vapor PMC performances and to anticipate the avenue of quantum devices based on it [139].

With respect to cold atom manipulation inside HCPCF, rapid progress was made in the last few years despite the experimental difficulties that loading cold atoms inside a micrometer scaled hollow-core entails. The first work on loading cold atoms inside a HCPCF was reported in 2009 by Bajcsy et al. where they demonstrated an optical switch [140,141]. A second work [142] concluded that the long-distance transportation of atoms is feasible which would be a route to guided matter wave interferometry, large-area Sagnac-type interferometers. Other works have recently involved loading Rydberg cold trapped atoms into the core of PBG fibers [21,143]. Finally, another seminal demonstration was motivated to explore quantum metrology and optical atomic clocks, and was reported in Reference [29]. Here, by designing an IC Kagome HCPCF with a sufficiently large core diameter (34 μm) and low residual birefringence, cold 88 Sr atoms from Magneto-optical trap (MOT) (see Figure 29) were successfully loaded into fiber by using optical lattice whose potential depth was 30 μK, whilst keeping the atoms trapped within the wells of an optical lattice. The main result of this work is the demonstration of an absorption line whose linewidth is set by the lifetime of the excited state instead of atom-wall collisional dephasing as one would expect given the microconfinement of the atoms.

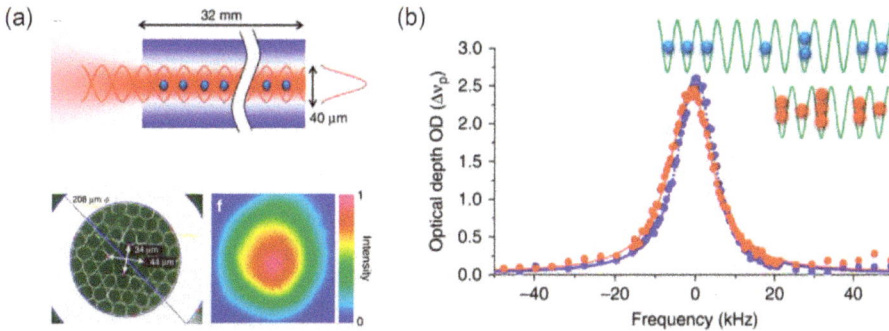

Figure 29. (**a**) Atoms in the HCPCF radially- and axially-confined by the optical lattice, preventing atoms from interacting with the fiber-wall (top). Microscope image of the Kagome fiber and Far-field pattern of the laser intensity passed through a 32-mm-long HCPCF (bottom). (**b**) Absorption spectra with and without atomic expansion over lattice sites, as illustrated in the inset, are displayed by blue and red symbols, respectively, corresponding to a mean atom occupation of m = 0.45 and m = 1.7 (Reprinted and adapted with permission from Reference [29], Nature, 2014).

5.3.2. Novel Stimulated Raman Scattering Configuration

Recently, an exotic configuration of SRS in H_2-filled PBG HCPCF was reported [28], which has the potential of offering both an alternative means for light control and trapping of molecules and for developing high-power and narrow linewidth exotic light sources. This configuration relies on a powerful CW excitation of hydrogen-filled PBG-HCPCF and the generation of forward and backward Stokes radiation. Here, molecular hydrogen is optically self-nanostructured into a periodic lattice whereby the Raman-active molecules are trapped within an array of potential wells with a nanometric width and a depth as high as 55 THz. Such a nanolocalization of the Raman-active molecules created a new Lamb–Dicke regime of SRS. The results and the theoretical model show an unconventional Stokes radiation with sub-recoil linewidth of ~3 kHz (four orders of magnitude lower than Doppler linewidth) (see Figure 30), and a rich spectral structure, which includes Rabi splitting sidebands, molecular motional sidebands, and inter-sideband four wave mixing. Finally, the results show a novel molecular acceleration.

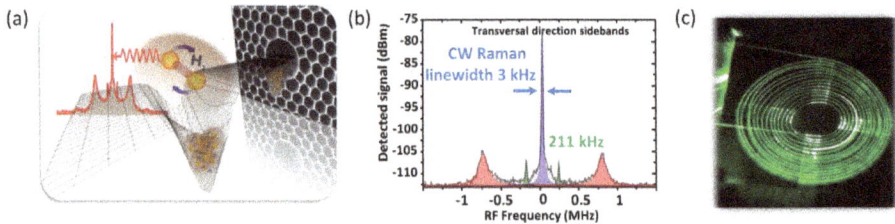

Figure 30. (**a**) Illustration of the molecular trapping experiment. (**b**) Stokes spectrum. (**c**) Moving scatter along the set fiber snapshot (reprinted with permission from Reference [28], Nature, 2016).

5.4. Quantum Information

One of the main driving subjects in the field of quantum information is the generation of single photon and/or entangled photon multiple sources. Among the recent avenues to generate such nonclassical photon sources are those based on spectrally-entangled photon generation. Here, promising progress in developing photonic devices that enable tuning photon pair phase-matched frequencies was demonstrated using photonic chip [144]. Within this context, recent results exploiting the gas-filled IC-HCPCF's dispersion spectral profile and its strong nonlinearity show that IC-HCPCF is an excellent platform to both generate and engineer spectrally-entangled photon pairs [145–147].

In order explain how the IC-HCPCF properties are relevant to spectrally-correlated photon-pair generation, we recall that the frequency correlation between signal and idler photon pair is described by the joint spectral amplitude function (JSA), which can be approximated as the product of the energy conservation function $\alpha(\omega s, \omega i)$ and the phase matching function $\varphi(\omega s, \omega i)$:$F(\omega s, \omega i) \approx \alpha(\omega s, \omega i)$ $\varphi(\omega s, \omega i)$. The profile of the JSA is a direct signature of the correlation nature between the photons of the photon pair; from separable pairs, which can be used for heralded single photon source, to entangle for quantum cryptography. The JSA profile can be engineered by either shaping the spectral/temporal form of the pump or by tuning the phase-matching of the generated single and idler.

In the work reported in Reference [148], the JSA was adjusted by chirp tuning of pump pulses and change of gas nature and pressure. Conversely, Cordier et al. [146] exploited the multidispersion band host to tune photon pair time-frequency correlations. Here, thanks to the IC HCPCFs dispersion profile made of different dispersion band with several zero dispersion wavelengths and S-shape curves, the authors demonstrated theoretically and experimentally the generation of photon pair with controllable joint spectral intensity (JSI = |JSA|^2), corresponding to different photon correlation. The photon pair was generated via FWM in different phase matching conditions using gas-filled IC-HCPCF with controllable dispersion and optical nonlinearity. The possibility is offered in such IC gas-filled fibers to position the pump, signal and idler photons in different transmission bands (as shown in Figure 31b). Several JSIs have been produced with different degrees of photons entanglement, and with an active control on the JSI by playing on gas pressure change [147]. In the results presented in Figure 31, the fiber exhibits strut thickness of 600 nm, core radius of 20 μm. The pump wavelength is fixed at 1030 nm and the seed laser is tunable around 1550 nm.

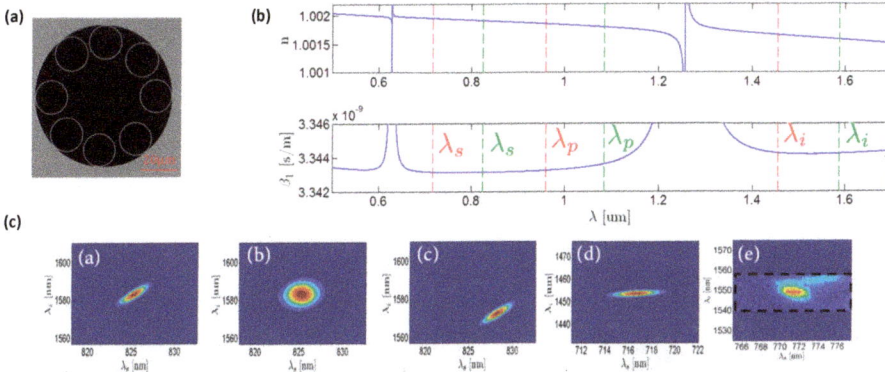

Figure 31. (**a**) SEM image of the 8 tubes IC fiber cross-section used for JSI experiments. (**b**) Calculated fundamental mode effective index and inverse group velocity in an 8 tubes IC fiber filled with 3-bar Xenon. Two specific four-wave-mixing configurations are identified with respectively the fulfillment of $\beta_1(\omega_p) = \beta_1(\omega_s)$ and $2\beta_1(\omega_p) = \beta_1(\omega_s) + \beta_1(\omega_i)$. (**a–d**) Simulated and (**e**) experimental JSI for different lengths, gas pressures, and pump wavelengths (reprinted with permission from Reference [146], OSA, 2018).

6. Conclusions and Future Trends

To conclude, we reviewed the recent advances in the design, fabrication, and application of hollow-core photonic crystal fibers by bringing to light a unique synergetic relationship between the fundamental ingredients in the rise of HCPCF-based "gas-photonics", which are glass, gas, and light.

Using notions from solid-state physics, we have given the underlying physical principles of the fiber's optical guidance mechanisms of the Photonic BandGap and the Inhibited Coupling. We emphasized the role of the modal spectrum of the fiber cladding structure in defining the HCPCF guidance properties. We have shown how an appropriate geometric glass structure of a fiber can structure the fiber modal spectrum. We gave a historical account to show how concepts of atomic physics and solid-state physics, such as tight-binding model or bound-state in a continuum, have been exploited in fiber–photonics to develop double photonic-bandgap HCPCF or hypocycloid core-contour (i.e., negative curvature core-contour) fibers. We listed the evolution of the dramatic drop of the attenuation in HCPCF that followed the introduction of the hypocycloid core-contour, along with the future prospect in fiber photonics that can be achieved with IC-HCPCF.

We finished the review by giving some examples of the results obtained in varied fields using HCPCF and its gas-filled form PMC. These include high energy ultrafast beam delivery, pulse compression with values down to the single-cycle regime, multi-octave fiber-based light sources based on stimulated Raman scattering, supercontinuum generation gases, UV-DUV radiation sources via by plasma generation, gas phase nanostructuring and molecular trapping, and finally, nonclassical light generation.

Author Contributions: Conceptualization, F.B.; Methodology, F.B.; Validation, F.B. and B.D.; Formal Analysis, F.B and B.D..; Investigation, F.B., B.D., F.G., F.A., L.V.; Resources, B.D., F.A., F.G., L.V., F.B.; Data Curation, B.D., F.A., F.G., L.V., F.B.; Writing-Original Draft Preparation, F.B., B.D.; Writing-Review & Editing, B.D., F.A., F.G., L.V., F.B.; Supervision, F.B.; Project Administration, F.B.

Funding: This research received no external funding.

Conflicts of Interest: The authors declare no conflict of interest.

References

1. Cregan, R.F.; Mangan, B.J.; Knight, J.C.; Birks, T.A.; Russell, P.S.J.; Roberts, P.J.; Allan, D.C. Single-Mode Photonic Band Gap Guidance of Light in Air. *Science* **1999**, *285*, 1537–1539. [CrossRef] [PubMed]
2. Couny, F.; Benabid, F.; Roberts, P.J.; Burnett, M.T.; Maier, S.A. Identification of Bloch-modes in hollow-core photonic crystal fiber cladding. *Opt. Express* **2007**, *15*, 325–338. [CrossRef] [PubMed]
3. Von Neumann, J.; Wigner, E. Über merkwürdige diskrete Eigenwerte. *Phys. Z.* **1929**, *30*, 465–467.
4. Couny, F.; Benabid, F.; Roberts, P.J.; Light, P.S.; Raymer, M.G. Generation and photonic guidance of multi-octave optical-frequency combs. *Science* **2007**, *318*, 1118–1121. [CrossRef]
5. Benabid, F. Hollow-core photonic bandgap fibre: New light guidance for new science and technology. *Philos. Trans. A Math. Phys. Eng. Sci.* **2006**, *364*, 3439–3462. [CrossRef]
6. Benabid, F.; Couny, F.; Knight, J.C.; Birks, T.A.; Russell, P.S.J. Compact, stable and efficient all-fibre gas cells using hollow-core photonic crystal fibres. *Nature* **2005**, *434*, 488. [CrossRef]
7. Benabid, F.; Roberts, P.J. Linear and nonlinear optical properties of hollow core photonic crystal fiber. *J. Mod. Opt.* **2011**, *58*, 87–124. [CrossRef]
8. Couny, F.; Benabid, F.; Roberts, P.J.; Light, P.S. Fresnel zone imaging of Bloch-modes from a Hollow-Core Photonic Crystal Fiber Cladding. In Proceedings of the 2007 Conference on Lasers and Electro-Optics (CLEO), Baltimore, MD, USA, 6–11 May 2007; pp. 1–2.
9. Couny, F.; Benabid, F.; Light, P.S. Subwatt threshold cw Raman fiber-gas laser based on H$_2$-filled hollow-core photonic crystal fiber. *Phys. Rev. Lett.* **2007**, *99*, 143903. [CrossRef]
10. Benoît, A.; Beaudou, B.; Alharbi, M.; Debord, B.; Gérôme, F.; Salin, F.; Benabid, F. Over-five octaves wide Raman combs in high-power picosecond-laser pumped H$_2$-filled inhibited coupling Kagome fiber. *Opt. Express* **2015**, *23*, 14002–14009. [CrossRef]
11. Heckl, O.H.; Baer, C.R.E.; Kränkel, C.; Marchese, S.V.; Schapper, F.; Holler, M.; Südmeyer, T.; Robinson, J.S.; Tisch, J.W.G.; Couny, F.; et al. High harmonic generation in a gas-filled hollow-core photonic crystal fiber. *Appl. Phys. B* **2009**, *97*, 369. [CrossRef]
12. Belli, F.; Abdolvand, A.; Chang, W.; Travers, J.C.; Russell, P.S.J. Vacuum-ultraviolet to infrared supercontinuum in hydrogen-filled photonic crystal fiber. *Optica* **2015**, *2*, 292–300. [CrossRef]
13. Debord, B.; Gérôme, F.; Honninger, C.; Mottay, E.; Husakou, A.; Benabid, F. Milli-Joule energy-level comb and supercontinuum generation in atmospheric air-filled inhibited coupling Kagome fiber. In Proceedings of the 2015 Conference on Lasers and Electro-Optics (CLEO), San Jose, CA, USA, 10–15 May 2015; pp. 1–2.
14. Light, P.S.; Couny, F.; Benabid, F. Low optical insertion-loss and vacuum-pressure all-fiber acetylene cell based on hollow-core photonic crystal fiber. *Opt. Lett.* **2006**, *31*, 2538–2540. [CrossRef] [PubMed]
15. Heckl, O.H.; Saraceno, C.J.; Baer, C.R.E.; Südmeyer, T.; Wang, Y.Y.; Cheng, Y.; Benabid, F.; Keller, U. Temporal pulse compression in a xenon-filled Kagome-type hollow-core photonic crystal fiber at high average power. *Opt. Express* **2011**, *19*, 19142–19149. [CrossRef] [PubMed]
16. Balciunas, T.; Fourcade-Dutin, C.; Fan, G.; Witting, T.; Voronin, A.A.; Zheltikov, A.M.; Gerome, F.; Paulus, G.G.; Baltuska, A.; Benabid, F. A strong-field driver in the single-cycle regime based on self-compression in a kagome fibre. *Nat. Commun.* **2015**, *6*, 6117. [CrossRef]
17. Debord, B.; Dontabactouny, M.; Alharbi, M.; Fourcade-Dutin, C.; Honninger, C.; Mottay, E.; Vincetti, L.; Gerome, F.; Benabid, F. Multi-meter fiber-delivery and compression of milli-Joule femtosecond laser and fiber-aided micromachining. In Proceedings of the Advanced Solid-State Lasers Congress, Paris, France, 27 October–1 November 2013.
18. Jones, A.M.; Nampoothiri, A.V.V.; Ratanavis, A.; Fiedler, T.; Wheeler, N.V.; Couny, F.; Kadel, R.; Benabid, F.; Washburn, B.R.; Corwin, K.L.; et al. Mid-infrared gas filled photonic crystal fiber laser based on population inversion. *Opt. Express* **2011**, *19*, 2309–2316. [CrossRef] [PubMed]
19. Nampoothiri, A.V.V.; Jones, A.M.; Fourcade-Dutin, C.; Mao, C.; Dadashzadeh, N.; Baumgart, B.; Wang, Y.Y.; Alharbi, M.; Bradley, T.; Campbell, N.; et al. Hollow-core Optical Fiber Gas Lasers (HOFGLAS): A review [Invited]. *Opt. Mater. Express* **2012**, *2*, 948–961. [CrossRef]
20. Ghosh, S.; Bhagwat, A.R.; Renshaw, C.K.; Goh, S.; Gaeta, A.L.; Kirby, B.J. Low-Light-Level Optical Interactions with Rubidium Vapor in a Photonic Band-Gap Fiber. *Phys. Rev. Lett.* **2006**, *97*, 23603. [CrossRef]
21. Epple, G.G.; Kleinbach, K.S.; Euser, T.G.; Joly, N.Y.; Pfau, T.; Russell, P.S.J.; Löw, R. Rydberg atoms in hollow-core photonic crystal fibres. *Nat. Commun.* **2014**, *5*, 4132. [CrossRef]

22. Perrella, C.; Light, P.S.; Anstie, J.D.; Benabid, F.; Stace, T.M.; White, A.G.; Luiten, A.N. High-efficiency cross-phase modulation in a gas-filled waveguide. *Phys. Rev. A* **2013**, *88*, 13819. [CrossRef]
23. Light, P.S.; Benabid, F.; Maric, M.; Luiten, A.N.; Couny, F. Electromagnetically induced transparency in rubidium-filled kagome HC-PCF. In Proceedings of the 2008 Conference on Lasers and Electro-Optics and 2008 Conference on Quantum Electronics and Laser Science, San Jose, CA, USA, 4–9 May 2008; pp. 1–2.
24. Knebl, A.; Yan, D.; Popp, J.; Frosch, T. Fiber enhanced Raman gas spectroscopy. *TrAC Trends Anal. Chem.* **2018**, *103*, 230–238. [CrossRef]
25. Wang, Y.Y.; Couny, F.; Roberts, P.J.; Benabid, F. Low loss broadband transmission in optimized core-shape Kagome hollow-core PCF. In Proceedings of the CLEO/QELS: 2010 Laser Science to Photonic Applications, San Jose, CA, USA, 16–21 May 2010; pp. 1–2.
26. Wang, Y.Y.; Wheeler, N.V.; Couny, F.; Roberts, P.J.; Benabid, F. Low loss broadband transmission in hypocycloid-core Kagome hollow-core photonic crystal fiber. *Opt. Lett.* **2011**, *36*, 669–671. [CrossRef] [PubMed]
27. Debord, B.; Jamier, R.; Gérôme, F.; Leroy, O.; Boisse-Laporte, C.; Leprince, P.; Alves, L.L.; Benabid, F. Generation and confinement of microwave gas-plasma in photonic dielectric microstructure. *Opt. Express* **2013**, *21*, 25509–25516. [CrossRef] [PubMed]
28. Alharbi, M.; Husakou, A.; Chafer, M.; Debord, B.; Gérôme, F.; Benabid, F. Raman gas self-organizing into deep nano-trap lattice. *Nat. Commun.* **2016**, *7*, 12779. [CrossRef] [PubMed]
29. Okaba, S.; Takano, T.; Benabid, F.; Bradley, T.; Vincetti, L.; Maizelis, Z.; Yampol'skii, V.; Nori, F.; Katori, H. Lamb-Dicke spectroscopy of atoms in a hollow-core photonic crystal fibre. *Nat. Commun.* **2014**, *5*, 4096. [CrossRef] [PubMed]
30. Debord, B.; Amsanpally, A.; Chafer, M.; Baz, A.; Maurel, M.; Blondy, J.M.; Hugonnot, E.; Scol, F.; Vincetti, L.; Gérôme, F.; Benabid, F. Ultralow transmission loss in inhibited-coupling guiding hollow fibers. *Optica* **2017**, *4*, 209–217. [CrossRef]
31. Debord, B.; Alharbi, M.; Vincetti, L.; Husakou, A.; Fourcade-Dutin, C.; Hoenninger, C.; Mottay, E.; Gérôme, F.; Benabid, F. Multi-meter fiber-delivery and pulse selfcompression of milli-Joule femtosecond laser and fiber-aided laser-micromachining. *Opt. Express* **2014**, *22*, 10735–10746. [CrossRef]
32. Birks, T.A.; Roberts, P.J.; Russell, P.S.J.; Atkin, D.M.; Shepherd, T.J. Full 2-D photonic bandgaps in silica/air structures. *Electron. Lett.* **1995**, *31*, 1941–1943. [CrossRef]
33. Birks, T.A.; Knight, J.C.; Russell, P.S.J. Endlessly single-mode photonic crystal fiber. *Opt. Lett.* **1997**, *22*, 961–963. [CrossRef]
34. Knight, J.C.; Birks, T.A.; Cregan, R.F.; Russell, P.S.J.; de Sandro, P.D. Large mode area photonic crystal fibre. *Electron. Lett.* **1998**, *34*, 1347–1348. [CrossRef]
35. Limpert, J.; Liem, A.; Reich, M.; Schreiber, T.; Nolte, S.; Zellmer, H.; Tünnermann, A.; Broeng, J.; Petersson, A.; Jakobsen, C. Low-nonlinearity single-transverse-mode ytterbium-doped photonic crystal fiber amplifier. *Opt. Express* **2004**, *12*, 1313–1319. [CrossRef]
36. Ortigosa-Blanch, A.; Knight, J.C.; Wadsworth, W.J.; Arriaga, J.; Mangan, B.J.; Birks, T.A.; Russell, P.S.J. Highly birefringent photonic crystal fibers. *Opt. Lett.* **2000**, *25*, 1325–1327. [CrossRef] [PubMed]
37. Mangan, B.J.; Couny, F.; Farr, L.; Langford, A.; Roberts, P.J.; Williams, D.P.; Banham, M.; Mason, M.W.; Murphy, D.F.; Brown, E.A.M.; et al. Slope-matched dispersion-compensating photonic crystal fibre. In Proceedings of the Conference on Lasers and Electro-Optics/International Quantum Electronics Conference and Photonic Applications Systems Technologies, San Francisco, CA, USA, 16–21 May 2014.
38. Argyros, A.; Birks, T.A.; Leon-Saval, S.G.; Cordeiro, C.M.B.; Russell, P.S.J. Guidance properties of low-contrast photonic bandgap fibres. *Opt. Express* **2005**, *13*, 2503–2511. [CrossRef] [PubMed]
39. Grujic, T.; Kuhlmey, B.T.; Argyros, A.; Coen, S.; de Sterke, C.M. Solid-core fiber with ultra-wide bandwidth transmission window due to inhibited coupling. *Opt. Express* **2010**, *18*, 25556–25566. [CrossRef]
40. Cerqueira, S.A.; Luan, F.; Cordeiro, C.M.B.; George, A.K.; Knight, J.C. Hybrid photonic crystal fiber. *Opt. Express* **2006**, *14*, 926–931. [CrossRef]
41. Monro, T.M.; West, Y.D.; Hewak, D.W.; Broderick, N.G.R.; Richardson, D.J. Chalcogenide holey fibres. *Electron. Lett.* **2000**, *36*, 1998–2000. [CrossRef]
42. Sanghera, J.S.; Shaw, L.B.; Pureza, P.; Nguyen, V.Q.; Gibson, D.; Busse, L.; Aggarwal, I.D.; Florea, C.M.; Kung, F.H. Nonlinear Properties of Chalcogenide Glass Fibers. *Int. J. Appl. Glas. Sci.* **2010**, *1*, 296–308. [CrossRef]
43. John, S. Strong localization of photons in certain disordered dielectric superlattices. *Phys. Rev. Lett.* **1987**, *58*, 2486–2489. [CrossRef]

44. Yablonovitch, E. Inhibited Spontaneous Emission in Solid-State Physics and Electronics. *Phys. Rev. Lett.* **1987**, *58*, 2059–2062. [CrossRef]
45. Yablonovitch, E. Photonic band-gap structures. *J. Opt. Soc. Am. B* **1993**, *10*, 283. [CrossRef]
46. Benabid, F.; Knight, J.C.; Antonopoulos, G.; Russell, P.S.J. Stimulated Raman scattering in hydrogen-filled hollow-core photonic crystal fiber. *Science* **2002**, *298*, 399–402. [CrossRef]
47. Venkataraman, N.; Gallagher, M.T.; Smith, C.M.; Muller, D.; West, J.A.; Koch, K.W.; Fajardo, J.C. Low Loss (13 dB/km) Air Core Photonic Band-Gap Fibre. In Proceedings of the 2002 28th European Conference on Optical Communication, Copenhagen, Denmark, 8–12 September 2002; Volume 5, pp. 1–2.
48. Russell, P. Photonic Crystal Fibers. *Science* **2003**, *299*, 358–362. [CrossRef]
49. Roberts, P.; Couny, F.; Sabert, H.; Mangan, B.J.; Williams, D.P.; Farr, L.; Mason, M.W.; Tomlinson, A.; Birks, T.A.; Knight, J.C.; et al. Ultimate low loss of hollow-core photonic crystal fibres. *Opt. Express* **2005**, *13*, 236–244. [CrossRef] [PubMed]
50. Marcatili, E.A.J.; Schmeltzer, R.A. Hollow metallic and dielectric waveguides for long distance optical transmission and lasers. *Bell Syst. Tech. J.* **1964**, *43*, 1783–1809. [CrossRef]
51. Duguay, M.A.; Kokubun, Y.; Koch, T.L.; Pfeiffer, L. Antiresonant reflecting optical waveguides in SiO_2-Si multilayer structures. *Appl. Phys. Lett.* **1986**, *49*, 13–15. [CrossRef]
52. Hsu, C.W.; Zhen, B.; Stone, A.D.; Joannopoulos, J.D.; Soljačić, M. Bound states in the continuum. *Nat. Rev. Mater.* **2016**, *1*, 16048. [CrossRef]
53. Pryamikov, A.D.; Biriukov, A.S.; Kosolapov, A.F.; Plotnichenko, V.G.; Semjonov, S.L.; Dianov, E.M. Demonstration of a waveguide regime for a silica hollow—Core microstructured optical fiber with a negative curvature of the core boundary in the spectral region > 3.5 μm. *Opt. Express* **2011**, *19*, 1441. [CrossRef] [PubMed]
54. Yu, F.; Wadsworth, W.J.; Knight, J.C. Low loss silica hollow core fibers for 3–4 μm spectral region. *Opt. Express* **2012**, *20*, 11153–11158. [CrossRef]
55. Vincetti, L.; Setti, V. Waveguiding mechanism in tube lattice fibers. *Opt. Express* **2010**, *18*, 23133. [CrossRef]
56. Poletti, F. Nested antiresonant nodeless hollow core fiber. *Opt. Express* **2014**, *22*, 23807. [CrossRef]
57. Habib, M.S.; Bang, O.; Bache, M. Low-loss hollow-core silica fibers with adjacent nested anti-resonant tubes. *Opt. Express* **2015**, *23*, 17394–17406. [CrossRef]
58. Gao, S.; Wang, Y.; Ding, W.; Jiang, D.; Gu, S.; Zhang, X.; Wang, P. Hollow-core conjoined-tube negative-curvature fibre with ultralow loss. *Nat. Commun.* **2018**, *9*, 2828. [CrossRef] [PubMed]
59. Belardi, W.; Knight, J.C. Hollow antiresonant fibers with reduced attenuation. *Opt. Lett.* **2014**, *39*, 1853. [CrossRef] [PubMed]
60. Belardi, W. Design and Properties of Hollow Antiresonant Fibers for the Visible and Near Infrared Spectral Range. *J. Light. Technol.* **2015**, *33*, 4497–4503. [CrossRef]
61. Wang, Y.Y.; Peng, X.; Alharbi, M.; Dutin, C.F.; Bradley, T.D.; Gérôme, F.; Mielke, M.; Booth, T.; Benabid, F. Design and fabrication of hollow-core photonic crystal fibers for high-power ultrashort pulse transportation and pulse compression. *Opt. Lett.* **2012**, *37*, 3111–3113. [CrossRef] [PubMed]
62. Bradley, T.D.; Wang, Y.; Alharbi, M.; Debord, B.; Fourcade-Dutin, C.; Beaudou, B.; Gerome, F.; Benabid, F. Optical properties of low loss (70 dB/km) hypocycloid-core kagome hollow core photonic crystal fiber for Rb and Cs based optical applications. *J. Light. Technol.* **2013**, *31*, 2752–2755. [CrossRef]
63. Debord, B.; Alharbi, M.; Bradley, T.; Fourcade-Dutin, C.; Wang, Y.Y.; Vincetti, L.; Gérôme, F.; Benabid, F. Hypocycloid-shaped hollow-core photonic crystal fiber Part I: Arc curvature effect on confinement loss. *Opt. Express* **2013**, *21*, 28597–28608. [CrossRef] [PubMed]
64. Debord, B.; Alharbi, M.; Benoît, A.; Ghosh, D.; Dontabactouny, M.; Vincetti, L.; Blondy, J.-M.; Gérôme, F.; Benabid, F. Ultra low-loss hypocycloid-core Kagome hollow-core photonic crystal fiber for green spectral-range applications. *Opt. Lett.* **2014**, *39*, 6245–6248. [CrossRef] [PubMed]
65. Maurel, M.; Chafer, M.; Amsanpally, A.; Adnan, M.; Amrani, F.; Debord, B.; Vincetti, L.; Gérôme, F.; Benabid, F. Optimized inhibited-coupling Kagome fibers at Yb-Nd:Yag (8.5 dB/km) and Ti:Sa (30 dB/km) ranges. *Opt. Lett.* **2018**, *43*, 1598–1601. [CrossRef] [PubMed]
66. Bagley, R.D. Extrusion Method for Forming thin-Walled Honeycomb Structures. U.S. Patent US3790654A, 5 February 1974.
67. Knight, J.C.; Birks, T.A.; Russell, P.S.J.; Atkin, D.M. All-silica single-mode optical fiber with photonic crystal cladding. *Opt. Lett.* **1996**, *21*, 1547. [CrossRef] [PubMed]

68. Edagawa, K. Photonic crystals, amorphous materials, and quasicrystals. *Sci. Technol. Adv. Mater.* **2014**, *15*, 34805. [CrossRef] [PubMed]
69. Snyder, A.W.; Love, J.D. *Optical Waveguide Theory*; Chapman & Hall: London, UK; New York, NY, USA, 1983.
70. Shechtman, D.; Blech, I.; Gratias, D.; Cahn, J.W. Metallic Phase with Long-Range Orientational Order and No Translational Symmetry. *Phys. Rev. Lett.* **1984**, *53*, 1951–1953. [CrossRef]
71. Couny, F.; Benabid, F.; Light, P.S. Large-pitch kagome-structured hollow-core photonic crystal fiber. *Opt. Lett.* **2006**, *31*, 3574–3576. [CrossRef] [PubMed]
72. Marzari, N.; Mostofi, A.A.; Yates, J.R.; Souza, I.; Vanderbilt, D. Maximally localized Wannier functions: Theory and applications. *Rev. Mod. Phys.* **2012**, *84*, 1419–1475. [CrossRef]
73. Argyros, A.; Birks, T.A.; Leon-Saval, S.G.; Cordeiro, C.M.B.; Luan, F.; Russell, P.S.J. Photonic bandgap with an index step of one percent. *Opt. Express* **2005**, *13*, 309–314. [CrossRef] [PubMed]
74. Light, P.S.; Couny, F.; Wang, Y.Y.; Wheeler, N.V.; Roberts, P.J.; Benabid, F. Double photonic bandgap hollow-core photonic crystal fiber. *Opt. Express* **2009**, *17*, 16238–16243. [CrossRef] [PubMed]
75. Hedley, T.D.; Bird, D.M.; Benabid, F.; Knight, J.C.; Russell, P.S.J. Modelling of a novel hollow-core photonic crystal fibre. In Proceedings of the Conference on Quantum Electronics and Laser Science (QELS), Baltimore, MD, USA, 6 June 2003.
76. Archambault, J.L.; Black, R.J.; Lacroix, S.; Bures, J. Loss calculations for antiresonant waveguides. *J. Light. Technol.* **1993**, *11*, 416–423. [CrossRef]
77. Tamir, T. Integrated Optics. In *Topics in Applied Physics*; Springer: Berlin, Germany, 1975; Volume 7.
78. Benabid, F.; Roberts, P.J. Photonic Crystal Hollow Waveguides. In *Handbook Of Optofluidics*; Hawkins, A.R., Schmidt, H., Eds.; Taylor & Francis Group: Abingdon, UK, 2010.
79. Birks, T.A.; Bird, D.M.; Benabid, F.; Roberts, P.J. Strictly-bound modes of an idealised hollow-core fibre without a photonic bandgap. In Proceedings of the European Conference on Optical Communication, ECOC, Torino, Italy, 19–23 September 2010; pp. 1–2.
80. Couny, F.; Benabid, F.; Light, P.S. Large Pitch Kagome-Structured Hollow-Core PCF. In Proceedings of the 2007 Conference on Lasers and Electro-Optics (CLEO), Baltimore, MD, USA, 6–11 May 2007; pp. 1–2.
81. Couny, F.; Roberts, P.J.; Birks, T.A.; Benabid, F. Square-lattice large-pitch hollow-core photonic crystal fiber. *Opt. Express* **2008**, *16*, 20626–20636. [CrossRef] [PubMed]
82. Beaudou, B.; Couny, F.; Benabid, F.; Roberts, P.J. Large Pitch Hollow Core Honeycomb Fiber. In Proceedings of the Conference on Lasers and Electro-Optics/Quantum Electronics and Laser Science Conference and Photonic Applications Systems Technologies, San Jose, CA, USA, 4–9 May 2008.
83. Alharbi, M.; Bradley, T.; Debord, B.; Fourcade-Dutin, C.; Ghosh, D.; Vincetti, L.; Gérôme, F.; Benabid, F. Hypocycloid-shaped hollow-core photonic crystal fiber Part II: Cladding effect on confinement and bend loss. *Opt. Express* **2013**, *21*, 28609–28616. [CrossRef]
84. Fokoua, E.N.; Poletti, F.; Richardson, D.J. Analysis of light scattering from surface roughness in hollow-core photonic bandgap fibers. *Opt. Express* **2012**, *20*, 20980. [CrossRef]
85. Kharadly, M.M.Z.; Lewis, J.E. Properties of dielectric-tube waveguides. *Proc. Inst. Electr. Eng.* **1969**, *116*, 214–224. [CrossRef]
86. Burns, W.K. Normal mode analysis of waveguide devices. I. Theory. *J. Light. Technol.* **1988**, *6*, 1051–1057. [CrossRef]
87. Uebel, P.; Günendi, M.C.; Frosz, M.H.; Ahmed, G.; Edavalath, N.N.; Ménard, J.; Russell, P.S.J. Broadband robustly single-mode hollow-core PCF by resonant filtering of higher-order modes. *Opt. Lett.* **2016**, *41*, 1961. [CrossRef] [PubMed]
88. Michieletto, M.; Lyngsø, J.K.; Jakobsen, C.; Lægsgaard, J.; Bang, O.; Alkeskjold, T.T. Hollow-core fibers for high power pulse delivery. *Opt. Express* **2016**, *24*, 7103. [CrossRef] [PubMed]
89. Hayes, J.R.; Sandoghchi, S.R.; Bradley, T.D.; Liu, Z.; Slavík, R.; Gouveia, M.A.; Wheeler, N.V.; Jasion, G.; Chen, Y.; Fokoua, E.N.; et al. Antiresonant Hollow Core Fiber With an Octave Spanning Bandwidth for Short Haul Data Communications. *J. Light. Technol.* **2017**, *35*, 437–442. [CrossRef]
90. Chafer, M.; Delahaye, F.; Amrani, F.; Debord, B.; Gérôme, F.; Benabid, F. 1 km long HC-PCF with losses at the fundamental Rayleigh scattering limit in the green wavelength range. In Proceedings of the Conference on Lasers and Electro-Optics, San Jose, CA, USA, 13–18 May 2018.
91. Gao, S.-F.; Wang, Y.-Y.; Ding, W.; Wang, P. Hollow-core negative-curvature fiber for UV guidance. *Opt. Lett.* **2018**, *43*, 1347. [CrossRef] [PubMed]

92. Litchinitser, N.M.; Abeeluck, A.K.; Headley, C.; Eggleton, B.J. Antiresonant reflecting photonic crystal optical waveguides. *Opt. Lett.* **2002**, *27*, 1592–1594. [CrossRef] [PubMed]

93. Miyagi, M.; Nishida, S. A Proposal of Low-Loss Leaky Waveguide for Submillimeter Waves Transmission. *IEEE Trans. Microw. Theory Tech.* **1980**, *28*, 398–401. [CrossRef]

94. Bornstein, A.; Croitoru, N. Experimental evaluation of a hollow glass fiber. *Appl. Opt.* **1986**, *25*, 355–358. [CrossRef]

95. Zeisberger, M.; Schmidt, M.A. Analytic model for the complex effective index of the leaky modes of tube-type anti-resonant hollow core fibers. *Sci. Rep.* **2017**, *7*, 11761. [CrossRef]

96. Bird, D. Attenuation of model hollow-core, anti-resonant fibres. *Opt. Express* **2017**, *25*, 23215–23237. [CrossRef] [PubMed]

97. Fini, J.M.; Nicholson, J.W.; Mangan, B.; Meng, L.; Windeler, R.S.; Monberg, E.M.; DeSantolo, A.; DiMarcello, F.V.; Mukasa, K. Polarization maintaining single-mode low-loss hollow-core fibres. *Nat. Commun.* **2014**, *5*, 5085. [CrossRef] [PubMed]

98. Nicholson, J.W.; Yablon, A.D.; Ramachandran, S.; Ghalmi, S. Spatially and spectrally resolved imaging of modal content in large-mode-area fibers. *Opt. Express* **2008**, *16*, 7233–7243. [CrossRef] [PubMed]

99. Debord, B.; Gérôme, F.; Paul, P.-M.; Husakou, A.; Benabid, F. 2.6 mJ energy and 81 GW peak power femtosecond laser-pulse delivery and spectral broadening in inhibited coupling Kagome fiber. In Proceedings of the CLEO: 2015, San Jose, CA, USA, 10–15 May 2015.

100. Beaudou, B.; Gerôme, F.; Wang, Y.Y.; Alharbi, M.; Bradley, T.D.; Humbert, G.; Auguste, J.-L.; Blondy, J.-M.; Benabid, F. Millijoule laser pulse delivery for spark ignition through kagome hollow-core fiber. *Opt. Lett.* **2012**, *37*, 1430. [CrossRef] [PubMed]

101. Peng, X.; Mielke, M.; Booth, T. High average power, high energy 155 μm ultra-short pulse laser beam delivery using large mode area hollow core photonic band-gap fiber. *Opt. Express* **2011**, *19*, 923. [CrossRef] [PubMed]

102. Shephard, J.D.; Couny, F.; Russell, P.S.J.; Jones, J.D.C.; Knight, J.C.; Hand, D.P. Improved hollow-core photonic crystal fiber design for delivery of nanosecond pulses in laser micromachining applications. *Appl. Opt.* **2005**, *44*, 4582. [CrossRef]

103. Tauer, J.; Orban, F.; Kofler, H.; Fedotov, A.B.; Fedotov, I.V.; Mitrokhin, V.P.; Zheltikov, A.M.; Wintner, E. High-throughput of single high-power laser pulses by hollow photonic band gap fibers. *Laser Phys. Lett.* **2007**, *4*, 444–448. [CrossRef]

104. Hädrich, S.; Rothhardt, J.; Demmler, S.; Tschernajew, M.; Hoffmann, A.; Krebs, M.; Liem, A.; de Vries, O.; Plötner, M.; Fabian, S.; et al. Scalability of components for kW-level average power few-cycle lasers. *Appl. Opt.* **2016**, *55*, 1636. [CrossRef]

105. Lee, E.; Luo, J.; Sun, B.; Ramalingam, V.L.; Yu, X.; Wang, Q.; Yu, F.; Knight, J.C. 45W 2 μm Nanosecond Pulse Delivery Using Antiresonant Hollow-Core Fiber. In Proceedings of the Conference on Lasers and Electro-Optics, San Jose, CA, USA, 13–18 May 2018.

106. Hädrich, S.; Krebs, M.; Hoffmann, A.; Klenke, A.; Rothhardt, J.; Limpert, J.; Tünnermann, A. Exploring new avenues in high repetition rate table-top coherent extreme ultraviolet sources. *Light Sci. Appl.* **2015**, *4*, e320. [CrossRef]

107. Jaworski, P.; Yu, F.; Carter, R.M.; Knight, J.C.; Shephard, J.D.; Hand, D.P. High energy green nanosecond and picosecond pulse delivery through a negative curvature fiber for precision micro-machining. *Opt. Express* **2015**, *23*, 8498. [CrossRef]

108. Gebert, F.; Frosz, M.H.; Weiss, T.; Wan, Y.; Ermolov, A.; Joly, N.Y.; Schmidt, P.O.; Russell, P.S.J. Damage-free single-mode transmission of deep-UV light in hollow-core PCF. *Opt. Express* **2014**, *22*, 15388–15396. [CrossRef] [PubMed]

109. Pricking, S.; Gebs, R.; Fleischhaker, R.; Kleinbauer, J.; Budnicki, A.; Sutter, D.H.; Killi, A.; Weiler, S.; Mielke, M.; Beaudou, B.; et al. Hollow core fiber delivery of sub-ps pulses from a TruMicro 5000 Femto edition thin disk amplifier. *Proc. SPIE* **2015**, *9356*, 935602.

110. Yalin, A.P. High power fiber delivery for laser ignition applications. *Opt. Express* **2013**, *21*, A1102. [CrossRef]

111. Subramanian, K.; Gabay, I.; Ferhanoğlu, O.; Shadfan, A.; Pawlowski, M.; Wang, Y.; Tkaczyk, T.; Ben-Yakar, A. Kagome fiber based ultrafast laser microsurgery probe delivering micro-Joule pulse energies. *Biomed. Opt. Express* **2016**, *7*, 4639. [CrossRef] [PubMed]

112. Ferhanoglu, O.; Yildirim, M.; Subramanian, K.; Ben-Yakar, A. A 5-mm piezo-scanning fiber device for high speed ultrafast laser microsurgery. *Biomed. Opt. Express* **2014**, *5*, 2023. [CrossRef] [PubMed]

113. Keller, U. Recent developments in compact ultrafast lasers. *Nature* **2003**, *424*, 831. [CrossRef] [PubMed]
114. Mak, K.F.; Travers, J.C.; Joly, N.Y.; Abdolvand, A.; Russell, P.S.J. Two techniques for temporal pulse compression in gas-filled hollow-core kagomé photonic crystal fiber. *Opt. Lett.* **2013**, *38*, 3592–3595. [CrossRef]
115. Guichard, F.; Giree, A.; Zaouter, Y.; Hanna, M.; Machinet, G.; Debord, B.; Gérôme, F.; Dupriez, P.; Druon, F.; Hönninger, C.; et al. Nonlinear compression of high energy fiber amplifier pulses in air-filled hypocycloid-core Kagome fiber. *Opt. Express* **2015**, *23*, 7416–7423. [CrossRef]
116. Emaury, F.; Dutin, C.F.; Saraceno, C.J.; Trant, M.; Heckl, O.H.; Wang, Y.Y.; Schriber, C.; Gerome, F.; Südmeyer, T.; Benabid, F.; Keller, U. Beam delivery and pulse compression to sub-50 fs of a modelocked thin-disk laser in a gas-filled Kagome-type HC-PCF fiber. *Opt. Express* **2013**, *21*, 4986. [CrossRef]
117. Gebhardt, M.; Gaida, C.; Heuermann, T.; Stutzki, F.; Jauregui, C.; Antonio-Lopez, J.; Schulzgen, A.; Amezcua-Correa, R.; Limpert, J.; Tünnermann, A. Nonlinear pulse compression to 43 W GW-class few-cycle pulses at 2 μm wavelength. *Opt. Lett.* **2017**, *42*, 4179–4182. [CrossRef]
118. Murari, K.; Stein, G.J.; Cankaya, H.; Debord, B.; Gérôme, F.; Cirmi, G.; Mücke, O.D.; Li, P.; Ruehl, A.; Hartl, I.; et al. Kagome-fiber-based pulse compression of mid-infrared picosecond pulses from a Ho:YLF amplifier. *Optica* **2016**, *3*, 816–822. [CrossRef]
119. Benabid, F.; Bouwmans, G.; Knight, J.C.; Russell, P.S.J.; Couny, F. Ultrahigh Efficiency Laser Wavelength Conversion in a Gas-Filled Hollow Core Photonic Crystal Fiber by Pure Stimulated Rotational Raman Scattering in Molecular Hydrogen. *Phys. Rev. Lett.* **2004**, *93*, 123903. [CrossRef] [PubMed]
120. Wang, Y.Y.; Wu, C.; Couny, F.; Raymer, M.G.; Benabid, F. Quantum-Fluctuation-Initiated Coherence in Multioctave Raman Optical Frequency Combs. *Phys. Rev. Lett.* **2010**, *105*, 123603. [CrossRef] [PubMed]
121. Hosseini, P.; Abdolvand, A.; Russell, P.S.J. Generation of spectral clusters in a mixture of noble and Raman-active gases. *Opt. Lett.* **2016**, *41*, 5543. [CrossRef] [PubMed]
122. Alharbi, M.; Debord, B.; Dontabactouny, M.; Gérôme, F.; Benabid, F. 17.6 THz waveform synthesis by phase-locked Raman sidebands generation in HC-PCF. In Proceedings of the CLEO: 2014, San Jose, CA, USA, 8–13 June 2014.
123. Cassataro, M.; Novoa, D.; Günendi, M.C.; Edavalath, N.N.; Frosz, M.H.; Travers, J.C.; Russell, P.S.J. Generation of broadband mid-IR and UV light in gas-filled single-ring hollow-core PCF. *Opt. Express* **2017**, *25*, 7637–7644. [CrossRef] [PubMed]
124. Raizer, Y.P. *Gas Discharge Physics*; Springer: Berlin, Ggrmany, 1991.
125. Moisan, M.; Beaudry, C.; Leprince, P. A Small Microwave Plasma Source for Long Column Production without Magnetic Field. *IEEE Trans. Plasma Sci.* **1975**, *3*, 55–59. [CrossRef]
126. Debord, B.; Alves, L.L.; Gérôme, F.; Jamier, R.; Leroy, O.; Boisse-Laporte, C.; Leprince, P.; Benabid, F. Microwave-driven plasmas in hollow-core photonic crystal fibres. *Plasma Sources Sci. Technol.* **2014**, *23*, 015022. [CrossRef]
127. Vial, F.; Gadonna, K.; Debord, B.; Delahaye, F.; Amrani, F.; Leroy, O.; Gérôme, F.; Benabid, F. Generation of surface-wave microwave microplasmas in hollow-core photonic crystal fiber based on a split-ring resonator. *Opt. Lett.* **2016**, *41*, 2286–2289. [CrossRef]
128. Amrani, F.; Delahaye, F.; Debord, B.; Alves, L.L.; Gerome, F.; Benabid, F. Gas mixture for deep-UV plasma emission in a hollow-core photonic crystal fiber. *Opt. Lett.* **2017**, *42*, 3363–3366. [CrossRef]
129. Chang, W.; Nazarkin, A.; Travers, J.C.; Nold, J.; Hölzer, P.; Joly, N.Y.; Russell, P.S.J. Influence of ionization on ultrafast gas-based nonlinear fiber optics. *Opt. Express* **2011**, *19*, 4856–4863. [CrossRef]
130. Saleh, M.F.; Biancalana, F. Understanding the dynamics of photoionization-induced nonlinear effects and solitons in gas-filled hollow-core photonic crystal fibers. *Phys. Rev. A* **2011**, *84*, 063838. [CrossRef]
131. Köttig, F.; Novoa, D.; Tani, F.; Günendi, M.C.; Cassataro, M.; Travers, J.C.; Russell, P.S.J. Mid-infrared dispersive wave generation in gas-filled photonic crystal fibre by transient ionization-driven changes in dispersion. *Nat. Commun.* **2017**, *8*, 813. [CrossRef] [PubMed]
132. Novoa, D.; Cassataro, M.; Travers, J.C.; Russell, P.S.J. Photoionization-Induced Emission of Tunable Few-Cycle Midinfrared Dispersive Waves in Gas-Filled Hollow-Core Photonic Crystal Fibers. *Phys. Rev. Lett.* **2015**, *115*, 033901. [CrossRef] [PubMed]
133. Hölzer, P.; Chang, W.; Travers, J.C.; Nazarkin, A.; Nold, J.; Joly, N.Y.; Saleh, M.F.; Biancalana, F.; Russell, P.S.J. Femtosecond Nonlinear Fiber Optics in the Ionization Regime. *Phys. Rev. Lett.* **2011**, *107*, 203901. [CrossRef] [PubMed]

134. Saleh, M.F.; Chang, W.; Hölzer, P.; Nazarkin, A.; Travers, J.C.; Joly, N.Y.; Russell, P.S.J.; Biancalana, F. Theory of Photoionization-Induced Blueshift of Ultrashort Solitons in Gas-Filled Hollow-Core Photonic Crystal Fibers. *Phys. Rev. Lett.* **2011**, *107*, 203902. [CrossRef]
135. Meng, F.; Liu, B.; Wang, S.; Liu, J.; Li, Y.; Wang, C.; Zheltikov, A.M.; Hu, M. Controllable two-color dispersive wave generation in argon-filled hypocycloid-core kagome fiber. *Opt. Express* **2017**, *25*, 32972. [CrossRef]
136. Joly, N.Y.; Nold, J.; Chang, W.; Hölzer, P.; Nazarkin, A.; Wong, G.K.L.; Biancalana, F.; Russell, P.S. Bright Spatially Coherent Wavelength-Tunable Deep-UV Laser Source Using an Ar-Filled Photonic Crystal Fiber. *Phys. Rev. Lett.* **2011**, *106*, 203901. [CrossRef] [PubMed]
137. Mak, K.F.; Travers, J.C.; Hölzer, P.; Joly, N.Y.; Russell, P.S.J. Tunable vacuum-UV to visible ultrafast pulse source based on gas-filled Kagome-PCF. *Opt. Express* **2013**, *21*, 10942. [CrossRef]
138. Light, P.S.; Benabid, F.; Couny, F.; Maric, M.; Luiten, A.N. Electromagnetically induced transparency in Rb-filled coated hollow-core photonic crystal fiber. *Opt. Lett.* **2007**, *32*, 1323–1325. [CrossRef]
139. Bradley, T.D.; Jouin, J.; McFerran, J.J.; Thomas, P.; Gerome, F.; Benabid, F. Extended duration of rubidium vapor in aluminosilicate ceramic coated hypocycloidal core Kagome HC-PCF. *J. Light. Technol.* **2014**, *32*, 2486–2491.
140. Bajcsy, M.; Hofferberth, S.; Balic, V.; Peyronel, T.; Hafezi, M.; Zibrov, A.S.; Vuletic, V.; Lukin, M.D. Efficient All-Optical Switching Using Slow Light within a Hollow Fiber. *Phys. Rev. Lett.* **2009**, *102*, 203902. [CrossRef] [PubMed]
141. Bajcsy, M.; Hofferberth, S.; Peyronel, T.; Balic, V.; Liang, Q.; Zibrov, A.S.; Vuletic, V.; Lukin, M.D. Laser-cooled atoms inside a hollow-core photonic-crystal fiber. *Phys. Rev. A* **2011**, *83*, 63830. [CrossRef]
142. Vorrath, S.; Möller, S.A.; Windpassinger, P.; Bongs, K.; Sengstock, K. Efficient guiding of cold atoms through a photonic band gap fiber. *New J. Phys.* **2010**, *12*, 123015. [CrossRef]
143. Langbecker, M.; Noaman, M.; Kjaergaard, N.; Benabid, F.; Windpassinger, P. Rydberg excitation of cold atoms inside a hollow-core fiber. *Phys. Rev. A* **2017**, *96*, 41402. [CrossRef]
144. Kumar, R.; Ong, J.R.; Savanier, M.; Mookherjea, S. Controlling the spectrum of photons generated on a silicon nanophotonic chip. *Nat. Commun.* **2014**, *5*, 5489. [CrossRef]
145. Ortiz-Ricardo, E.; Bertoni-Ocampo, C.; Ibarra-Borja, Z.; Ramirez-Alarcon, R.; Cruz-Delgado, D.; Cruz-Ramirez, H.; Garay-Palmett, K.; U'Ren, A.B. Spectral tunability of two-photon states generated by spontaneous four-wave mixing: Fibre tapering, temperature variation and longitudinal stress. *Quantum Sci. Technol.* **2017**, *2*, 034015. [CrossRef]
146. Cordier, M.; Orieux, A.; Debord, B.; Gérome, F.; Gorse, A.; Chafer, M.; Diamanti, E.; Delaye, P.; Benabid, F.; Zaquine, I. Shaping photon-pairs time-frequency correlations in inhibited-coupling hollow-core fibers. In Proceedings of the Conference on Lasers and Electro-Optics, San Jose, CA, USA, 13–18 May 2018.
147. Cordier, M.; Orieux, A.; Debord, B.; Gérome, F.; Gorse, A.; Chafer, M.; Diamanti, E.; Delaye, P.; Benabid, F.; Zaquine, I. Active engineering of four-wave mixing spectral entanglement in hollow-core fibers. *ArXiv*, 2018; arXiv:1807.11402.
148. Finger, M.A.; Joly, N.Y.; Russell, P.S.J.; Chekhova, M.V. Characterization and shaping of the time-frequency Schmidt mode spectrum of bright twin beams generated in gas-filled hollow-core photonic crystal fibers. *Phys. Rev. A* **2017**, *95*, 053814. [CrossRef]

fibers

MDPI

Review

Revolver Hollow Core Optical Fibers

Igor A. Bufetov, Alexey F. Kosolapov *, Andrey D. Pryamikov, Alexey V. Gladyshev, Anton N. Kolyadin, Alexander A. Krylov, Yury P. Yatsenko and Alexander S. Biriukov

Fiber Optics Research Center, Russian Academy of Sciences, Moscow 119333, Russia; iabuf@fo.gpi.ru(I.A.B.); pryamikov@fo.gpi.ru (A.D.P.); alexglad@fo.gpi.ru (A.V.G.); kolyadin@fo.gpi.ru (A.N.K.); krylov@fo.gpi.ru (A.A.K.); yuriya@fo.gpi.ru (Y.P.Y.); biriukov@fo.gpi.ru (A.S.B.)
* Correspondence: kaf@fo.gpi.ru; Tel.: +7-499-503-8207

Received: 15 May 2018; Accepted: 5 June 2018; Published: 7 June 2018

Abstract: Revolver optical fibers (RF) are special type of hollow-core optical fibers with negative curvature of the core-cladding boundary and with cladding that is formed by a one ring layer of capillaries. The physical mechanisms contributing to the waveguiding parameters of RFs are discussed. The optical properties and possible applications of RFs are reviewed. Special attention is paid to the mid-IR hydrogen Raman lasers that are based on RFs and generating in the wavelength region from 2.9 to 4.4 µm.

Keywords: hollow-core fibers; Raman lasers; negative curvature fibers; microstructured optical fibers

1. Introduction

Revolver fibers (RF), which are a special type of hollow-core fibers (HCF), were proposed and experimentally realized for the first time in FORC RAS in 2011 [1]. Since then, various RF designs have been demonstrated (Figure 1a–c). The key concept behind RFs is the negative curvature of the core-cladding interface. For the first time, this concept was introduced in [1]; where it was clearly shown that it is negative curvature that reduces the optical losses in the HCF significantly.

The importance of this concept was highlighted by authors [1] who initially suggested referring to such fibers as negative-curvature hollow-core fibers (NC HCF). Later, the abbreviation "NC HCF" became also to be applied to other fiber designs, such as Kagome HCFs with hypocycloid core-cladding boundary [2], and HCFs with "ice-cream-cone" shaped cladding [3]. However, RFs should be distinguished among other types of NC HCFs, because (1) the RFs provide an extremely low overlap of the optical mode with the cladding material and (2) RFs have very simple design of the cladding, which is based on a single layer of cylindrical or elliptical capillaries, and this fact enables new possibilities to optimize the fiber performance [4]. The design simplicity makes RFs a new starting point for the further development of low loss HCFs (Figure 1a–c). To distinguish such fibers from other NC HCFs, a separate name "revolver fibers" was proposed in [5].

It should be noted that RF scarcely could be attributed to photonic crystal fibers (PCF). The key feature of PCFs is the structure of fiber cladding that can be described as some unit cell linearly translated with some period in two dimensions of the fiber cross section. This complex structured cladding defines optical bandgaps, which correspond to the transmission spectral bands of the PCF, and thus, governs the optical properties of fiber. Alternatively, the cladding of RFs is not structured as a crystal. The optical properties of RFs are defined by reflection of light on the elements of the core-cladding interface only.

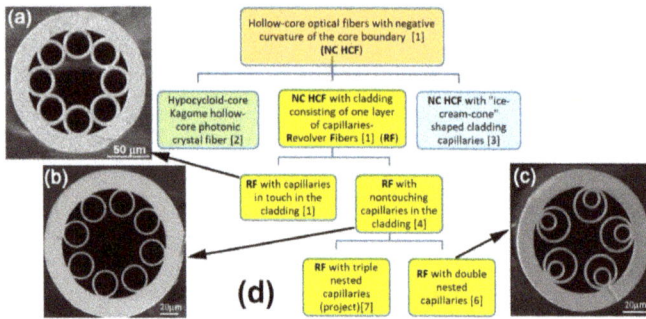

Figure 1. Cross section images of revolver fibers (RF): (**a**) RF with single touching capillaries in the cladding [1]; (**b**) RF with single non-touching capillaries in the cladding [4]; (**c**) RF with double nested non-touching capillaries [6]; and, (**d**) The family of hollow-core optical fibers with negative curvature of the core boundary [7].

This fact was confirmed by numerical simulations in many works (see, e.g., [8]). So, it seems not appropriate to refer RF as PCFs.

This review is organized as follows. Section 2 describes the general properties of RFs by simple analytical models. In Section 3, the properties of experimentally realized RFs are reviewed and supported by numerical simulations. Section 4 is devoted to a fabrication technology of the revolver fibers. Then, some applications of the RFs are discussed in Section 5, which reviews the recent advances in mid-infrared Raman lasers based on gas-filled revolver fibers, and in Section 6, where short pulse propagation in air-filled RFs is considered.

2. Physical Demonstrative Approach to the Waveguiding Properties of RF

Unlike PCFs that use the phenomenon of the energy band gaps formation to limit the propagation of light in a direction perpendicular to the axis of the fiber, the waveguiding properties of RFs are due to the reflection of radiation from structures that are located at the core-cladding interface. This interface influences the optical properties of RFs by means of many interrelated geometrical parameters, such as diameter of the hollow core, the shape and number of capillaries in the cladding, capillary diameter and wall thickness, etc. Usually, the effects of all geometrical parameters are precisely taken into account via time consuming numerical simulations (see, e.g., recent extensive review [9] and references therein). This Section, however, highlights the fact that general waveguiding properties of revolver fibers can be understood on the basis of simple analytical models using a kind of method of successive approximations.

As an initial approximation to RF, one can take the simplest model of optical waveguide in the form of an opening in a dielectric (its scheme is depicted in Figure 2a). Such hollow waveguide (HW) was considered in detail in [10]. In this case, the Fresnel reflection from the surface separating the hollow core with the dielectric determines the optical loss level of this fiber (Figure 2d, line 1). It is possible to significantly reduce the optical losses of such a fiber by increasing the reflection coefficient from the core-cladding interface e.g., by reflection from two surfaces, using as a waveguide, a capillary with a thin glass wall (tube waveguide—TW) and constructive interference of radiation that is reflected from both surfaces of the capillary. Such a fiber was considered in [11], and it can be taken as a second approximation to RF (model TW). In this case, the capillary wall serves as a Fabry-Perot interferometer, the transmission spectrum of the optical fiber, respectively, acquires a band structure. When the resonance condition for the radiation incident on the wall at an incidence angle of almost $\pi/2$ is satisfied, the reflection coefficient decreases, which leads to large optical losses in the optical fiber. If the resonance condition is violated (or the antiresonance condition is met), then the reflection coefficient from the capillary wall increases significantly and the optical fiber transparency zones are formed (see Figure 2d, line 2). Later, such a mechanism was actually re-considered in [12], and it was given the abbreviated name ARROW (AntiResonant Reflecting Optical Waveguide). Note that it is possible to further develop the resonantly reflecting structure of the fiber cladding (see, for example, [13]). In the case of RF, such a development leads to a structure with double nested capillaries [6,7] (see Figure 1c). Finally, an introduction of negative curvature at the core-cladding interface by forming a reflective cladding as a capillary layer (RF, [1]) preserves the band structure of the light transmission spectrum, but it leads to a further increase in the reflection coefficient of radiation at the core-cladding interface. As a consequence, the optical fiber losses are significantly reduced. This can be explained qualitatively, as follows. The parts of the capillaries walls that are located closer to the center of the core act as parts of the cladding in the TW model. Parts of the capillary walls, which deviate significantly from the circle inscribed into the RF core, interact with electromagnetic radiation as the sides of the corner with highly reflective coating (see Figure 2c, 2). In the ray approximation, it can be said that the light rays are reflected from these corner structures (see Figure 2c, 3), and the decrease in the glancing angle of the rays and the decrease in the radiation intensity as we approach the angle vertex [14] leads to a significant decrease in optical losses in comparison with the model TW. In addition, when the capillaries in the shell are separated by a distance $d \ll (2 \cdot \pi)/k_\perp$ from each other, where $k_\perp$ is the component of the wave vector perpendicular to the axis of the fiber, the propagation conditions of the radiation along the core practically do not change. This roughly corresponds to the removal of a part of the "mirror" angle at its vertex, which is indicated by the dotted line in Figure 2c, 3. However, in this case, the excitation of the cladding modes that are associated with the areas of contact of capillaries with each other will be substantially reduced. All of this leads as a result to the further effective reduction of optical losses in the RFs. This was confirmed experimentally in [4].

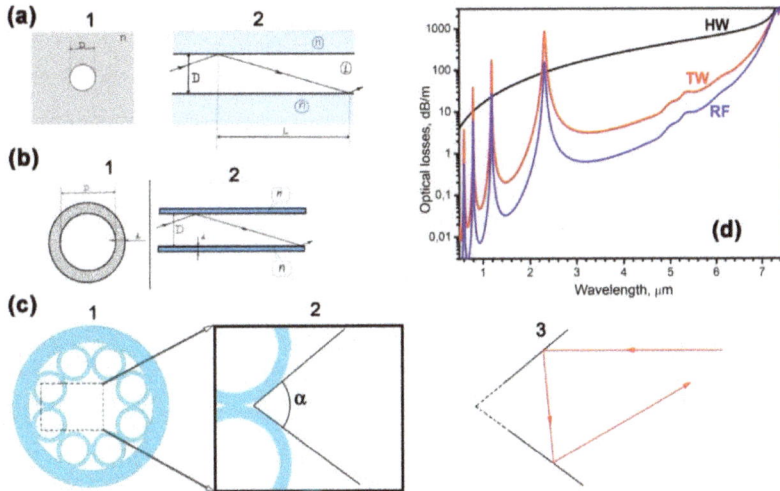

Figure 2. (a) Cross-(1) and longitudinal (2) sections of the hollow waveguide [10]; (b) Cross-(1) and longitudinal (2) sections of the tube waveguide [11]; (c) Cross-section (1) of the RF [1], approximation of the part of the walls of the reflecting capillaries by the mirror sides of the angle α (2), a reflection scheme of a ray propagating along the RF, from a corner with mirror sides (a projection onto the cross section of the optical fiber) (3); and, (d) Calculated optical loss spectrum for silica fibers: hollow waveguide (HW), tube waveguide (TW) and RF. For all waveguides, the hollow core diameter is assumed to be 77 μm, and the thickness of the capillary wall (for TW and RF) is 1.15 μm.

Already in the case of HW, the high reflection coefficient at the core-cladding interface leads to the fact that the modes in HW are similar to the modes in perfect conducting metallic waveguides, when operating far from cutoff. This feature was mentioned in [10]. RFs, as any other NC HCFs, have an even higher reflection at the core-cladding boundary. For this reason, the radiation power in RF is even more concentrated in the hollow fiber core. In these circumstances, the optical absorption of the material, from which the RF is made, recedes into the background, while the optical losses of the RF become mostly defined by the geometric parameters of the optical fiber design and by the conditions for Fresnel reflection at the interfaces. This means that RF can be used to exploit a variety of optical phenomena, even in those spectral regions where the fiber material (e.g., silica) is opaque.

As shown by the simple RF models that are discussed above, the possibility of waveguiding the laser radiation with low losses in RFs in the UV and mid-IR spectral ranges is mainly due to dispersion of complex refractive index $n(\lambda) = Re(n(\lambda)) + i \cdot Im(n(\lambda))$ of a fiber material. Note, the choice of materials for RFs is largely limited: up to now, RFs have been made of silica glass, chalcogenide glass [15], and organic glass (polymethylmethacrylate) [16]. Waveguiding properties are also influenced by the ratio of the wavelength to the basic geometric dimensions of RF, such as the diameter of the hollow core D_{core} and the thickness of the capillary walls d. However, this ratio can be optimized for a wavelength of interest during the fiber manufacturing process.

For RFs that are made of silica glass, the optical loss increases with wavelength in the mid-IR range up to a wavelength of 7.3 µm (see Figure 2d). This occurs for two reasons: $Re(n_{SiO_2})$ decreases [17], which leads to a decrease in the Fresnel reflection coefficient from air-glass surfaces (here the Fresnel reflection coefficient is mainly determined by the $Re(n_{SiO_2})$ value, since the value of $Im(n_{SiO_2})$ is small in comparison with unity). In addition, radiation absorption in the silica capillary wall begins to reveal itself in the wavelength region of about 5 µm. As a result, the efficiency of the ARROW mechanism decreases and the value of the reflection coefficient from the capillary wall starts to decrease further, approaching the values that are characteristic for the HW model. This is also true for longer wavelengths of the mid-IR range, with the exception of small regions around 7.3 µm, and possibly around 9 µm and 20 µm. Note, that around those wavelengths the value of $Re(n_{SiO_2})$ is close to unity, and reflection at the air-silica boundary is practically absent so that silica hollow core microstructured fibers (HCMFs) cannot demonstrate any waveguide properties.

In the near-IR and visible ranges, RF from silica, like silica glass itself, exhibit their best properties. While shifting along the wavelengths towards the UV band, the value of $Re(n_{SiO_2})$ increases, thus reducing the optical losses in RFs. However, at a wavelength of about 150 nm, the value of $Im(n_{SiO_2})$ has sharp increase, which, like in the mid-IR range, leads to the "shutdown" of the ARROW interference mechanism. As a result, the value of optical losses increases up to the level that is determined by only one reflection at the core-cladding interface (HW). Nevertheless, the results that were obtained show that silica-based RFs can be used up to vacuum ultraviolet (124 nm) [18]. In the case of RFs that are made of chalcogenide glass, the waveguiding properties of RFs have been demonstrated up to wavelengths ~10 µm [15]. THz radiation can also be transmitted in polymethylmethacrylate waveguides that are similar to RFs [16].

3. RF with Various Cladding Structures

Spectral properties of RFs were investigated in detail by numerical simulation and experimentally. Optical loss is one of the main parameters of RFs. Figure 3 shows the most available now experimental data on optical losses in silica glass RFs of various types. For comparison, the optical losses in different types of silica glass HCFs and the absorption spectrum for pure silica glass are also shown. As can be seen in Figure 3, in the UV range the optical losses of RFs (Figure 3, data 15) are approaching to the absorption level of a pure silica glass (Figure 3, data 2), while HCFs that have square (data 14) and hexagonal (data 11) cores without negative curvature demonstrate properties that are similar to RFs. In the near-IR range, the photonic bandgap HCFs (data 6) have the lowest level of optical losses. At wavelengths 3–4.4 µm in the mid-IR both RFs (data 5, 16, 17) and "ice-cream-cone" shaped HCFs (data 8) show similar optical losses, which outperform the attenuation in pure silica glass (data 3).

Note, the RFs (Figure 3, data 4, 16) are the only hollow-core silica fibers that have demonstrated optical transmission at wavelengths above ~4.4 μm.

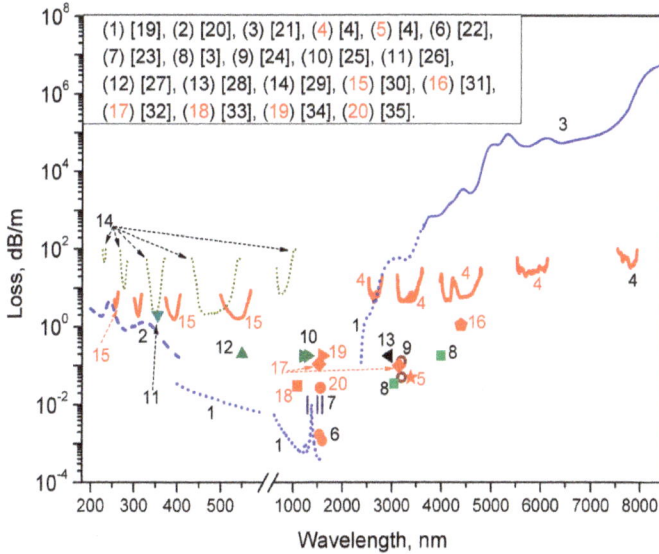

Figure 3. Optical loss in pure silica glass (F300, Heraeus, Hanau, Germany) (blue curves, 1–3) and the minimal optical loss obtained to date in different types of silica glass hollow-core fibers (HCFs). Data on RFs is highlighted in red. In the figure legend, the plotted experimental data are indexed by numbers in round brackets followed by references to literature in square brackets. More details are given in the text [19–35].

Figure 3 indicates that RFs extend the applicability of silica glass technology into the mid-IR spectral range (above ~3 μm). For example, at the wavelength of 4.4 μm, where absorption coefficient of the silica glass is about 4000 dB/m (Figure 3, data 3), the RFs that are made of silica glass allowed for the demonstration of optical losses as low as 1 dB/m (Figure 3, data 16). One should note, that compared with RFs made of silica glass, the solid-core non-silica fibers have lower optical losses in the mid-IR. For example, optical losses less than 0.1 dB/m were demonstrated in fluoroindate fibers in the 2.0–4.5 μm spectral range [36] and in chalcogenide fibers at a wavelength of up to 6.5 μm [37,38]. At an even longer wavelength (8–16 μm), the silver halides fibers with optical losses below 1 dB/m are available [39]. In general, non-silica solid-core mid-IR fibers provide the level of optical losses that is 10–1000 times lower than the optical losses of the silica RFs at the same wavelengths. Nevertheless, the usage of RFs that are made of silica glass can be advantageous at wavelengths of up to ~5 μm, as reasonable level of optical attenuation can be achieved using well-developed silica glass technology. Moreover, a damage threshold of RFs is much higher when compared with solid-core non-silica mid-IR fibers. Thus, hollow-core silica RFs are indispensable for high power applications.

3.1. RFs with Touching and Non-Touching Capillaries in a Cladding

After the first RFs that had touching capillaries in the cladding (Figure 1a) [1], a modified RF with non-contacting capillaries in the cladding was proposed (Figure 1b) [4]. It turned out that this RF structure has lower optical losses than the previous one. Subsequently, RFs with non-contacting capillaries in the cladding were used in many works (see, e.g., [32,34,40,41]).

It was numerically demonstrated that the absence of touching points between capillaries removes the additional resonances in the transmission bands (Figure 4). In simulations, the two models of RFs were analyzed and compared. All of the geometrical parameters of the RFs were identical, except that gaps between the capillaries in one of the fibers (red line in Figure 4) were filled with glass (see inset in Figure 4). So, all of the difference between two curves in Figure 4a is defined by the presence of the nodes between the capillaries in the cladding. The cladding of each virtual fiber consisted of eight capillaries that had the outer and inner diameters of 63 μm and 51 μm, respectively. The minimal distance between the non-touching capillaries was 1.3 μm. The loss spectra were calculated in the spectral range of 3–6.5 μm for both RF models. It was clearly shown that the presence of nodes between the capillaries leads to an increase of optical losses due to the occurrence of resonances between the core and cladding modes.

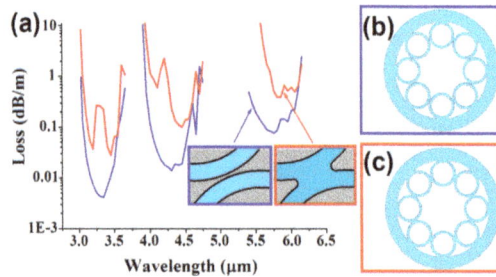

Figure 4. (a) The calculated fundamental mode loss for a silica RF with capillaries in touch and non-touching capillaries in the cladding; (b,c) Schemes of cross sections of these RFs.

For the first time, a revolver fiber with separate capillaries in the cladding was fabricated in [4]. The real fiber had outer diameter of 290 μm, the core diameter of 110 μm and the capillary wall thickness of 6 μm. The fiber cross section and the measured spectrum of optical losses are shown in Figure 5. One can see that the average level of optical losses measured in the spectral range of 2.5–5 μm was about 4–5 dB/m. At longer wavelengths of around 5.8 and 7.7 μm the losses were measured to be 30 and 50 dB/m, respectively. Transmission bands at 3.3 and 4.3 μm have a number of absorption peaks, which are related to the absorption lines of HCl (similar to work [3]) and atmospheric CO_2, respectively. Optical absorption of fused silica glass is also shown for comparison (Figure 5, black curve).

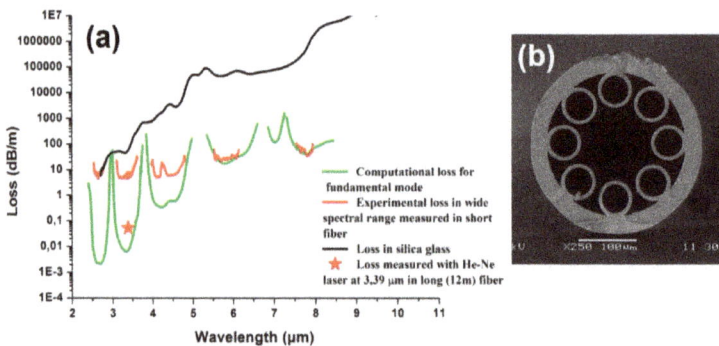

Figure 5. (a) The measured loss (red); the loss measured with He-Ne laser at 3.39 μm (red asterisk); the material loss in fused silica (black); the calculated loss of the fundamental mode (green) of the RF; (b) the micrograph image of the RF cross section, D_{core} = 110 μm and capillary wall thickness d = 6 μm.

To analyze the experimental results, a numerical modeling of the optical losses for the fundamental mode of the fiber was carried out (Figure 5, green curves). As can be seen from the Figure 5, the calculated and the experimental band edges superpose very well. On the other hand, the minimum loss level in transmission bands differs significantly. It occurs mainly due to the presence of higher order modes in the process of loss measurement (multimode light source was used to excite a short fiber section during the experiment). This idea was confirmed by loss measurement that was carried out by cut-back technique in 11-m-long fiber and a few-mode 3.39 μm He-Ne laser as a light source. This experiment showed that at this wavelength the mode content in the fiber was stabilized when the fiber is longer than 3 m. When only first several modes are present in the fiber, the loss level reaches 50 dB/km (red asterisk in Figure 5), which is much closer to the loss level that was calculated for fundamental mode at 3.39 μm. Thus, real loss level in the fiber is low enough and can be estimated using the calculated loss spectrum (Figure 5, green curves).

3.2. RF with a Cladding of Single and Double Nested Capillaries

Material loss of silica glass changes from 0.1 dB/m to 10^5 dB/m in the wavelength range from 2 μm to 6 μm [42]. Thus, starting at wavelengths >2 μm, the total losses in silica RF begin to be increasingly determined by the material losses of silica glass. It turns out that the optical loss behavior for RFs with simple capillaries in the cladding ((Figure 1a,b) and for RFs with nested capillaries (Figure 1c) is different in the region of high material losses [42]. The optical losses in the nested RF are lower than in RFs with one row of cladding capillaries up to a certain wavelength in the mid IR spectral range. In this case, the nested capillaries work as additional reflectors and an increase in reflection coefficient gives a win in comparison with RFs with one row of cladding capillaries (Figure 1b,c). Under a further increase in the wavelength, the reflection from the nested cladding capillaries cannot compensate for the growth of material loss and the total losses, correspondingly. That is why the first hydrogen Raman laser with a generation wavelength of 4.4 μm [31,43,44] was built using RF with one row of cladding capillaries. In [4], it was shown that RFs made of silica glass could transmit light up to wavelength of 8 μm with losses of about several tens of dB/m (Figure 5). This level of optical losses is too high for practical use. Nevertheless, silica glass RFs with one row of cladding capillaries can be used in practice up to wavelength of 5 μm. It is possible because material loss of silica glass increases by approximately an order of magnitude in comparison with previous values (Figure 5a). In this way, nested RFs have lower losses in comparison with RFs with one row of cladding capillaries in the transparency region of silica glass. RFs with one row of cladding capillaries have an advantage in the region of high loss of silica glass. Also, it is necessary to take into account bend loss.

RF bending naturally results in an increase of fiber losses. However, fiber bending reveals another important feature of RF: the resonance coupling of the hollow core modes with the cladding capillary modes (Figure 6). For the first time such, the resonant coupling was found by numerical simulation in [15], and then this effect was experimentally investigated in [8,16,45].

As can be seen from Figure 6d, with a decrease in the bending diameter, the RF transmission decreases non-monotonically, but high loss peaks due to the resonant coupling between the core mode and the cladding capillary modes are observed. Depending on the bending radius, the resonant coupling can occur both in the same capillary for modes of different orders, and in different capillaries of the cladding Figure 6a–c. In both cases, the resonant coupling increases the losses in hollow core mode dramatically. Similarly, higher order core modes can be resonantly coupled to capillary modes. This RF feature can be used for the filtering of the hollow core modes. For example, if the ratio of the inner diameter of the capillary to the hollow core diameter is equal to 0.68, a resonant coupling of the core mode LP11 to the capillary mode LP01 occurs, i.e., the RF becomes quasi-single-mode [33,34].

To transmit light in RF made of silica glass in the mid IR spectral range in the vicinity of wavelength of 5 μm with losses about 1 dB/m, it is necessary to carry out optimization of the cladding geometry. The optimization parameters are thickness of the capillary wall to obtain the transmission band at the desired wavelength, the number of the cladding capillaries and diameter of the hollow core. An alternative way of solving the problem of light transmission in the mid IR spectral range with low loss is that to use soft glasses (chalcogenide, tellurite). They have low material losses in the mid IR spectral range.

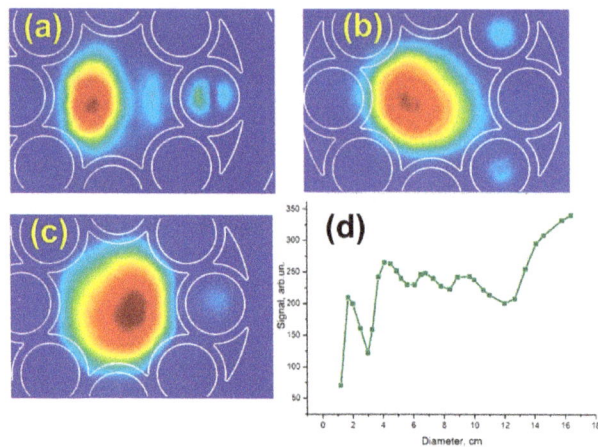

Figure 6. The experimentally measured intensity distributions of the hollow core modes arising at the bending of the RF; white lines show a RF cross-section contour. Intensity distributions are shown for: (**a**) RF bend diameter of 1.6 cm; (**b**) bend diameter of 3.6 cm; (**c**) bend diameter 5.6 cm; (**d**) The experimental dependence of the light intensity transmitted through the RF vs. bend diameter of the fiber [15].

3.3. Optical Properties of RFs in UV Spectral Range

A different situation is observed under light transmission in the UV spectral range. Measured losses of silica glass in the spectral range from 200 nm to 400 nm are not as high as in the mid IR spectral range. They vary from a few tenth to 10 dB/m [20]. Therefore, when comparing waveguide losses and material losses one can conclude that the former play the main role in this case (for wavelengths $\lambda > 150$ nm). In this way, the main mechanism allowing for one to decrease the level of total losses is an optimization of geometric structure of silica glass RF. On the one hand, low waveguide losses can be obtained by increasing the air core diameter. On the other hand, it leads to excitation of many air core modes due to inhomogeneous construction of the fiber occurring under the drawing. It also leads to a narrowing of the transmission bands [30]. Therefore, it is necessary to carry out the optimization of the RF geometric structure as in the case of light transmission in the mid IR spectral range. The usage of nested RFs is unlikely due to complexity of their fabrication, for example, it is very difficult to keep the sizes and shapes of the cladding capillaries under the drawing. Besides, it is important to choose a number of the cladding capillaries. For example, in [46], the waveguide regime was in silica glass RF with four capillaries in the cladding was demonstrated at loss level about 0.5 dB/m at wavelength of 350 nm. The sizes of the cladding capillaries were comparable with the air core diameter, which should lead to high bend losses. In work [30], silica glass RF with eight capillaries in the cladding was demonstrated. It transmitted light in the spectral region from 350 nm to 200 nm and the loss level was about 1–2 dB/m. The authors explained such rather high loss level by imperfect construction of the drawing fiber. In conclusion, it is worth saying about polygonal fiber, which localizes light at the expense of double antiresonant mechanism [29]. The authors of [29] proposed to use the fiber with the square core-cladding boundary. Such fiber allowed for transmitting light up to wavelength of 241 nm with optical losses that are comparable with those that were reported in work [30].

4. Technology of the RF

Until now, most of the revolver-type optical fibers had been made of silica glass. Significantly greater difficulties arise with the manufacturing of RF from glasses of other types, for example, chalcogenide glass. Let us consider the manufacturing processes of these fibers one after another.

4.1. Technology of the Silica Glass Based RF

The significant advantages of RFs over, e.g., Kagome type HCMFs, is a simpler waveguide structure. Accordingly, RFs have a simpler manufacturing technique, since the reflecting cladding contains only one layer of capillaries.

Usually, RFs are manufactured by the 'stack and draw' technique. This technology consists in stacking the prefabricated capillaries inside the support tube (Figure 7). In doing so, the capillaries can either touch each other (Figure 7a), or between the capillaries can be inserted additional elements at the beginning and at the end of the preform (Figure 7b). In the first case, a RF with touched capillaries [1] or an "ice-cream-cone" structure [3] is obtained. In the second case, it is possible to obtain a RF with non-touching capillaries [4]. Furthermore, simple single capillaries (Figure 7a), or capillaries of a more complex structure, for example, double nested capillaries (Figure 7b) can be used to fabricate the RF. The result is either RF with single capillaries that works best in spectral areas where silica glass has high absorption, or RF with double nested capillaries that provides less optical loss in spectral regions with low silica glass absorption.

After the stacking step, the preform is usually treated by flame to weld the capillaries with the supporting tube. However, this procedure is not necessary in some cases. For example, in [15], the RF was made from chalcogenide glass, which has a high temperature expansion coefficient, and the heat treatment was not carried out to avoid the possible cracking of the preform.

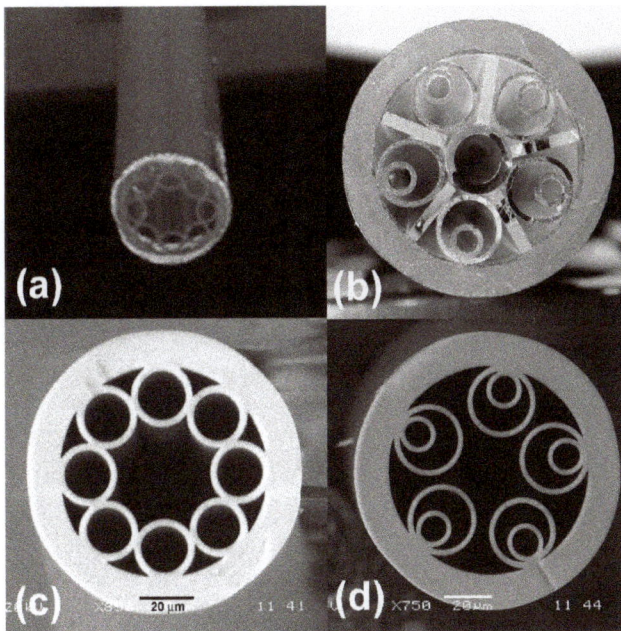

Figure 7. Cross sections of optical elements at the main stages of RF production: (**a**) Picture of the RF preform with single capillaries touched to each other; (**b**) Cross-section of the RF preform with silica elements between the double nested capillaries, Ø 25 mm; (**c**) SEM image of the RF cross section (drawn from the preform in Figure 7a); and, (**d**) SEM image of the RF cross section with double non-touching capillaries, Ø 110 μm (from the preform in Figure 7b).

During the fiber drawing process, an excess gas pressure is applied to the capillaries, in order to prevent their collapse under the action of surface tension forces. Usually, a gas pressure regulator is used for this purpose. The regulator is connected to all capillaries and it provides the necessary overpressure. This solution has a significant drawback, which does not allow for the fabrication of RF with strictly identical capillaries. Obviously, the capillaries have dimensional deviations, and therefore the surface tension force, which tends to collapse the capillary, will be different for each of the capillaries. So, for each capillary, it is necessary to use an individual pressure regulation system, which in practice is hard to realize. As a rule, in practice, the same pressure is applied to all capillaries, while the capillaries dimensional deviations increase during the drawing process (see Figure 8, upper way). This happens because the capillaries of a smaller diameter tend to collapse during the drawing process, while capillaries of larger diameter tend to blow up. Nevertheless, the using of high-quality tubes (for example silica tubes by "Heraeus") for the capillaries production allows for achieving good results even in technology with the same pressure being applied to all capillaries.

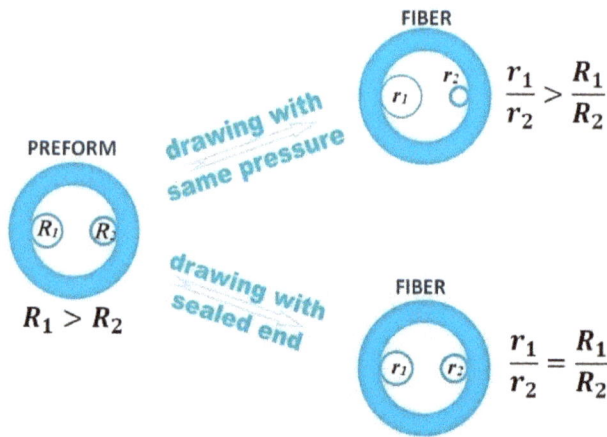

Figure 8. The scheme of RF drawing. Drawing with same pressure in all capillaries (on the top), drawing with sealed end (on the bottom).

There is an alternative approach, which is usually applied to drawing micro-structured fibers, but is also applicable to RF production. The method is called "fiber drawing with sealed upper end" (Figure 8, bottom). The key point of this technology is that the holes of the preform are sealed at the top end of the preform. In such a way, each hole initially contains a certain gas volume and all of the holes are isolated from each other. During the drawing process, the lower part of the preform is heated, the pressure in the holes increases, and the holes expand. Therewith, the increase of the hole volume is determined by the ratio of temperature on top and bottom of the preform. In the first approximation, the increase of the hole volume does not depend on the diameter of the hole. In other words, all of the holes of the preform expand in the same way [47]. Thus, the hole size distortion, which is present in the perform, remains in the fiber. While using the technology with the same pressure being applied to all holes, the hole size deviation from the mean value increases when the fiber is drawn.

4.2. Technology and Properties of the RF Based on Chalcogenide Glasses

Although the waveguide regime in RF made of silica glass was demonstrated up to a wavelength of 7.9 μm [4], the optical losses in RFs are higher than 10 dB/m at wavelengths longer than 4.5 μm. Apparently, it is impossible to construct a silica glass hollow fiber of a reasonable core diameter, which would have optical losses at a level of ~1 dB/m at wavelengths that are greater than 5 μm. Therefore, to work in the longer wavelength range, it is necessary to develop hollow fibers that are made of other materials. Tellurite glasses have a wider transparency region [48], so, they can enable extension of the RFs low-loss region to the wavelength of 5.5–6 μm. But, a more radical benefit is the use of chalcogenide glass.

However, from the technology point of view, the chalcogenide glass is substantially less manufacturable. Both the quality of the chalcogenide glass tubes and their physical properties lead to the fact that production chalcogenide RF with perfect geometric parameters is a much more complicated task (compare the cross sections of quartz RF in Figure 7c,d and chalcogenide RF in Figure 9b,d). For the first time, a chalcogenide RF was made in 2011 [15]. To fabricate the optical fiber, a high-purity $As_{30}Se_{50}Te_{20}$ glass was used. The fiber was made by the "stack and draw" technique. The support tube was made by the centrifugal casting method inside an evacuated silica tube. Capillaries were produced by the double crucible method from the melt of chalcogenide glass. The stacked preform is shown in Figure 9a. The preform was drawn on standard drawing tower using special low-temperature furnace. The obtained fiber had outer diameter of 750 um, core diameter of 260 μm, and capillary wall thickness of 13 μm (see Figure 9b).

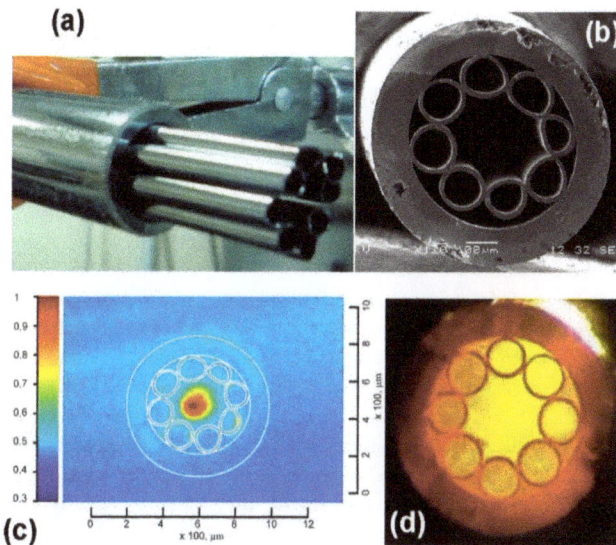

Figure 9. (a) The stacked perform of RF made of $As_{30}Se_{50}Te_{20}$ glass, the tube outer diameter—16 mm, the tube inner diameter—11 mm, the capillaries outer diameter—3 mm, the capillaries inner diameter—2.4 mm; (b) SEM image of the $As_{30}Se_{50}Te_{20}$ RF cleaved end face, the fiber diameter is 750 μm; (c) the CO_2-laser radiation intensity (in a.u.) distribution over the $As_{30}Se_{50}Te_{20}$ RF cross section. White lines show the cross section of the RF; and, (d) optical microscope image of the As_2S_3 RF cleaved end face, the fiber diameter is 820 μm.

Optical loss of his fiber was 11 dB/m at the wavelength of 10.6 μm. Also, the propagation of CO_2 laser radiation along the hollow core has been detected by the thermal imaging camera (Figure 9c).

Then, in 2014, a similar method was used to fabricate a fiber from more technologically simple glass $As_2S_3As_2S_3$ [49] (Figure 9d). The minimum optical losses in that fiber were 3 dB/m at the wavelength of 4.8 μm. Also, in the loss spectrum of that fiber, there were significant absorption bands of typical impurities, for example: S-H bonds at 6.8; 4.1; 3.7; 3.1 μm; CO_2 impurity at 4.31 and 4.34 μm; OH groups at 2.92 μm; and, molecular H_2O at 6.33 μm. This indicates that a significant part of optical power in that fiber propagates through the glass in the process of loss measurements.

In 2015, the Gattass et al. [50] used the extrusion technology to obtain a preform of a RF of As_2S_3 glass. The preform had a diameter of 18 mm and a length of 135 mm (Figure 10a). The fiber drawn from that preform had the following geometric dimensions: the core diameter was 172 μm, the size of the oval capillaries was 70 by 75 μm, and the thickness of the capillary walls was 7 μm (Figure 10b). Optical loss turned out to be 2.1 dB/m at the wavelength of 10 μm.

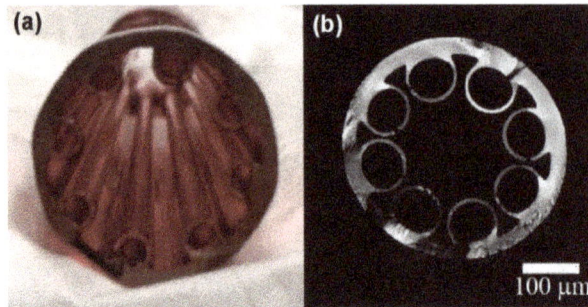

Figure 10. (a) The preform made by extrusion technique of As_2S_3 glass; and, (b) The drawn As_2S_3 RF [50].

In all works concerned with the chalcogenide RF, the theoretical calculations of optical losses using the finite element method were carried out. Actually, the experimentally measured loss is almost always several orders of magnitude higher than the theoretically calculated loss. The observed difference is explained by the strong sensitivity of the fiber optical properties to the fiber geometric deviations, and also by the imperfection of the obtained fiber structures. The increased sensitivity of chalcogenide fibers to the geometric deviations in comparison with silica fibers is due to the high refractive index of chalcogenide glasses, i.e., with the same absolute variations of the geometric thicknesses, the variations in the optical thicknesses in chalcogenide fibers are 3 to 4 times greater than in silica fibers. In addition, the dependence of the viscosity of chalcogenide glasses on temperature is several times stronger than the dependence of the viscosity of silica glass on temperature (chalcogenide glasses are "short"), i.e., at identical temperature gradients and temperature fluctuations in the fiber drawing process, the quality of chalcogenide microstructured fibers is lower than the quality of silica fibers. It is also clear that the purity and homogeneity of the chalcogenide glasses is worse than the purity and homogeneity of the high quality silica glass that was used for HCMF fabrication.

In all studies of chalcogenide RF, the loss spectra that were obtained by theoretical simulation contain a lot of resonant loss peaks. This fact is explained by the resonant coupling of hollow core modes with special type cladding modes. Similar resonance peaks are sometimes observed in the theoretically calculated loss spectra of silica RFs at the long-wavelength edge of the transmission bands [8] (see [4], Figure 4). However, in the case of chalcogenide RFs, the irregularity of the simulated spectra is observed practically always. This is because the density of states in the cladding of chalcogenide RF is much higher than in case of silica RFs due to the high refractive index of chalcogenide glasses. However, in the experiment, narrow peaks of optical losses in the transparency regions of chalcogenide RFs have never been observed. Apparently, because of the faulty geometry of the produced RFs, the real loss spectra are broadened.

5. Mid-Infrared Raman Laser Based on Revolver Fibers

What makes revolver fibers attractive is their ability to provide an extremely low overlap of an optical mode with a cladding material. As a result, the RFs can have low optical losses even in those spectral regions where the cladding material has strong fundamental absorption. In particular, silica glass can be used to fabricate revolver fibers for UV and mid-IR spectral ranges. It is instructive to compare the measured optical losses of RFs with those of pure silica glass (see Figure 3). Although, in the UV range, the RFs do not yet outperform silica glass in terms of optical losses, in the mid-IR range, the RFs optical losses are orders of magnitude less than the optical losses of silica glass. Thus, current state-of-the-art RFs fabrication technology enables the development of silica fiber devices, including fiber lasers, for the mid-IR spectral range.

HCFs paved the way to a new class of lasers—the gas fiber lasers (GFL) [51]. Such lasers combine the advantages of both fiber lasers (compactness, reliability and excellent beam quality) and gas lasers (wide range of lasing wavelengths, high output power and narrow linewidth). The gain medium of GFLs is a gas, which fills the hollow core and has dipole-active or Raman-active transitions. The hollow-core fiber ensures a small mode field diameter and a long interaction length of the light and gain medium. As a result, thresholds for nonlinear processes, such as stimulated Raman scattering (SRS), can be reduced by several orders of magnitude with respect to non-guiding schemes.

Currently, GFLs development for spectral range of 3–5 μm is an area of active research. In particular, stimulated Raman scattering (SRS) in gas-filled HCFs is used to generate mid-IR radiation. For example, 2.9–4.4 μm Raman lasers that are based on gas-filled revolver silica fibers have been recently demonstrated [43,44,52]. Such lasers in the NIR range have been implemented with fewer difficulties, since RF and HCF of other types have significantly lower optical losses in this region [51,53].

In most studies, gas fiber lasers are constructed in a cavity-free, single-pass scheme [5,51,53–55]. Due to strong localization of light in their core (MFD ~5 ÷ 50 μm) along the entire length of the fiber (~1 ÷ 10 m) active gas-filled HCFs provide a single-pass gain that is sufficient for laser radiation build-up from quantum noise. Thus, a single-pass scheme allows for one to realize efficient GFLs that are based on both SRS [51–54] and population inversion [55]. Designing a cavity for GFLs remains a challenging problem because there are neither fiber couplers nor analogues of fiber Bragg gratings for hollow-core fibers. Nevertheless, a few studies addressed cavity-based GFL schemes using a ring cavity that was made from bulk elements [40] and a Fabry-Perot cavity formed by Bragg gratings spliced to the end faces of an active hollow-core fiber [56].

An active medium of gas fiber Raman lasers can be formed with light molecular gases, such as light hydrogen ($^1H_2$), deuterium (D_2), methane (CH_4), and ethane (C_2H_6). These gases have a Raman shift (4155, 2987, 2917, and 2954 cm^{-1}, respectively) that is large enough to enable single-stage Raman conversion of 1.5 μm radiation, generated by well-developed pulsed erbium-doped fiber lasers, into mid-IR spectral range.

A key component for making efficient mid-IR Raman fiber lasers is a hollow-core fiber, whose characteristics should satisfy certain conditions. A necessary condition is that the optical loss in the fiber should not exceed the Raman gain of an active medium. Previously [57], a quality parameter (P_F) was introduced to characterize an optical fiber as a Raman-active medium:

$$P_F = \left(\sqrt{\frac{\alpha(\lambda_p)}{g_0}} + \sqrt{\frac{\alpha(\lambda_s)}{g_0}} \right)$$

where g_0 is the Raman gain coefficient (in units of dB/(m*W)) of the active fiber for a particular Raman conversion $\lambda \to \lambda_S$; $\alpha(\lambda_P)$ and $\alpha(\lambda_S)$ are optical losses of the fiber at pump and Stokes wavelengths, respectively. The parameter P_F has the same dimensions as power and is measured in watts. By its physical meaning, P_F is the threshold pump power of a CW Raman laser that is based on the fiber under consideration placed in some high-Q cavity [57]. Thus, the less the value of P_F, the better the fiber is for Raman conversion $\lambda_P \to \lambda_S$, provided the pump pulse duration is sufficiently long.

The parameter P_F provides a convenient tool to optimize fiber characteristics for Raman conversion. Let us consider the dependence of P_F on the diameter of the hollow core using simplified analytical models of hollow waveguide (HW) and tube waveguide (TW) (see Section 2). It is known that for a straight (i.e., non-bent) HW and TW, the optical losses depend on the diameter of the hollow core D as $1/D^3$ and $1/D^4$, respectively [13]. Therefore, the models predict that for a straight fiber the figure of merit P_F is proportional to $1/D$ (for the HW) and $1/D^2$ (for the TW), because g_0 is proportional to $1/D^2$ in a case when an effective area of a fiber is proportional to D^2. Consequently, the larger D, the smaller the P_F. Thus, there is no optimum for the straight optical fibers with respect to the diameter of the hollow core. However, bent-induced losses must be also considered. The ability to bend is one of the main advantages of optical fibers. Assuming that we are working with HW and TW optical fibers that are coiled to a certain radius R, the bent-induced losses in such fibers are proportional to the diameter of the hollow core D. As a result, the optical loss of such fibers has a minimum at some value of the hollow core diameter at any definite wavelength. Correspondingly, the figure of merit P_F reaches its minimal value at some hollow core diameter D_{min}, which determines the optimal diameter of the fiber core for Raman fiber. If we choose a typical bending radius of R = 15 cm and assume that the revolver fiber is filled with hydrogen at a pressure of 30 bar, then the estimates based on both HW and TW models give rise to the value of $D_{min} \approx 75$ μm. Much more rigorous and complicated numerical modeling of real RFs gives approximately the same value [58].

It is important for P_F to be much lower than the pump power that is achievable in the experiment. Let us consider a model fiber with a hollow-core diameter of ~75 μm filled with molecular hydrogen at room temperature and a pressure above 10 atm. Assuming that such fiber may in principle have optical losses $\alpha(\lambda_P)$ ~0.1 dB/m (in the near-IR) and $\alpha(\lambda_S)$ ~1 dB/m (in the mid-IR), we obtain for the quality parameter P_F ~100 W. Much higher peak pump power can be reached using existing nanosecond solid-state and fiber lasers. However, it is worth noting that the fabrication of hollow-core silica fibers with mid-IR losses within 1 dB/m is a nontrivial task, because, in the wavelength range from 3 to 5 μm, the material absorption in silica glass rises sharply, from ~50 to ~50,000 dB/m (see Figure 3). Nevertheless, revolver silica fibers, which have characteristics that are mentioned above, were fabricated and mid-IR gas fiber Raman lasers were demonstrated using such fibers.

A typical single-pass scheme of the mid-IR Raman GFLs is illustrated in Figure 11. Lens system is used to couple pump radiation at the wavelength near 1.5 μm into the gas-filled revolver fiber. Both ends of the RF are hermetically sealed into miniature gas cells, which had inlets for gas injection and sapphire windows to couple/decouple the radiation. Radiation at the RF output is collimated by ZnSe lens, passes through a set of optional optical filters, and is then analyzed by spectrum analyzer and/or powermeter.

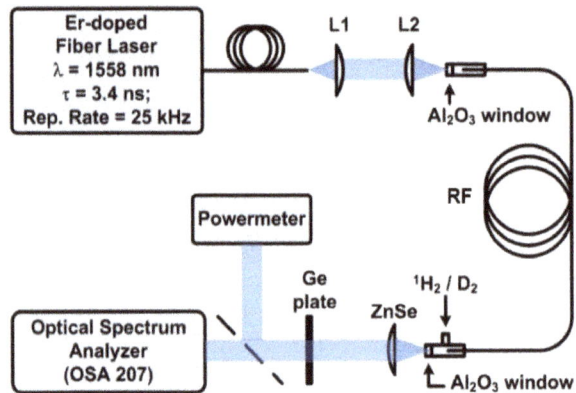

Figure 11. Scheme of the experimental setup: L1 and L2—aspheric fused silica lenses; RF—revolver hollow-core fiber; Al$_2$O$_3$—sapphire windows of the gas cells at the HCF ends; ZnSe—collimating lens made of zinc selenide; and, Ge—2-mm-thick germanium plate.

The first mid-IR Raman GFL generating at wavelengths around 3 μm was demonstrated in [59,60], where silica glass RF was used with the calculated transmission spectrum that is shown in Figure 12a. The mode field diameter of 11-m-long fiber was 45 μm. Filling the hollow core with D_2 molecular deuterium (partial pressure of 28 atm) containing $^1H_2$ molecular light hydrogen impurities (partial pressure of 2 atm) made it possible to obtain Raman lasing at wavelengths of 2.9, 3.3, and 3.5 μm

(Figure 12b). The peak power in the mid-IR spectral region was about 400 W, which corresponded to an average power of about 40 mW. Conversion quantum efficiency was 10% (at λ = 2.9 μm) and 6% (at λ = 3.5 μm), with the possibility of further optimization. Note that adjusting the $^1H_2$ and D_2 partial pressures and pump power enabled predominant lasing at a wavelength of 2.9 or 3.5 μm to be obtained.

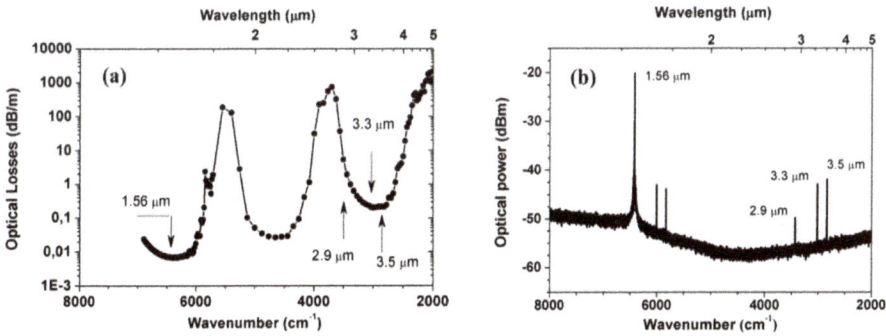

Figure 12. (a) Calculatedspectrum of optical losses for RF that was developed in [59,60] to realize Raman gas fiber lasers (GFL) for 2.9–3.5 μm spectral region; (b) Output spectrum of the Raman gas fiber laser [60]. Peak pump power at λ = 1.56 μm coupled into the RF was 14 kW. A mixture of light hydrogen and deuterium at partial pressures of 2 and 28 atm, respectively, was used as an active medium.

Later, in studies [43,52,61], the design of silica revolver fiber was modified so as to shift the transmission spectrum of the fiber to the ~4 μm range (Figure 13a). The mode field diameter was 56 μm. Filling the hollow core with $^1H_2$ molecular hydrogen at a pressure of 30 atm, the first Raman lasing at a wavelength of 4.4 μm was demonstrated (Figure 13b) [43,52]. Using single-mode output of the Raman laser, the loss in the revolver fiber at this wavelength was measured to be 1.13 dB/m, being in good agreement with numerical simulation results (0.92 dB/m) [61]. Note for comparison, that the material absorption in silica glass at this wavelength is ~4000 dB/m. The use of 15-m-long RF ensured Raman lasing with a quantum efficiency of ~15% and the average power of 30 mW at the generation wavelength of 4.4 μm [52].

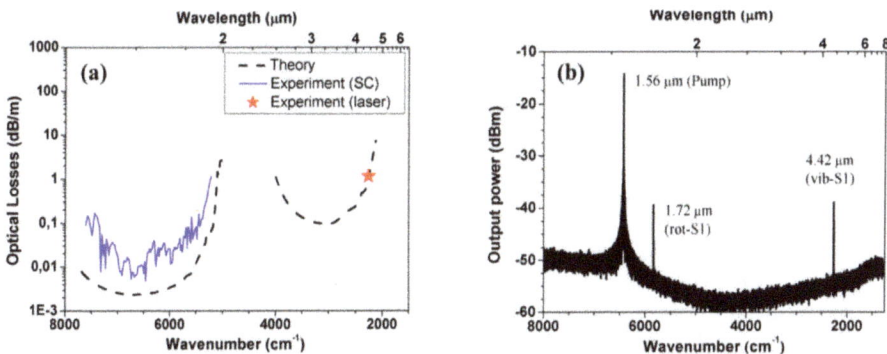

Figure 13. (a) Calculated optical loss spectrum of a hollow-core revolver fiber (black dashed lines) [52]. Experimentally measured losses are also shown: near-IR loss spectrum was measured using a supercontinuum source (blue solid line), and optical losses at wavelength of 4.4 μm (asterisks) was measured using narrow-band laser source [61]; and, (b) Output emission spectrum of an RF filled with $^1H_2$ at room temperature and a pressure of 30 atm. The launched peak pump power was 18 kW [43,52].

To improve efficiency and output power of the 4.4 μm Raman laser, it was analyzed theoretically by numerically solving a system of coupled wave equations for vibrational SRS in $^1H_2$ molecular hydrogen [61]. For this purpose, measured optical losses at wavelengths of 1.56 and 4.4 μm (Figure 13a) were taken into account. Raman gain coefficient $g_R = 0.43$ cm/GW was calculated for $1.56 \rightarrow 4.4$ μm conversion using the available data on the linewidth and scattering cross section of the Q(1) vibrational transition of molecular hydrogen [62–64]. The theoretically evaluated optimal Raman laser length was found to be ~3.5 m, which is substantially shorter than the hollow-core fiber length (15 m) that was used in previous experiments.

One interesting result that was obtained in [61] is the possibility of maintaining steady-state SRS when GFLs are pumped by nanosecond pulses. It is known [62] that, if the pump pulse duration (τ_p) and the dephasing time of optical phonons (T_2) meet the relationship $\tau_p \leq 20 \cdot T_2$, SRS conversion is a transient process, in which the Raman gain coefficient decreases. However, T_2 can be easily controlled by varying the pressure of a gas, filling RF core, since the collision frequency of molecules grows with gas pressure, which leads to more frequent changes in the phase of molecular vibrations, and, therefore, to T_2 reduction. This effect was observed in [61], as the hydrogen pressure was varied in the range from 10 to 70 atm and ensured an increase in the output power of a Raman laser (Figure 14a) pumped by 3.5-ns pulses.

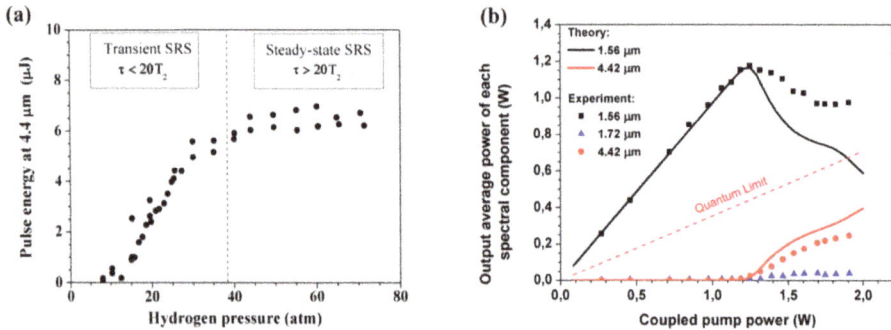

Figure 14. (**a**) Experimentally determined output pulse energy at 4.4 μm as a function of $^1H_2$ pressure in the hollow core [61]. The pump pulse duration was $\tau_p = 3.5$ ns. The vertical dashed line represents a pressure at which the dephasing time of molecular hydrogen vibrations satisfies the relation $\tau_p = 20 \cdot T_2$; and, (**b**) Calculated (lines) and measured (data points) average output power of spectral components as a function of average launched pump power for the Raman GFL. Different colors correspond to different spectral components: 1.56 μm (black), 1.72 μm (blue), and 4.42 μm (red). Dashed red line represents a quantum limit for 4.42 μm generation. The measurements and calculations were made at the optimal fiber length (3.2 m) and hydrogen pressure (50 atm) [61].

As a result of optimization of the fiber length and the hydrogen pressure in the hollow core, 4.4-μm Raman generation of nanosecond pulses was demonstrated with an average power that was as high as ~250 mW and quantum efficiency as high as 36%. In this process, the rotational component at wavelengths of 1.72 μm was significantly suppressed (Figure 14b) [61].

To date, the efficiency of the mid-IR Raman GFLs is limited by the level of optical losses at the Stokes wavelength (~1 dB/m). At the same time, it is seen in Figure 12a that the level of losses at wavelengths below 4 μm is an order of magnitude lower (0.1–0.2 dB/m). This spectral range is suitable for making more efficient Raman gas fiber lasers that are based on the already existing silica revolver fibers. The rise in optical losses at wavelengths above 4 μm is caused by the sharp increase in material absorption in silica glass (as well as by the reduction in its refractive index). New solutions, which are capable of further minimizing the overlap of the optical mode field with the silica cladding, are needed to reduce the optical loss in this spectral region. Note also that the use of higher peak power pump lasers may improve the efficiency of the mid-IR Raman GFLs because a shorter length of the hollow-core fibers can be used, thus reducing the detrimental effect of optical losses.

To date, the peak power of pulsed nanosecond Raman GFLs emitting in the range 3–5 μm has been demonstrated to reach ~2 kW [61]. This parameter is rather limited by the pump power that is achievable with erbium-doped fiber lasers than by any characteristics of a hollow-core revolver fiber. Recent works [65,66] have demonstrated Raman gas fiber lasers with an output peak power of 400 and 150 kW at wavelengths of 1.55 and 1.9 μm, respectively. The use of such lasers as pump sources for gas-filled hollow-core silica fibers paves the way to efficient Raman GFLs generating nanosecond pulses with a peak power of ~100 kW in the spectral range 3–5 μm. Moreover, such mid-IR lasers can be realized by means of two-stage SRS in a given revolver fiber segment that is filled with one or a few gases. First experimental demonstration of two-stage SRS (1.06 → 1.54 → 2.81 μm) in a revolver fiber has been recently demonstrated in [67], where picoseconds pulses at the wavelength of 2.81 μm was generated with a peak power of about 10 MW.

The near-IR (λ = 1.56 μm) to mid-IR (λ = 3–5 μm) SRS conversion is known to be accompanied by a large quantum defect, which may hinder obtaining a high average power at the Stokes wavelength. At the same time, in a recent study [55] that was related to the gas fiber lasers based on population inversion, efficient lasing at a wavelength of 3.1 μm was demonstrated under pumping at λ_p = 1.53 μm. Despite the large quantum defect, which is comparable to that in Raman lasers, they reached a CW output power above 1 W. This result suggests the possibility of high average power of gas fiber lasers, including Raman lasers, in the mid-IR spectral range.

6. High-Power Femtosecond Pulse Propagation in Air-Filled RF

High power ultra-short optical pulses (USP) of pico- and femtosecond durations are powerful tools for high precision material processing applications, such as micromachining, laser surgery, and micro-modification [68]. Moreover, a performance of USP source can be noticeably improved by means of a flexible and robust high power USP delivery option due to specially developed fiber with extremely low nonlinearity and dispersion also with acceptable attenuation and beam quality. It is evident that HCMF being capable of light localization in the large air-filled core is the best candidate for this purpose [69], since the Kerr nonlinear refractive index of air is three orders of magnitude lower than that of silica glass [70].

Moreover, HCMFs that are filled with air at atmospheric pressure are of particular interest for the creation of all-fiber systems for the transmission of high-power femtosecond pulses, since they do not have a sophisticated technology for pumping gas. Nevertheless, in the absence of evacuation at sufficiently long interaction lengths of the radiation with the gaseous medium, the nonlinear properties of the gas can exert a strong influence on the spectral-temporal characteristics of powerful ultrashort pulses [18,71–75].

The propagation of femtosecond pulses in a photonic crystal fiber with a hollow core filled with atmospheric air was studied in [74]. The 2.4 MW pulse with a spectrum shifted to the long-wavelength edge of the band was obtained at the fiber output, when a 110-fs pulse with 900-nJ energy at a wavelength of 1470 nm, was launched into a three-m-long fiber. In [75], the pulses were transmitted through a photonic crystal fiber with a hollow core filled with air at a wavelength of 800 nm. The Raman solitons at the output of the 5-m-long fiber had a peak power of 208 kW and 290 fs duration.

Pulses with 105 μJ energy and 844 fs duration were transmitted at a wavelength of 1550 nm through a Kagome-type HCF with 70 μm core diameter [76]. At the end of a 2.3-m-long fiber, the 300 fs pulses of 78-μJ energy (240 MW peak power) were obtained due to soliton compression. In the work [77], the Kagome HCF had a transmission band in the 900–1300 nm region, and loss figure of 200 dB/km at 1030 nm. A solitonic propagation regime was observed in a three-m-long air-filled fiber for pulse energies higher than 100 μJ.

Thanks to relatively simple design, the RF offers remarkable possibilities by means of accurately maintaining relationships between effective mode field diameter, dispersion, and transmission bandwidth in any wavelength region that is required for pulse delivery. In the RF, the propagation of femtosecond pulses was studied for a few wavelength bands [35,78–80].

6.1. Linear and Nonlinear Pulse Propagation Regimes

The linear propagation regime (without distortion of pulse spectrum) of femtosecond pulses in RF at a wavelength of 0.748 μm was demonstrated in [78]. A fiber with eight separate capillaries and a core diameter of 21 μm (Figure 15a) had a transmission band in the range 700–800 nm. Input pulses with an average power of 1.3 W, a repetition rate of 76 MHz, and a duration of 180 fs (95 kW peak power) passed through 10-m-long fiber without distortions of the spectrum and with a dispersion induced temporal broadening (≈2 times), in accordance with measured group velocity dispersion (GVD) value of 7.7 ps/nm/km.

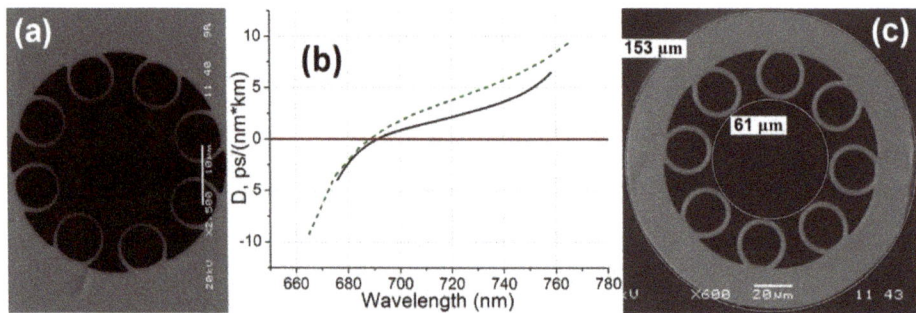

Figure 15. (**a**) RF SEM cross section image [78]; (**b**) Measured (solid black line) and simulated (dashed green line) group velocity dispersion(GVD) of the fiber used in experiments at 748 nm wavelength [78]; and, (**c**) Cross-section image of RF used in experiments at 1560 nm [35].

Long-distance delivery through a low-loss RF of ~1 MW sub-picosecond pulses in the telecom spectral band was experimentally demonstrated in [35].

Sub-picosecond pulses with up to 530 nJ energy and 1.42 W average power at 1.56 μm central wavelength from the all-fiber erbium CPA (chirped pulse amplification) source [35] were launched to the 11.7-m-long air-filled RF coiled to a diameter of ≈30 cm, with ≈80% coupling efficiency.

The cross-section image of the RF is shown in Figure 15c. The RF cladding was formed from eight silica glass capillaries with ≈2.6 μm wall thickness. Core and outer cladding diameters are 61 μm and 153 μm, respectively. Optical loss measurement by means of the careful fundamental mode excitation at 1560 nm wavelength yields the attenuation of ≈27 dB/km being one of the best results being obtained for revolver HCFs. GVD for fundamental mode amounts to $\beta_2 = -1.42$ ps^2/km (D = 1.1 ps/nm/km), while the fiber nonlinearity coefficient γ has been estimated taking into account Kerr nonlinear refractive index of air $n_2 = 3 \times 10^{-23}$ m^2/W [70], as $\gamma \approx 10^{-7}$ m^{-1}·W^{-1} ($\gamma = 2\pi n_{2K}/\lambda A_{eff}$) at a wavelength of 1.56 μm.

If pulse energy at CPA source output was less than ≈380 nJ, the linear pulse propagation regime through RF was realized [79], with a spectrum at HCF output being almost the same as the corresponding spectrum of the CPA source.

The CPA source and RF output pulse-widths together with CPA source pulse peak power are plotted on Figure 16 as a function of CPA source pulse energy. The highest CPA source pulse energy reaches 530 nJ at 1.42 W average power, while the shortest pulses (360 fs) are obtained at ≈380 nJ energy resulting in the ≈1 MW peak power. However, strong self-phase modulation (SPM) effect in the high-power amplification stage of the CPA source results in the broad spectrum wings origination after the pulse energy reaches ≈380 nJ (≈1.0 W average power).

Furthermore, the CPA source pulse-width has a clear minimum of 360 fs at 381 nJ energy and rapidly grows at higher pulse energies with simultaneous saturation of the peak power. As it has been mentioned above, strong SPM action in the amplification stage of the CPA source prevents further pulse shortening due to the excessive nonlinear pulse chirping that cannot be compensated by the grating pair compressor.

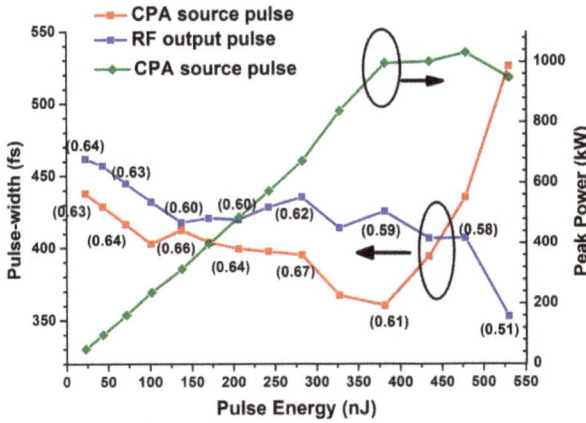

Figure 16. Chirped pulse amplification(CPA) source (red) and RF output (blue) pulse-width also with CPA source pulse peak power (green) versus CPA source pulse energy. (TBP values are given in parentheses) [35].

RF output pulse-width is almost monotonically decreased during pulse energy growth, as seen in Figure 16. Thus, as short as 353 fs Gaussian-type pulses have been obtained at 0.94 W maximum average power at RF output. Hence, nonlinear spectral wings inherent to higher pulse energies are filtered out at RF output, which is also accompanied by simultaneous pulse shortening. Here we suggest nonlinearity influence (SPM) in RF on high peak power pulse propagation that depends on the chirp sign and value of the input pulse, since pulse spectrum undergoes either broadening or narrowing under SPM influence depending on the initial chirp sign [81].

Finally, RF output beam quality have been examined by means of the beam profile scanning (in the X and Y planes) in the far field (at a distance of z = 50 mm from RF end face) with 105/125 μm multimode fiber. Taking into account a RF fundamental mode-field diameter of 45 μm, the M^2 values have been estimated to be $M^2 \approx 1.3$ and $M^2 \approx 1.4$ at low and high output average power, respectively.

In work [79], the propagation of high-power 100-fs Gaussian pulses in RF with a transmission band in the region of 1.56 μm was investigated in a broad range of pump powers. The fiber had eight capillaries with a wall thickness of 2.5 μm, a core diameter of 55 μm, and an outer diameter of 140 μm, while the power attenuation in the fiber was 0.175 dB/m. The dispersion length for 100 fs Gaussian pulses at a wavelength of 1.56 μm was calculated to be $L_d = t_0^2 / \beta_2 = 2.54$ m. The calculated fundamental mode field diameter at a wavelength of 1.56 μm was 40 μm. Taking into account the nonlinearity coefficient for air at 1 atm, $\gamma = 9.65 \times 10^{-8}$ m^{-1}W^{-1}, the linear propagation regime in this RF is limited to a peak power of about 300 ÷ 400 kW, in the case when a nonlinear length of a pulse $L_{nl} = 1/\gamma P$ exceeds the effective absorption length $L_{eff} \approx 25$ m.

Propagation of pulses through the fiber at pulse powers such, that a nonlinear propagation regime is realized ($L_{nl} < L_d$), was investigated numerically using the generalized nonlinear Schrödinger equation for the complex spectral envelope of a pulse [82], taking into account the higher-order dispersion, the Kerr nonlinearity, and stimulated Raman scattering by rotational transitions of nitrogen [83,84].

Figure 17 shows the results that were obtained for a pulse peak power of 10 MW, at which the Kerr nonlinear length is 2.5 times smaller than dispersion one for a 100-fs bandwidth-limited Gaussian pulse. The density plot (Figure 17a) demonstrates a shift of the spectrum at 54 nm to the Stokes region, in which the structure is much weaker than that in the anti-Stokes region, where several characteristic bands are clearly distinguishable. Figure 17b demonstrates the dependence of the spectral shift on the fiber length. The highest shift rate occurs at the first 4 m of the fiber, where the spectrum is strongly broadened due to self-phase modulation.

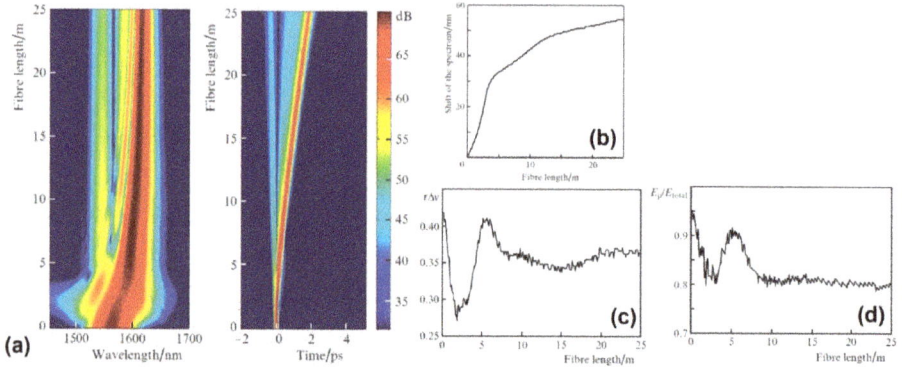

Figure 17. Propagation of a 100-fs transform-limited Gaussian pulse with an input power of 10 MW through the RF: (**a**) color density plot of pulse evolution; (**b**) red-shift of the pulse spectrum; and, (**c**,**d**) time-bandwidth product and pulse energy as functions of fiber length [79].

As follows from Figure 17c,d, the pulse preserves the time-bandwidth product and the ratio of the pulse energy at the half-maximum level to the total energy (within 1%) at fiber lengths between 8 m and 25 m. This behavior may be attributed to Raman soliton that sustains its shape when propagating along the fiber (in this case, the soliton order is $N = 1.6$).

In the nonlinear propagation regime ($L_{nl} < L_{eff}$), 160-fs pulse with a peak power of more than 12 MW (1.92 µJ pulse energy) can be retrieved from the 5-m-long RF, when 100-fs pulse with 40 MW peak power (4 µJ pulse energy) is launched into the fiber. Transmission of radiation in the form of Raman solitons with megawatt-level peak powers, without spreading into a supercontinuum, is possible within the entire effective absorption length of 25 m.

It is known that the creation and amplification of dispersion waves leads to the instability of a multisoliton pulse, transforming its spectrum into the supercontinuum. The aforementioned RF has a relatively narrow transmission window in the range 1450 ÷ 1700 nm, with a dispersion zero at 1.514 µm shifted to the short-wavelength edge of the RF transmission band. For a pulse at a wavelength of 1560 nm, phase matching, which ensures the efficient transfer of soliton energy to dispersion waves, is possible at wavelengths that are around 1442 nm located outside the transmission band. Thus, they cannot be amplified. The dominant process forming the structure of the spectrum is SRS, which can provide a power-dependent spectral shift of up to 130 nm under appropriate peak power of the input 100-fs pulse.

6.2. Multi-Band Supercontinuum Generation in RF

The most impressive results on supercontinuum generation were obtained in Kagome-type holy core fibers [18,71]. By varying gas pressure, the supercontinuum spanning more than three octaves from 124 to 1200 nm was obtained in [18]. The Kagome HCF used for supercontinuum generation in these studies had one or a few broad transmission bands, which, however, limited supercontinuum span.

In the work [80], the possibility of a multi-band supercontinuum generation in the RF with separated capillaries and a core filled with atmospheric air was demonstrated. For these studies, eight-capillary RF with a core diameter of 61.5 μm, a capillary wall thickness of 2.7 μm, and an outer capillary diameter of 25 μm was fabricated. With this capillary thickness, the RF had a large number of transmission bands with slightly different spectral widths, extending from UV to middle IR spectral range (Figure 18a). Due to the low GVD of the atmospheric air, the dispersion characteristics for a fundamental mode in slightly different neighboring transmission bands also had a small difference. In particular, the presence of GVD zeros near the center of each band created favorable conditions for the efficient band-to-band transfer of radiation due to nonlinear effects.

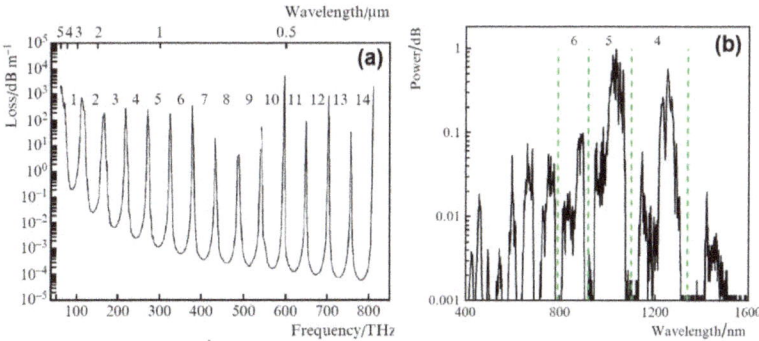

Figure 18. (a) Calculated losses for the fundamental mode in the spectral range corresponding to 14 transmission bands; and, (b) Laser emission spectrum measured at the output of the 3-m-long fiber at the input pulse energy of 110 μJ [80].

A powerful femtosecond solid-state laser was used as a pump source emitting 205 fs pulses with up to 130 μJ energy at a central wavelength of 1028 nm. It should be noted that the pump laser generation wavelength lies in the center of the 5th transmission band, in accordance with the ARROW model, as it is depicted in Figure 18a. Figure 18b shows an experimentally obtained supercontinuum with a spectral range extending from 415 to 1593 nm wavelength that overlaps 11 transmission bands, when 205 fs pulses with 110 μJ energy are launched into the RF.

The multimode nature of the light propagation in this RF reduces the efficiency of nonlinear processes responsible for the supercontinuum generation, owing to a redistribution of energy between higher-order RF modes. Numerical analysis proved (Figure 19) that at a comparable input pulse power and a single-mode propagation, the expected supercontinuum can overlap 14 transmission bands, extending from 370 nm to 4200 nm. Such a spectral width (exceeding three octaves) can be obtained at fiber lengths of ~50 cm, while the pulse retains more than 50% of its energy.

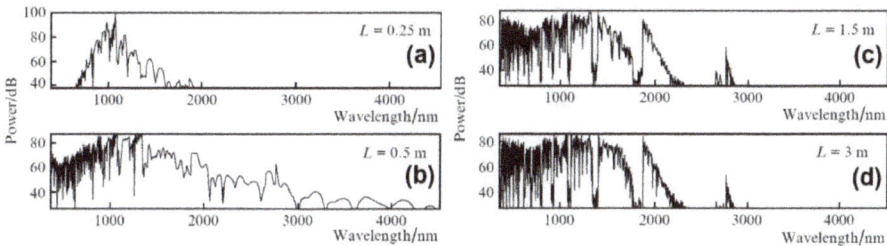

Figure 19. (a–d) Calculated supercontinuum spectra at different RF lengths for pure single-mode propagation at an input pulse energy of 110 μJ for various fiber lengths from 0.25 to 3 m [80].

Detailed analysis of the spectrum structure in various transmission bands made it possible to establish the main nonlinear processes that are responsible for transferring energy from one band to another, such as degenerate and non-degenerate four-wave mixing, and the generation of dispersion waves. It is necessary to emphasize that high efficiency of cascaded nonlinear processes in the RF filled with atmospheric air is accounted for the uniform distribution of zero dispersion wavelengths over a wide spectral range.

7. Conclusions

To conclude, revolver fibers are a versatile tool for transmission, generation, and nonlinear conversion of light in regimes that are not possible in solid-core fibers. Moreover, the design simplicity of the revolver fibers distinguishes them among other types of hollow-core fibers. At the same time, the RFs provide an extremely low overlap of an optical mode with a cladding material. As a result, the RFs can have low optical losses even in those spectral regions where the cladding material has strong fundamental absorption. In particular, silica glass can be applied to fabricate revolver fibers for UV and mid-IR spectral ranges. It was experimentally demonstrated that optical losses of the revolver silica fibers can be as low as ~1 dB/m at wavelength up to 200 nm in the UV and up to 4.4 µm in the mid-IR.

Recently, the implementation of the gas-filled revolver fibers enabled one to demonstrate mid-IR gas fiber lasers that are based on stimulated Raman scattering. Pumping molecular gases, such as light hydrogen $^1H_2$ and deuterium D_2, by nanosecond pulses of a 1.56 µm Er-doped fiber laser, Raman generation in the wavelength range of 2.9–4.4 µm has been demonstrated. In spite of high quantum defect for $1.56 \rightarrow 4.4$ µm conversion, the average output power as high as 250 mW was generated at the wavelength of 4.4 µm with quantum a conversion efficiency as high as 36%. We believe that the efficiency and output power of the Raman gas fiber lasers can be dramatically improved.

An extremely small overlap of the optical mode with the cladding material is also responsible for another useful property of the revolver fibers: such fibers can transmit intense ultrashort pulses without distortion. This fact is advantageous for pulse delivery in material processing applications. Nonlinear and dispersive properties of the cladding material have a limited effect on pulse propagation, as the intensity of light in the cladding is rather low. On the other hand, if optical pulses have high enough intensity, the nonlinearity of a gas inside the hollow core comes into play, enabling various nonlinear phenomena, such as Raman soliton propagation and supercontinuum generation. In particular, multiband supercontinuum generation has been demonstrated in an air-filled revolver fiber in the spectral range of 400–1500 nm. According to numerical simulations, such multiband supercontinuum can be extended up to ~4.5 µm in revolver fibers that are made of silica glass. Extension towards even longer wavelengths can be achieved using chalcogenide revolver fibers, which have already demonstrated optical transmission at the wavelength as long as 10.6 µm.

Finally, we believe that further development of the revolver fibers and devices based on them can make valuable contribution to numerous applications in biomedicine, spectroscopy, and material processing.

Author Contributions: A.F.K. fabricated revolver fibers, A.D.P., Y.P.Y., A.S.B. and I.A.B. made theoretical analysis and calculations, A.V.G., A.N.K. and A.A.K. carried out experiments. All the authors participated in discussion of the results and manuscript preparation.

Acknowledgments: This work was supported by the Presidium of the Russian Academy of Sciences (Program No 1.7: Topical Problems of Photonics, Probing of inhomogeneous Media and Materials).

Conflicts of Interest: The authors declare no conflict of interest.

References

1.   Pryamikov, A.D.; Biriukov, A.S.; Kosolapov, A.F.; Plotnichenko, V.G.; Semjonov, S.L.; Dianov, E.M. Demonstration of a waveguide regime for a silica hollow—Core microstructured optical fiber with a negative curvature of the core boundary in the spectral region > 3.5 µm. *Opt. Express* **2011**, *19*, 1441–1448. [CrossRef] [PubMed]

2. Wang, Y.Y.; Couny, F.; Roberts, P.J.; Benabid, F. Low Loss Broadband Transmission in Optimized Core-Shape Kagome Hollow-Core PCF. In Proceedings of the CLEO'2010, San Jose, CA, USA, 16–21 May 2010.

3. Yu, F.; Wadsworth, W.J.; Knight, J.C. Low loss silica hollow core fibers for 3–4 μm spectral region. *Opt. Express* **2012**, *20*, 11153–11158. [CrossRef] [PubMed]

4. Kolyadin, A.N.; Kosolapov, A.F.; Pryamikov, A.D.; Biriukov, A.S.; Plotnichenko, V.G.; Dianov, E.M. Light transmission in negative curvature hollow core fiber in extremely high material loss region. *Opt. Express* **2013**, *21*, 9514–9519. [CrossRef] [PubMed]

5. Gladyshev, A.V.; Kolyadin, A.N.; Kosolapov, A.F.; Yatsenko, Y.P.; Pryamikov, A.D.; Biryukov, A.S.; Bufetov, I.A.; Dianov, E.M. Efficient 1.9-μm Raman generation in a hydrogen-filled hollow-core fibre. *Quantum Electron.* **2015**, *45*, 807–812. [CrossRef]

6. Kosolapov, A.F.; Alagashev, G.K.; Kolyadin, A.N.; Pryamikov, A.D.; Biryukov, A.S.; Bufetov, I.A.; Dianov, E.M. Hollow-core revolver fibre with a reflecting cladding consisting of double capillaries. *Quantum Electron.* **2016**, *46*, 10–14. [CrossRef]

7. Belardi, W.; Knight, J.C. Hollow antiresonant fibers with reduced attenuation. *Opt. Lett.* **2014**, *39*, 1853–1856. [CrossRef] [PubMed]

8. Alagashev, G.K.; Pryamikov, A.D.; Kosolapov, A.F.; Kolyadin, A.N.; Lukovkin, A.Y.; Biriukov, A.S. Impact of geometrical parameters on the optical properties of negative curvature hollow-core fibers. *Laser Phys.* **2015**, *25*, 055101. [CrossRef]

9. Wei, C.; Weiblen, R.J.; Menyuk, C.R.; Hu, J. Negative curvature fibers. *Adv. Opt. Photonics* **2017**, *9*, 504–561. [CrossRef]

10. Marcatili, E.A.J.; Schmeltzer, R.A. Hollow metallic and dielectric waveguides for long distance optical transmission and lasers. *Bell Syst. Tech. J.* **1964**, *43*, 1783–1809. [CrossRef]

11. Miyagi, M.; Nishida, S. Transmission characteristics of dielectric tube leaky waveguide. *IEEE Trans. Microw. Theory Tech.* **1980**, *28*, 536–541. [CrossRef]

12. Litchinitser, N.M.; Abeeluck, A.K.; Headley, C.; Eggleton, B.J. Antiresonant reflecting photonic crystal optical waveguides. *Opt. Lett.* **2002**, *27*, 1592–1594. [CrossRef] [PubMed]

13. Zheltikov, A.M. Colors of thin films, antiresonant phenomena in optical systems, and the limiting loss of modes in hollow optical waveguides. *Uspekhi Fiz. Nauk* **2008**, *178*, 619–629.

14. Landau, L.D.; Lifshitz, E.M. *Electrodynamics of Continuous Media*, 2nd ed.; Pergamon Press Ltd.: New York, NY, USA, 1984.

15. Kosolapov, A.F.; Pryamikov, A.D.; Biriukov, A.S.; Vladimir, S.; Astapovich, M.S.; Snopatin, G.E.; Plotnichenko, V.G.; Churbanov, M.F.; Dianov, E.M. Demonstration of CO_2-laser power delivery through chalcogenide-glass fiber with negative-curvature hollow core. *Opt. Express* **2011**, *19*, 25723–25728. [CrossRef] [PubMed]

16. Setti, V.; Vincetti, L.; Argyros, A. Flexible tube lattice fibers for terahertz applications. *Opt. Express* **2013**, *23*, 3388–3399. [CrossRef] [PubMed]

17. Kitamura, R.; Pilon, L.; Jonasz, M. Optical constants of silica glass from extreme ultraviolet to far infrared at near room temperature. *Appl. Opt.* **2007**, *46*, 8118–8133. [CrossRef] [PubMed]

18. Belli, F.; Abdolvand, A.; Chang, W.; Travers, J.C.; Russell, P.S.J. Vacuum-ultraviolet to infrared supercontinuum in hydrogen-filled photonic crystal fiber. *Optica* **2015**, *2*, 292–300. [CrossRef]

19. Humbach, O.; Fabian, H.; Grzesik, U.; Haken, U.; Heitmann, W. Analysis of OH absorption bands in synthetic silica. *J. Non-Cryst. Solids* **1996**, *203*, 19–26. [CrossRef]

20. Tomashuk, A.L.; Golant, K.M. Radiation-resistant and radiation-sensitive silica optical fibers. In Proceedings of the SPIE, Moscow, Russia, 17 May 2000.

21. Kryukova, E.B.; Plotnichenko, V.G.; Dianov, E.M. IR absorption spectra in high-purity silica glasses fabricated by different technologies. In Proceedings of the SPIE, Moscow, Russia, 17 May 2000.

22. Roberts, P.; Couny, F.; Sabert, H.; Mangan, B.; Williams, D.; Farr, L.; Mason, M.; Tomlinson, A.; Birks, T.; Knight, J.; et al. Ultimate low loss of hollow-core photonic crystal fibres. *Opt. Express* **2005**, *13*, 236–244. [CrossRef] [PubMed]

23. Fini, J.M.; Nicholson, J.W.; Windeler, R.S.; Monberg, E.M.; Meng, L.; Mangan, B.; DeSantolo, A.; DiMarcello, F.V. Low-loss hollow-core fibers with improved single-modedness. *Opt. Express* **2013**, *21*, 6233–6242. [CrossRef] [PubMed]

24. Wheeler, N.; Heidt, A.; Petrovich, M.; Baddela, N.; Numkam-fokoua, A.; Hayes, J.; Sandoghchi, S.R.; Poletti, F.; Wheeler, N.V.; Heidt, A.M. Low-loss and low-bend-sensitivity mid-infrared guidance in a hollow-core-photonic-bandgap fiber. *Opt. Lett.* **2014**, *39*, 295–298. [CrossRef] [PubMed]

25. Wang, Y.Y.; Wheeler, N.V.; Couny, F.; Roberts, P.J.; Benabid, F. Low loss broadband transmission in hypocycloid-core Kagome hollow-core photonic crystal fiber. *Opt. Lett.* **2011**, *36*, 669–671. [CrossRef] [PubMed]

26. Février, S.; Beaudou, B.; Viale, P. Understanding origin of loss in large pitch hollow-core photonic crystal fibers and their design simplification. *Opt. Express* **2010**, *18*, 5142–5150. [CrossRef] [PubMed]

27. Gérôme, F.; Jamier, R.; Auguste, J.-L.; Humbert, G.; Blondy, J.-M. Simplified hollow-core photonic crystal fiber. *Opt. Lett.* **2010**, *35*, 1157–1159. [CrossRef] [PubMed]

28. Urich, A.; Maier, R.R.J.; Yu, F.; Knight, J.C.; Hand, D.P.; Shephard, J.D. Flexible delivery of Er: YAG radiation at 2. 94 μm with negative curvature silica glass fibers: A new solution for minimally invasive surgical procedures. *Biomed. Opt. Express* **2013**, *4*, 7139–7144. [CrossRef] [PubMed]

29. Hartung, A.; Kobelke, J.; Schwuchow, A.; Wondraczek, K.; Bierlich, J.; Popp, J.; Frosch, T.; Schmidt, M.A. Double antiresonant hollow core fiber—Guidance in the deep ultraviolet by modified tunneling leaky modes. *Opt. Express* **2014**, *22*. [CrossRef] [PubMed]

30. Pryamikov, A.D.; Kosolapov, A.F.; Alagashev, G.K.; Kolyadin, A.N.; Vel'miskin, V.V.; Biriukov, A.S.; Bufetov, I.A. Hollow-core microstructured "revolver" fibre for the UV spectral range. *Quantum Electron.* **2016**, *46*, 1129–1133. [CrossRef]

31. Gladyshev, A.V.; Kosolapov, A.F.; Kolyadin, A.N.; Astapovich, M.S.; Pryamikov, A.D.; Likhachev, M.E.; Bufetov, I.A. Mid-IR hollow-core silica fibre Raman lasers. *Quantum Electron.* **2017**, *47*, 1078–1082. [CrossRef]

32. Wang, Z.; Belardi, W.; Yu, F.; Wadsworth, W.J.; Knight, J.C. Efficient diode-pumped mid-infrared emission from acetylene-filled hollow-core fiber. *Opt. Express* **2014**, *22*, 21872–21878. [CrossRef] [PubMed]

33. Michieletto, M.; Lyngsø, J.K.; Jakobsen, C.; Lægsgaard, J.; Bang, O.; Alkeskjold, T.T. Hollow-core fibers for high power pulse delivery. *Opt. Express* **2016**, *24*, 7103–7119. [CrossRef] [PubMed]

34. Uebel, P.; Günendi, M.C.; Frosz, M.H.; Ahmed, G.; Edavalath, N.N.; Ménard, J.-M.; Russell, P.S.J. A broad-band robustly single-mode hollow-core PCF by resonant filtering of higher order modes. In Proceedings of the Frontiers in Optics 2015, San Jose, CA, USA, 18–22 October 2015.

35. Krylov, A.A.; Senatorov, A.K.; Pryamikov, A.D.; Kosolapov, A.F.; Kolyadin, A.N.; Alagashev, G.K.; Gladyshev, A.V.; Bufetov, I.A. 1.56 μm sub-microjoule femtosecond pulse delivery through low-loss microstructured revolver hollow-core fiber. *Laser Phys. Lett.* **2017**, *14*, 035104. [CrossRef]

36. Gauthier, J.-C.; Fortin, V.; Carrée, J.-Y.; Poulain, S.; Poulain, M.; Vallée, R.; Bernier, M. Mid-IR supercontinuum from 2.4 to 5.4 μm in a low-loss fluoroindate fiber. *Opt. Lett.* **2016**, *41*, 1756–1759. [CrossRef] [PubMed]

37. Tang, Z.; Shiryaev, V.S.; Furniss, D.; Sojka, L.; Sujecki, S.; Benson, T.M.; Seddon, A.B.; Churbanov, M.F. Low loss Ge-As-Se chalcogenide glass fiber, fabricated using extruded preform, for mid-infrared photonics. *Opt. Mater. Express* **2015**, *5*, 1722–1737. [CrossRef]

38. Sanghera, J.; Aggarwal, I.D. *Infrared Fiber Optics*; CRC Press: Boca Raton, FL, USA, 1998.

39. Artyushenko, V.; Bocharnikov, A.; Sakharova, T.; Usenov, I. Mid-infrared fiber optics for 1–18 μm range. IR-fibers and waveguides for laser power delivery and spectral sensing. *Opt. Photonik* **2014**, *4*, 35–39. [CrossRef]

40. Hassan Muhammad Rosdi, A.; Yu, F.; Wadsworth, J.W.; Knight, J.C. Cavity-based mid-IR fiber gas laser pumped by a diode laser. *Optica* **2016**, *3*, 218–221. [CrossRef]

41. Uebel, P.; Günendi, M.C.; Frosz, M.H.; Ahmed, G.; Edavalath, N.N.; Ménard, J.-M.; Russell, P.S.J. Broadband robustly single-mode hollow-core PCF by resonant filtering of higher-order modes. *Opt. Lett.* **2016**, *41*, 1961–1964. [CrossRef] [PubMed]

42. Wei, C.; Hu, J.; Menyuk, C.R. Comparison of loss in silica and chalcogenide negative curvature fibers as the wavelength varies. *Front. Phys.* **2016**, *4*, 1–10. [CrossRef]

43. Gladyshev, A.V.; Kosolapov, A.F.; Khudyakov, M.M.; Yatsenko, Y.P.; Kolyadin, A.N.; Krylov, A.A. 4.4 μm raman laser based on hydrogen-filled hollow-core silica fiber. In Proceedings of the CLEO'2017, San Jose, CA, USA, 14–19 May 2017.

44. Gladyshev, A.V.; Kosolapov, A.F.; Khudyakov, M.M.; Yatsenko, Y.P.; Kolyadin, A.N.; Krylov, A.A.; Pryamikov, A.D.; Biriukov, A.S.; Likhachev, M.E.; Bufetov, I.A.; Dianov, E.M. 4.4-μm Raman laser based on hollow-core silica fibre. *Quantum Electron.* **2017**, *47*, 491–494. [CrossRef]

45. Alharbi, M.; Bradley, T.; Debord, B.; Fourcade-Dutin, C.; Ghosh, D.; Vincetti, L.; Gérôme, F.; Benabid, F. Hypocycloid-shaped hollow-core photonic crystal fiber Part II: Cladding effect on confinement and bend loss. *Opt. Express* **2013**, *21*, 28609–28616. [CrossRef] [PubMed]

46. Gao, S.-F.; Wang, Y.-Y.; Ding, W.; Wang, P. Hollow-core negative-curvature fiber for UV guidance. *Opt. Lett.* **2018**, *43*, 1347. [CrossRef] [PubMed]

47. Denisov, A.N.; Kosolapov, A.F.; Senatorov, A.K.; Pal'tsev, P.E.; Semjonov, S.L. Fabrication of microstructured optical fibres by drawing preforms sealed at their top end. *Quantum Electron.* **2016**, *46*, 1031–1039. [CrossRef]

48. Yakovlev, A.I.; Snetkov, I.L.; Dorofeev, V.V.; Motorin, S.E. Magneto-optical properties of high-purity zinc-tellurite glasses. *J. Non. Cryst. Solids* **2018**, *480*, 90–94. [CrossRef]

49. Shiryaev, V.S.; Kosolapov, A.F.; Pryamikov, A.D.; Snopatin, G.E.; Churbanov, M.F.; Biriukov, A.S.; Kotereva, T.V.; Mishinov, S.V.; Alagashev, G.K.; Kolyadin, A.N. Development of technique for preparation of As2S3 glass preforms for hollow core microstructured optical fibers. *J. Optoelectron. Adv. Mater.* **2014**, *16*, 1020–1025.

50. Gattass, R.R.; Rhonehouse, D.; Gibson, D.; McClain, C.C.; Thapa, R.; Nguyen, V.Q.; Bayya, S.S.; Weiblen, R.J.; Menyuk, C.R.; Shaw, L.B.; et al. Infrared glass-based negative-curvature anti-resonant fibers fabricated through extrusion. *Opt. Express* **2016**, *24*, 25697–25703. [CrossRef] [PubMed]

51. Benabid, F.; Knight, J.C.; Antonopoulos, G.; Russell, P.S.J. Stimulated raman scattering in hydrogen-filled hollow-core photonic crystal fiber. *Science* **2002**, *298*, 399–402. [CrossRef] [PubMed]

52. Gladyshev, A.V.; Kosolapov, A.F.; Astapovich, M.S.; Kolyadin, A.N.; Pryamikov, A.D.; Khudyakov, M.M.; Likhachev, M.E.; Bufetov, I.A. Revolver Hollow-Core Fibers and Raman Fiber Lasers. In Proceedings of the OFC'2018, San Diego, CA, USA, 11–15 March 2018.

53. Wang, Z.; Yu, F.; Wadsworth, W.J.; Knight, J.C. Efficient 1.9 μm emission in H$_2$-filled hollow core fiber by pure stimulated vibrational Raman scattering. *Laser Phys. Lett.* **2014**, *11*, 105807. [CrossRef]

54. Benoit, A.; Beaudou, B.; Debord, B.; Gerome, F.; Benabid, F. High power Raman-converter based on H$_2$-filled inhibited coupling HC-PCF. In Proceedings of the SPIE 10088, San Francisco, CA, USA, 28 January–2 February 2017.

55. Xu, M.; Yu, F.; Knight, J. Mid-infrared 1W hollow-core fiber gas laser source. *Opt. Lett.* **2017**, *42*, 4055. [CrossRef] [PubMed]

56. Couny, F.; Benabid, F.; Light, P.S. Subwatt threshold cw raman fiber-gas laser based on H$_2$-filled hollow-core photonic crystal fiber. *Phys. Rev. Lett.* **2007**, *99*, 143903. [CrossRef] [PubMed]

57. Bufetov, I.A.; Dianov, E.M. A simple analytic model of a cw multicascade fibre Raman laser. *Quantum Electron.* **2007**, *30*, 873–877. [CrossRef]

58. Kolyadin, A.N.; Astapovich, M.S.; Gladyshev, A.V.; Kosolapov, A.F. The design optimization and experimental investigation of the 4.4 μm raman laser basedon hydrogen-filled revolver silica fiber. In *VII International Conference on Photonics and Information Optics*; KnE Energy & Physics: Moscow, Russia, 2018; Volume 2018, pp. 47–64.

59. Gladyshev, A.V.; Kosolapov, A.F.; Khudyakov, M.M.; Yatsenko, Y.P.; Kolyadin, A.N.; Krylov, A.A. Raman generation in 2.9–3.5 μm spectral range in revolver hollow-core silica fiber filled by H$_2$/D$_2$ mixture. In Proceedings of the CLEO'2017, San Jose, CA, USA, 14–19 May 2017.

60. Gladyshev, A.V.; Kosolapov, A.F.; Khudyakov, M.M.; Yatsenko, Y.P.; Kolyadin, A.N.; Krylov, A.A.; Pryamikov, A.D.; Biriukov, A.S.; Likhachev, M.E.; Bufetov, I.A.; et al. 2.9, 3.3, and 3.5 μm Raman lasers based on revolver hollow-core silica fiber filled by 1H2/D2 gas mixture. *IEEE J. Sel. Top. Quantum Electron.* **2018**, *24*, 0903008. [CrossRef]

61. Astapovich, M.S.; Kolyadin, A.N.; Gladyshev, A.; Kosolapov, A.F.; Pryamikov, A.D.; Khudyakov, M.; Likhachev, M.E.; Bufetov, I.A. Efficient 4.4 μm Raman laser based on hydrogen-filled hollow-core silica fiber. *arXiv*, **2018**, arXiv:1801.01729.

62. Hanna, D.C.; Pointer, D.J.; Pratt, D.J. Stimulated raman scattering of picosecond light pulses in hydrogen, deuterium, and methane. *IEEE J. Quantum Electron.* **1986**, *22*, 332–336. [CrossRef]

63. Weber, M.J. *CRC Handbook of Laser Science and Technology Supplement 2: Optical Materials*; CRC Press: Boca Raton, FL, USA, 1994.

64. Bischel, W.K.; Black, G. Wavelength dependence of raman scattering cross sections from 200–600 nm. In *AIP Conference Proceedings*; American Institute of Physics: College Park, MD, USA, 1983.

65. Chen, Y.; Wang, Z.; Gu, B.; Yu, F.; Lu, Q. Achieving a 1.5 μm fiber gas Raman laser source with about 400 kW of peak power and a 6.3 GHz linewidth. *Opt. Lett.* **2016**, *41*, 5118–5121. [CrossRef] [PubMed]

66. Wang, Z.; Gu, B.; Chen, Y.; Li, Z.; Xi, X. Demonstration of a 150-kW-peak-power, 2-GHz-linewidth, 1.9-μm fiber gas Raman source. *Appl. Opt.* **2017**, *56*, 7657–7661. [CrossRef] [PubMed]

67. Cao, L.; Gao, S.-F.; Peng, Z.-G.; Wang, X.-C.; Wang, Y.-Y.; Wang, P. High peak power 2.8 μm Raman laser in a methane-filled negative-curvature fiber. *Opt. Express* **2018**, *26*, 5609–5615. [CrossRef] [PubMed]

68. Fermann, M.E.; Galvanauskas, A.; Sucha, G. *Ultrafast Lasers: Technology and Applications*; CRC Press: Boca Raton, FL, USA, 2002.

69. Funck, M.; Wedel, B. Industrial fiber beam delivery system for ultrafast lasers. *Laser Tech. J.* **2016**, *13*, 42–44. [CrossRef]

70. Nibbering, E.T.J.; Grillon, G.; Franco, M.A.; Prade, B.S.; Mysyrowicz, A. Determination of the inertial contribution to the nonlinear refractive index of air, N_2, and O_2 by use of unfocused high-intensity femtosecond laser pulses. *J. Opt. Soc. Am. B* **1997**, *14*, 650. [CrossRef]

71. Travers, J.C.; Chang, W.; Nold, J.; Joly, N.Y.; Russell, P.S.J. Ultrafast nonlinear optics in gas-filled hollow-core photonic crystal fibers. *J. Opt. Soc. Am. B* **2011**, *28*, A11–A26. [CrossRef]

72. Saleh, M.F.; Chang, W.; Hölzer, P.; Nazarkin, A.; Travers, J.C.; Joly, N.Y.; Russell, P.S.J.; Biancalana, F. Theory of photoionization-induced blueshift of ultrashort solitons in gas-filled hollow-core photonic crystal fibers. *Phys. Rev. Lett.* **2011**, *107*, 203902. [CrossRef] [PubMed]

73. Saleh, M.F.; Biancalana, F. Tunable frequency-up/down conversion in gas-filled hollow-core photonic crystal fibers. *J. Opt.* **2015**, *18*, 13002. [CrossRef]

74. Ouzounov, D.G.; Ahmad, F.R.; Müller, D.; Venkataraman, N.; Gallagher, M.T.; Thomas, M.G.; Silcox, J.; Koch, K.W.; Gaeta, A.L. Generation of megawatt optical solitons in hollow-core photonic band-gap fibers. *Science* **2003**, *301*, 1702–1704. [CrossRef] [PubMed]

75. Luan, F.; Knight, J.C.; Russell, P.S.J.; Campbell, S.; Xiao, D.; Reid, D.T.; Mangan, B.J.; Williams, D.P.; Roberts, P.J. Femtosecond soliton pulse delivery at 800 nm wavelength in hollow-core photonic bandgap fibers. *Opt. Express* **2004**, *12*, 835–840. [CrossRef] [PubMed]

76. Wang, Y.Y.; Peng, X.; Alharbi, M.; Dutin, C.F.; Bradley, T.D.; Gérôme, F.; Mielke, M.; Booth, T.; Benabid, F. Design and fabrication of hollow-core photonic crystal fibers for high-power ultrashort pulse transportation and pulse compression. *Opt. Lett.* **2012**, *37*, 3111–3113. [CrossRef] [PubMed]

77. Debord, B.; Alharbi, M.; Vincetti, L.; Husakou, A.; Fourcade-Dutin, C.; Hoenninger, C.; Mottay, E.; Gérôme, F.; Benabid, F. Multi-meter fiber-delivery and pulse self-compression of milli-Joule femtosecond laser and fiber-aided laser-micromachining. *Opt. Express* **2014**, *22*, 10735–10746. [CrossRef] [PubMed]

78. Kolyadin, A.N.; Alagashev, G.K.; Pryamikov, A.D.; Mouradian, L.; Zeytunyan, A.; Toneyan, H.; Kosolapov, A.F.; Bufetov, I.A. Negative curvature hollow-core fibers: Dispersion properties and femtosecond pulse delivery. *Phys. Procedia* **2015**, *73*, 59–66. [CrossRef]

79. Yatsenko, Y.P.; Krylov, A.A.; Pryamikov, A.D.; Kosolapov, A.F.; Kolyadin, A.N.; Gladyshev, A.V.; Bufetov, I.A. Propagation of femtosecond pulses in a hollow-core revolver fibre. *Quantum Electron.* **2016**, *46*, 617–626. [CrossRef]

80. Yatsenko, Y.P.; Pleteneva, E.N.; Okhrimchuk, A.G.; Gladyshev, A.V.; Kosolapov, A.F.; Kolyadin, A.N.; Bufetov, I.A. Multiband supercontinuum generation in an air-core revolver fibre. *Quantum Electron.* **2017**, *47*, 553. [CrossRef]

81. Agrawal, G. *Nonlinear Fiber Optics*, 5th ed.; Academic Press: Cambridge, MA, USA, 2012.

82. Dudley, J.M.; Taylor, J.R. *Supercontinuum Generation in Optical Fibers*; Cambridge University Press: Cambridge, MA, USA, 2010.

83. Sprangle, P.; Peñano, J.R.; Hafizi, B. Propagation of intense short laser pulses in the atmosphere. *Phys. Rev. E* **2002**, *66*, 046418. [CrossRef] [PubMed]

84. Peñano, J.R.; Sprangle, P.; Serafim, P.; Hafizi, B.; Ting, A. Stimulated Raman scattering of intense laser pulses in air. *Phys. Rev. E* **2003**, *68*, 056502. [CrossRef] [PubMed]

fibers

MDPI

Review

3D Printed Hollow-Core Terahertz Fibers

Alice L. S. Cruz [1,2,]*, **Cristiano M. B. Cordeiro [3]** and **Marcos A. R. Franco [1,2]**

[1] Instituto Tecnológico de Aeronáutica—ITA, São José dos Campos 12228-900, Brazil;
marcos.a.r.franco@gmail.com

[2] Instituto de Estudos Avançados—IEAv, São José dos Campos 12.228-001, Brazil

[3] Instituto de Física Gleb Wataghin, Universidade Estadual de Campinas—UNICAMP,
Campinas 13083-970, Brazil; cmbc@ifi.unicamp.br

* Correspondence: alicelscruz@gmail.com; Tel.: +55-11-99631-2736

Received: 22 May 2018; Accepted: 15 June 2018; Published: 21 June 2018

Abstract: This paper reviews the subject of 3D printed hollow-core fibers for the propagation of terahertz (THz) waves. Several hollow and microstructured core fibers have been proposed in the literature as candidates for low-loss terahertz guidance. In this review, we focus on 3D printed hollow-core fibers with designs that cannot be easily created by conventional fiber fabrication techniques. We first review the fibers according to their guiding mechanism: photonic bandgap, antiresonant effect, and Bragg effect. We then present the modeling, fabrication, and characterization of a 3D printed Bragg and two antiresonant fibers, highlighting the advantages of using 3D printers as a path to make the fabrication of complex 3D fiber structures fast and cost-effective.

Keywords: terahertz; THz; 3D printing; addictive manufacturing; waveguide; optical fiber

1. Introduction

The terahertz (THz) spectral range is the part of the electromagnetic spectrum between 0.1–10 THz or 0.03–3 mm wavelength. For a long period, the terahertz band was relatively unexplored due to the unavailability of cost-effective and powerful sources. Due to the evolution of these devices in the mid-1980s, however, terahertz radiation has attracted much more attention. Since this part of the spectrum is between the infrared (IR) and microwave frequency ranges, the development of waveguides [1–6], filters [7,8], polarizers [9,10], lenses [11,12], and other optical components benefits from the well-established technologies [13–15]. The characteristic of terahertz waves to penetrate most dielectric materials offers the possibility of many applications. The shorter wavelengths than microwave and millimeter waves allow much greater resolution in imaging, making it suitable for security scanning, imaging, and non-destructive testing [16,17]. Because of the non-ionizing characteristic of terahertz, it can pass through organic tissue without causing damage, and it can be safely applied in biomedical sensing [18,19]. In addition, it is possible to detect many chemicals and biological agents because they exhibit well defined spectral signatures in the terahertz range [20,21]. Radio astronomy and wireless communication are also fields with great interest in this spectral range. For example, terahertz waves could be used to detect cold bodies and debris in space or to increase data transmission using the larger bandwidth of the terahertz band [22–25].

Most terahertz systems are based on free-space propagation, which can control the high losses that occur as a result of absorption by water vapor. However, most terahertz sources and detectors are power inefficient and, in a free-space configuration, path power loss is a significant limitation. Moreover, free-space systems handicap integration with other components. In order to upgrade these systems to use guided waves one needs low-loss and low dispersion propagation waveguides as basic components. These waveguides can provide the transference of electromagnetic waves/information between two points and interconnect systems [4,5,26–31]. Furthermore, they can also be explored as sensors and imaging probes [21,32,33].

Over the last decade, a substantial amount of effort has been directed towards achieving significant low-loss terahertz fibers and waveguides. Some works show metal rods being used as terahertz waveguides, but finite conductivity limits their applications [34,35]. An alternative is to fabricate dielectric waveguides. Polymer optical fiber technology and simple designs, such as a rod or a dielectric tube, were initially investigated [36–38]. However, dielectric waveguides are lossy due to the bulk material absorption. Polymers, such as Zeonex® and Topas® (Cyclic Olefin Polymers), have losses with typical values of approximately 1 dB/cm [36] while silica, an usual glass used in optical fibers, has a typical loss of approximately 9 dB/cm [39,40]. The first terahertz dielectric waveguide designs tried to explore the concept of reducing the losses by increasing the air filling fraction of porous polymer fibers [41–44]. Many different configurations have been demonstrated: periodically microstructured fibers [36,37]; bandgap fibers [43]; fibers with elliptical air-holes; and fibers with rectangular slot air-holes to increase the birefringence [5,26,27]. Some waveguide designs have shown interesting results in terms of low-loss and low dispersion over certain frequency ranges [4,6,30,33]. For example, in [4] the authors achieved an effective material loss of 0.034 cm^{-1} at 1.0 THz. In spite of these results, issues such as broadband transmission, lower losses, low bending losses, easier cutting and splicing procedures, and availability in long lengths are still a challenge [1,30,41,43,45].

However, even with these achievements, the material losses are still high in porous terahertz fibers and the best option to overcome this issue is to move on to hollow-core fibers. Hollow-core fibers are good candidates for low-loss guidance because the material absorption loss can be significantly minimized. This reduction is mostly due to the modal energy being located within the cladding air-holes or air-core, reducing the effective material loss to less than 1/20th of the characteristic loss of the host material. The mentioned fibers and fibers' preforms can be fabricated via extrusion, stack-and-draw, and drilling and molding, but the fabrication of more complex structures, with higher air filling fraction, can be greatly simplified with more advanced manufacturing techniques.

The recent developments in rapid prototyping, from jewelry to food, have been shown as a path to meet the fabrication of complex 3D structures quickly and cost-effectively. Not only fibers but antennas, couplers, and metallic waveguides have been investigated and fabricated for GHz and THz frequencies [46,47]. The additive manufacturing technique creates structures layer by layer. Among the different additive manufacturing methods, polymer jetting (Polyjet) is the most commonly applied for the fabrication of millimetric and sub-millimetric components due to its superior spatial resolution around 100 μm.

This paper reviews the evolution of 3D printed hollow-core terahertz fibers, from the first terahertz fiber fabricated using knowledge from photonic crystal fibers (PCF) to the most recent achievements using additive manufacturing (3D printing). The paper is organized as follows: Section 2 outlines the evolution of additive manufacturing and its challenges. Section 3 relies on 3D printed hollow-core terahertz fibers. Section 4 focuses on numerical modeling and experimental characterization of a hollow-core terahertz Bragg fiber and two antiresonant fibers and, in Section 5, concluding remarks are presented as well as a brief discussion on the future.

2. Additive Manufacturing Technology

The first three-dimensional object created layer by layer via additive manufacturing (or 3D printing) was in the 1980's on the rapid prototyping field. Since then, this technology has revolutionized the manufacturing industry as well as research. Now, cost-effective, customizable, and quick fabrication is enabling the creation of prototypes or finished products with more efficiency. Additive manufacturing builds these objects by adding layers of material instead of removing material from a bulk, as in the milling process for example. Many different materials can be used in additive manufacturing such as polymers, metal [48], biocompatible material [49], ceramic [50] and organic compounds. Therefore, many different industries such as food [51], medical [52], pharmaceutical [53], mechanical [54], and microwaves [46] benefit from the technology.

Additive manufacturing can be split into several branches depending on the fabrication method. These branches include Fused Deposition Modeling (FDM), Stereolithography (SLA), Electron Beam Melting (EBM), Selective Laser Sintering (SLS), Polymer Jetting (Polyjet), and so on [55]. The common process of these methods is the model design, generally drawn in CAD software, converted to a STL file, and sent to the printer.

In the microwave and sub-millimetric wave fields, the use of additive manufacturing has grown. Recent works report the fabrication of waveguides, beam splitters, plasmonic devices, lenses, and antennas [56–59]. This great interest is due to the compatibility of the fabrication scale, the availability of several materials, fast processes, reproducibility, and low cost. For terahertz devices, the most common methods are fused deposition modelling (FDM), stereolithography apparatus (SLA) and Polyjet. In the FDM process, thermoplastic filaments are heated, extruded through a nozzle and subsequently deposited on the building bed. Its spatial resolution is given by the nozzle opening. The common materials are acrylonitrile butadiene styrene (ABS), polylactic acid (PLA), and polycarbonate (PC). In the SLA process a UV laser beam scans the surface of a photo-resin tank to form each layer of the object. In the Polyjet technique, a print head deposits thin layers of a UV-curable resin onto a construction tray. UV lamps cure the material as it is being deposited. After finishing one cross-section sheet another top layer is built. The advantage of Polyjet over the other methods is its superior spatial resolution of about 100 μm, which depends on the laser spot size.

One of the actual challenges for additive manufacturing is to produce complex components with high density ceramics. The ceramics are generally processed as powders and present high melting temperatures. Also, they are not resistant to thermal shocks. The most recent advance on this technology shows the application of SLA with ceramic suspension as the way to fabricate dense ceramics. Some commercial solutions are available, such as Admatec Europe. For terahertz devices, the main challenges of using these techniques are: building long length structures; high absorption losses of the available materials; surface finish; and the spatial resolution. Some authors have shown the fabrication of fibers' preforms with 3D printers and following that the fiber drawing [60] (what improves the finishing), terahertz optics devices printed with Topas (low-loss polymer) [61], and extremely high resolution fabrication (around 1 μm) [62]. These recent researches and innovations shown the great scientific interest in using additive manufacturing as a fabrication method. These achievements can lead the technology to become the main fabrication method of terahertz passive devices, keeping in mind the cost-efficiency of the technology.

3. Terahertz 3D Printed Waveguides

The terahertz waveguides should be able to promote propagation of the waves in dry air to decrease the material absorption contribution. To achieve this goal one of these three physical phenomena must occur: the photonic bandgap; the antiresonant effect; or the Bragg reflection. The photonic bandgap effect occurs in hollow-core fibers whose microstructured cladding has an appropriate distribution of air holes. In the bandgap condition, the terahertz modes cannot be guided in certain frequency ranges. The antiresonant effect occurs when the light launched in the fiber core is reflected on both interfaces of the core wall and a constructive interference occurs within the hollow-core. The transmission spectrum of such fibers can be easily obtained by knowing the contrast refractive indexes between clad and core as well as the capillary wall thickness, which is similar to a Fabry–Pérot cavity. Usually these fibers have a far simpler geometrical design than an ordinary tube (capillary). Another class of hollow-core fibers is based on structures with a cladding formed by a succession of material layers with low and high refractive indexes, giving rise to a kind of Bragg reflector known as OmniGuide or Bragg fibers [63].

Based on these physical phenomena, since 2011 researchers have been proposing new designs of air core terahertz fibers using 3D printing as a fabrication method. The first reported 3D printed fiber (Figure 1b) was based on a hollow-core PCF-like structure that was fabricated using the Polyjet technique [64]. In this case, it was possible to achieve a propagation loss of 0.03 dB/mm (0.3 dB/cm) at 105 GHz by applying a UV-resin with a dielectric constant of 2.75. In 2016, another terahertz hollow-core fiber based on photonic bandgap propagation and fabricated via Polyjet was proposed, see Figure 1c [65]. The fiber was printed using a 3D printer with a resolution of 600 dpi and a UV-curable polymer. One of the challenges using 3D prototyping is to build longer length structures since the currently available printers have a strict work-volume limitation. In order to overcome this issue, the authors printed two fibers and connected them mechanically, obtaining an average power propagation loss of 0.02 cm^{-1} (0.08 dB/cm) over 0.2–1.0 THz. Other authors are investigating the possibility of fabricating the preform of the terahertz fibers directly by 3D printing [60].

The next category of hollow-core printed fibers is based on the antiresonant effect with negative curvature in the core. In a negative curvature fiber, we have the surface normal vector of the core boundary directed towards the fiber's center [66]. This negative curvature helps to inhibit coupling between the fundamental core mode and the cladding modes, which considerably decreases the propagation losses. In Figure 1d, one can see the cross-section of a fiber fabricated via FDM. This fiber, built with ABS, was able to guide with low-loss in the transmission windows between 0.10–0.21, 0.30–0.40, and 0.5–1.1 THz [67]. The fiber whose cross section is shown in Figure 1e was fabricated using PC via the FDM technique and guides terahertz radiation with losses around 10's dB/cm over a 150 to 600 GHz range [68].

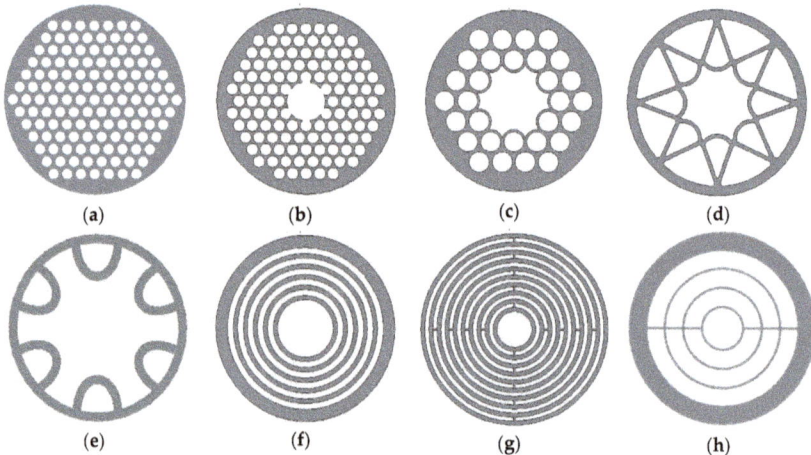

Figure 1. (a) Porous polymer terahertz fiber design [38] (b) First all-dielectric 3D printed terahertz waveguide [64]; (c) 3D printed terahertz waveguide based on Kagome photonic crystal structure [65]; (d) Hollow-core with negative curvature [67]; (e) 3D-printed polymer antiresonant waveguide [68]; (f) 3D printed terahertz Bragg [69]; (g) Bragg waveguide with defect layers [70]; (h) Single-mode Bragg waveguide [71].

The last group of fibers (Figure 1f–h) is based on the Bragg reflection. The characteristics of the first 3D printed Bragg fiber is all detailed in [69], see Figure 1f. Using the FDM technology and an ABS polymer, the authors were able to demonstrate low-loss propagation in a 93 mm long fiber. The authors in [70] showed the application of a 3D printed terahertz Bragg fiber as a powder and thin film sensor with sensitivity close to 0.1 GHz/μm (Figure 1g). The fiber was built using an SLA system, which has a transverse resolution of 50 μm and a longitudinal resolution of 1 μm. The printing resin has a refractive index and an absorption coefficient of around 1.64 and, 1.1 cm^{-1} at 0.2 THz, respectively. This fiber can propagate with low-loss propagation of 0.15 cm^{-1} (0.65 dB/cm) over frequencies higher than 0.35 THz. The cross-section of the single-mode and low-loss terahertz Bragg fiber presented in [71] can be seen in

Figure 1h. The authors reached single mode propagation and an average propagation loss of around 3 dB/m (0.03 dB/cm) at 0.27 THz. Table 1 summarizes the main characteristics of the cited fibers.

The porous fiber (Figure 1a) is easily obtained by drilling and drawing a plastic preform, but the design is limited by how thin the wall thickness can be during the drilling process. However, some energy will still overlap the lossy material leading to high absorption loss. The hollow-core bandgap fibers (photonic crystal and Bragg fibers) may decrease the propagation losses by guiding the wave in the air core, but normally this mechanism works in quite limited wavelength range (Figure 1b,c,f–h). This wavelength range can be broadened by using antiresonant hollow-core fibers (Figure 1d,e). In addition, negative curvature structures can avoid/reduce the coupling between the core/cladding modes, thereby decreasing the propagation loss.

Table 1. Summarized fibers characteristics.

Fiber	Guiding Method	Printing Method	Material	Loss (dB/cm)	Year
Figure 1b	Photonic Bandgap	Polyjet	UV-resin	0.3 @105 GHz	2011 [64]
Figure 1c	Photonic Bandgap	Polyjet	UV-resin	0.08 @1 THz	2016 [65]
Figure 1d	Antiresonant effect	FDM	ABS	0.3 @0.47 THz	2015 [67]
Figure 1e	Antiresonant effect	FDM	PC	10 @0.3 THz	2018 [68]
Figure 1f	Antiresonant effect	FDM	ABS	0.1 @0.4 THz	2015 [69]
Figure 1g	Bragg Reflection	SLA	UV-resin	0.65 @0.35 THz	2017 [70]
Figure 1h	Bragg Reflection	SLA	UV-resin	0.03 @0.27 THz	2018 [71]

4. Optical Characterization: Numerical Modeling and Experimental Data

In this section, we will present data from some of the 3D printed terahertz hollow-core fibers produced and studied by our research group in the last few years. We will focus on antiresonant and Bragg fibers, once they may present lower absorption and confinement losses. The Finite Element and Beam Propagation Method (FEM and BPM) were used to numerically model the transmittance spectrum of those waveguides. They were manufactured using a desktop 3D printer based on FDM as well as SLA [55].

The FDM printer used, Orion Delta (SeeME CNC), has an approximate resolution of 400 μm defined by an opening of the extruder nozzle that deposits polymer layers with thickness varying from 50 μm to 100 μm. The polymer used in this case was ABS. Also, the SLA printer Form 1+® (Formlabs) was used in this paper. The printer resolution depends on the laser spot size on the printer plane and on the displacement along the z-axis, being around 150 μm and 50 μm respectively.

4.1. Numerical Modeling

The simulated and fabricated fibers have the geometrical parameters described in Table 2, where D_{core} is the internal core diameter, D_{ext} is the external fiber diameter, e_h is the thickness of the high refractive index layer, e_l is the thickness of the low refractive index layer, and L is the fiber length. The antiresonant waveguide A (ARROW A) and the Bragg fiber were built via the FDM technique using ABS—which has a real refractive index around 1.6 at 1.0 THz and an imaginary part presented in [72] (material loss from 21 to 78 dB/cm in the 0.1–1.0 THz range). ARROW B was created via SLA [73]. The host material has a refractive index around 1.65 and an absorption coefficient of about 11 cm^{-1} at 1.0 THz (material loss of 47 dB/cm) [74].

The Bragg fiber design is based on five concentric polymer rings (e_h) separated by air layers (e_l). The ARROWs have negative curvature in the core. The first is based on the design of a silica hollow-core fiber [75] and the second is inspired by a core surrounded by nested capillaries.

Figure 2a shows the core mode effective refractive index calculated with the commercial software COMSOL® in the range of 0.1 to 1.0 THz. The fluctuation in the dispersion curves are related to the coupling between core and cladding modes. When an effective index phase match occurs, a resonant condition is reached. In those frequencies there is a strong exchange of energy between both core and cladding modes, leading to a higher confinement loss and oscillations showing in the dispersion curve. Fibers fabricated with thinner polymer web structures could reduce this undesirable mode coupling condition.

Table 2. Parameters of the 3D printed fibers [67,69,73,74].

Parameters	Arrow A [67]	Arrow B [73]	Bragg [69]
D_{core} (mm)	8.2	7.2	7.2
D_{ext} (mm)	55	18	22.8
e_h (mm)	0.54	0.4	0.55
e_l (mm)	-	-	0.75
L (mm)	67	93	93
Polymer	ABS	UV-resin	ABS
Material loss @1 THz	78 dB/cm	47 dB/cm [74]	78 dB/cm

Figure 2b shows the spectral transmission calculated for the three fiber samples in the same frequency range. Windows of low-loss propagation for all fibers can be observed. These windows are mainly related to the antiresonant propagation condition between two consecutive high loss resonances, where the fibers guide with losses smaller than 3 dB/m.

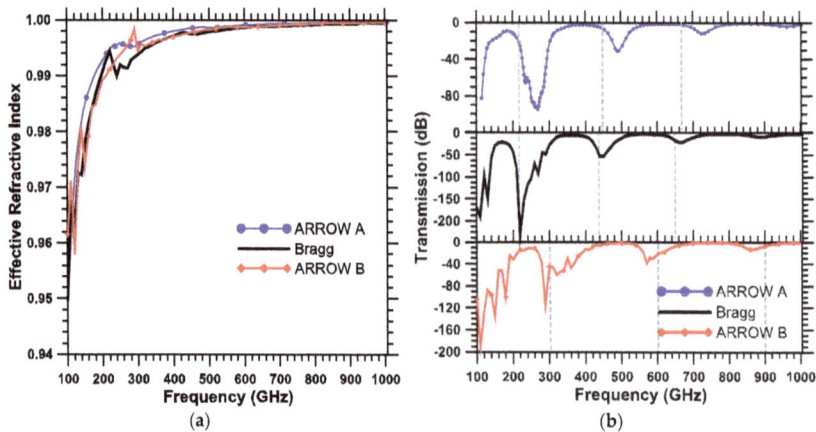

Figure 2. (a) Effective refractive index of ARROW A, ARROW B, and Bragg fiber; (b) Spectral transmission with high losses dips due to the resonant effect. The numerical simulations considered 1 m long fibers to the transmission analyses.

A simple analytical equation can predict the resonant frequency, taking into account the refractive index and thickness of the solid ring around the air core (dashed gray line in Figure 2b) [67]. It can be seen that these frequencies match the numerical data for the Bragg Fiber well. For the ARROW fibers, however, the analytical equation cannot predict precisely the resonances since the cores are not a perfect ring. Also, we see low-loss propagation windows in the Bragg fiber that match the calculated resonant frequencies. Consequently, it is possible to affirm that the main phenomena supporting the terahertz propagation is also the antiresonant effect at the first polymeric ring. However, using other materials with a lower refractive index contrast than polymer and air can allow the fiber to guide waves via Bragg reflections. Note that, the main geometrical parameter that affects the 3D printed fiber loss is the polymer thickness (e_h) [38]. Decreasing e_h shifts the transmission peaks to higher frequencies, and reduces the number of polymer/core mode couplings.

4.2. Experimental Characterization

The most common method used to characterize terahertz waveguides is the measurement of the transmission mode using a time domain spectrometer (TDS). Two terahertz electric pulses are measured, as shown in the Figure 3 inset. The first one, named reference pulse, is measured with all optics described in [1] except the waveguide. After that, a second pulse with the waveguide in the sampling area is taken, called a sample pulse. As demonstrated in [1], the loss and dispersion parameters can be calculated from these pulses.

Figure 3a shows the numerically and experimentally obtained spectral transmission of ARROW A (red and blue curves). During the numerical analyses, the polymer absorption was not considered. It was observed that the numerical data can predict the regions with high and low transmission, such as the frequency around 0.24 THz where core and polymer modes couple. As can be observed, the numerical and experimental spectral transmission data exhibit a difference in both frequency and amplitude. This mismatch can be attributed to the fact that the numerical data considered an idealized fiber with no absorption losses and no scattering due to the imperfect dielectric surfaces.

Figure 3b shows the normalized transmission spectrum to the Bragg fiber. Transmission bands were observed between 0.12–0.26 THz, 0.32–0.48 THz, and 0.50–1.00 THz. For lower frequencies there is good agreement between the bandgap regions. At high frequencies, however, the dips are shifted in frequency. Inaccuracy in the printing fabrication, such as roughness or deformation, could lead to this mismatch and should be further investigated. Moreover, the fiber length (93 mm) might not be long enough to establish the bandgaps for some frequency ranges.

It is important to note that these fibers guide with propagation losses significantly lower (around 0.3 dB/cm) in comparison with the absorption losses of the bulk material (about to 78 dB/cm).

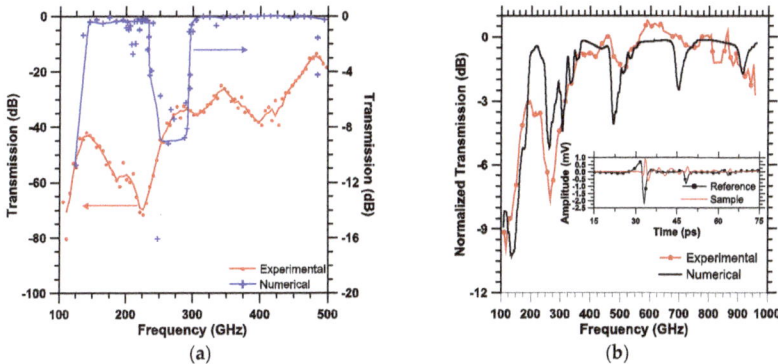

Figure 3. (a) Experimental and numerical transmission of ARROW A (93 mm long); (b) Experimental and numerical transmission of the Bragg fiber (100 mm long). Inset the reference and sample electrical pulse.

5. Discussions

We have reviewed different 3D printed hollow-core terahertz fibers focusing on low-loss propagation, breaking down the results from the literature according to their guidance phenomena. Terahertz hollow-core fibers, fabricated by additive manufacturing, are an attractive option to overcome the losses in terahertz waveguides.

Such manufacturing technology has experienced significant advances in recent years, providing a good solution on the fabrication of devices with complex geometries and low volume. It opens new opportunities to explore very complex fiber designs that are impossible to fabricate using conventional fiber optic manufacturing techniques, such as the ARROW B. Furthermore, we can consider the following advantages: the 3D CAD modeling provides many freedom degrees to design structures; final parts with low porosity; low material waste; availability to work with different materials such as food, ceramics, metal, and polymers, etc.; and the availability of a large number of commercial printers.

Despite the mentioned advantages, new research must increase the printing speed, develop and standardize the available materials, validate the materials thermal, mechanical, and optical properties; as well as increase the printers' spatial resolution. In addition, new means to overcome the short length print and the surface finish should be explored.

The great potential of this technology and the solution of the issues discussed above will likely lead 3D printing to be the fabrication method for millimetric terahertz components and waveguides, as recent works have shown.

Author Contributions: A.L.S.C., C.M.B.C. and M.A.R.F. designed the fiber samples and conceived the experiments. M.A.R.F. realized the numerical simulations. A.L.S.C. performed the experiments and drafted the paper and figures with input from all authors. All authors revised the manuscript. M.A.R.F. supervised the project.

Funding: This work was partially supported by CNPq and FAPESPA by the project INCT-Sensors and Optical Network, and CAPES.

Acknowledgments: The authors thank Gildo Rodrigues for fabricating the 'Arrow B' terahertz fiber.

Conflicts of Interest: The authors declare no conflicts of interest.

References

1. Atakaramians, S.; Afshar, S.V.; Monro, T.M.; Abbott, D. Terahertz dielectric waveguides. *Adv. Opt. Photonics* **2013**, *5*, 169–215. [CrossRef]
2. Markov, A.; Guerboukha, H.; Skorobogatiy, M. Hybrid metal wire–dielectric terahertz waveguides: Challenges and opportunities. *J. Opt. Soc. Am. B* **2014**, *31*, 2587–2600. [CrossRef]
3. Islam, M.; Chowdhury, D.R.; Ahmad, A.; Kumar, G. Terahertz Plasmonic Waveguide Based Thin Film Sensor. *J. Lightw. Technol.* **2017**, *35*, 5215–5221. [CrossRef]
4. Islam, M.S.; Sultana, J.; Atai, J.; Islam, M.R.; Abbott, D. Design and characterization of a low-loss, dispersion-flattened photonic crystal fiber for terahertz wave propagation. *Opt. Int. J. Light Electron Opt.* **2017**, *145*, 398–406. [CrossRef]
5. Islam, R.; Habib, M.S.; Hasanuzzaman, G.K.M.; Ahmad, R.; Rana, S.; Kaijage, S.F. Extremely High-Birefringent Asymmetric Slotted-Core Photonic Crystal Fiber in THz Regime. *IEEE Photonics Technol. Lett.* **2015**, *27*, 2222–2225. [CrossRef]
6. Islam, M.S.; Sultana, J.; Rana, S.; Islam, M.R.; Faisal, M.; Kaijage, S.F.; Abbott, D. Extremely low material loss and dispersion flattened TOPAS based circular porous fiber for long distance terahertz wave transmission. *Opt. Fiber Technol.* **2017**, *34*, 6–11. [CrossRef]
7. Fu, W.; Han, Y.; Li, J.; Wang, H.; Li, H.; Han, K.; Shen, X.; Cui, T. Polarization insensitive wide-angle triple-band metamaterial bandpass filter. *J. Phys. Appl. Phys.* **2016**, *49*, 285110. [CrossRef]
8. Jiang, L.-H.; Wang, F.; Liang, R.; Wei, Z.; Meng, H.; Dong, H.; Cen, H.; Wang, L.; Qin, S. Tunable Terahertz Filters Based on Graphene Plasmonic All-Dielectric Metasurfaces. *Plasmonics* **2018**, *13*, 525–530. [CrossRef]
9. Huang, Z.; Park, H.; Parrott, E.P.J.; Chan, H.P.; Pickwell-MacPherson, E. Robust Thin-Film Wire-Grid THz Polarizer Fabricated via a Low-Cost Approach. *IEEE Photonics Technol. Lett.* **2013**, *25*, 81–84. [CrossRef]
10. Cetnar, J.S.; Vangala, S.; Zhang, W.; Pfeiffer, C.; Brown, E.R.; Guo, J. High extinction ratio terahertz wire-grid polarizers with connecting bridges on quartz substrates. *Opt. Lett.* **2017**, *42*, 955–958. [CrossRef] [PubMed]
11. Machado, F.; Zagrajek, P.; Monsoriu, J.A.; Furlan, W.D. Terahertz Sieves. *IEEE Trans. Terahertz Sci. Technol.* **2017**, *8*, 140–143. [CrossRef]
12. Scherger, B.; Jördens, C.; Koch, M. Variable-focus terahertz lens. *Opt. Express* **2011**, *19*, 4528–4535. [CrossRef] [PubMed]

13. Tao, H.; Bingham, C.M.; Pilon, D.; Fan, K.; Strikwerda, A.C.; Shrekenhamer, D.; Padilla, W.J.; Zhang, X.; Averitt, R.D. A dual band terahertz metamaterial absorber. *J. Phys. Appl. Phys.* **2010**, *43*, 225102. [CrossRef]

14. Ma, Z.; Hanham, S.M.; Albella, P.; Ng, B.; Lu, H.T.; Gong, Y.; Maier, S.A.; Hong, M. Terahertz All-Dielectric Magnetic Mirror Metasurfaces. *ACS Photonics* **2016**, *3*, 1010–1018. [CrossRef]

15. Niu, T.; Withayachumnankul, W.; Upadhyay, A.; Gutruf, P.; Abbott, D.; Bhaskaran, M.; Sriram, S.; Fumeaux, C. Terahertz reflectarray as a polarizing beam splitter. *Opt. Express* **2014**, *22*, 16148–16160. [CrossRef] [PubMed]

16. Yakovlev, E.V.; Zaytsev, K.I.; Dolganova, I.N.; Yurchenko, S.O. Non-Destructive Evaluation of Polymer Composite Materials at the Manufacturing Stage Using Terahertz Pulsed Spectroscopy. *IEEE Trans. Terahertz Sci. Technol.* **2015**, *5*, 810–816. [CrossRef]

17. Mittleman, D.M. Twenty years of terahertz imaging. *Opt. Express* **2018**, *26*, 9417–9431. [CrossRef] [PubMed]

18. Zhang, M.; Yeow, J.T.W. Nanotechnology-Based Terahertz Biological Sensing: A review of its current state and things to come. *IEEE Nanotechnol. Mag.* **2016**, *10*, 30–38. [CrossRef]

19. Borovkova, M.; Khodzitsky, M.; Demchenko, P.; Cherkasova, O.; Popov, A.; Meglinski, I. Terahertz time-domain spectroscopy for non-invasive assessment of water content in biological samples. *Biomed. Opt. Express* **2018**, *9*, 2266–2276. [CrossRef] [PubMed]

20. Swearer, D.F.; Gottheim, S.; Simmons, J.G.; Phillips, D.J.; Kale, M.J.; McClain, M.J.; Christopher, P.; Halas, N.J.; Everitt, H.O. Monitoring Chemical Reactions with Terahertz Rotational Spectroscopy. *ACS Photonics* **2018**. [CrossRef]

21. Islam, M.S.; Sultana, J.; Ahmed, K.; Islam, M.R.; Dinovitser, A.; Ng, B.W.H.; Abbott, D. A Novel Approach for Spectroscopic Chemical Identification Using Photonic Crystal Fiber in the Terahertz Regime. *IEEE Sens. J.* **2018**, *18*, 575–582. [CrossRef]

22. Koch, M. Terahertz Communications: A 2020 vision. In *Terahertz Frequency Detection and Identification of Materials and Objects*; Miles, R.E., Zhang, X.C., Eisele, H., Krotkus, A., Eds.; Springer: Dordrecht, The Netherlands, 2007; pp. 325–338.

23. Hwu, S.U.; deSilva, K.B.; Jih, C.T. Terahertz (THz) wireless systems for space applications. In Sensors Applications Symposium. In Proceedings of the IEEE 2015 Sensors Applications Symposium, Galveston, TX, USA, 19–21 February 2013; pp. 171–175.

24. Akyildiz, I.F.; Jornet, J.M.; Han, C. Terahertz band: Next frontier for wireless communications. *Phys. Commun.* **2014**, *12*, 16–32. [CrossRef]

25. Yang, X.; Pi, Y.; Liu, T.; Wang, H. Three-Dimensional Imaging of Space Debris with Space-Based Terahertz Radar. *IEEE Sens. J.* **2018**, *18*, 1063–1072. [CrossRef]

26. Hasan, M.R.; Akter, S.; Khatun, T.; Rifat, A.A.; Anower, M.S. Dual-hole unit-based kagome lattice microstructure fiber for low-loss and highly birefringent terahertz guidance. *Opt. Eng.* **2017**, *56*, 043108. [CrossRef]

27. Faisal, M.; Shariful Islam, M. Extremely high birefringent terahertz fiber using a suspended elliptic core with slotted airholes. *Appl. Opt.* **2018**, *57*, 3340–3347. [CrossRef] [PubMed]

28. Wu, Z.; Shi, Z.; Xia, H.; Zhou, X.; Deng, Q.; Huang, J.; Jiang, X.; Wu, W. Design of Highly Birefringent and Low-Loss Oligoporous-Core THz Photonic Crystal Fiber with Single Circular Air-Hole Unit. *IEEE Photonics J.* **2016**, *8*, 1–11. [CrossRef]

29. Islam, M.S.; Faisal, M.; Razzak, S.M.A. Dispersion Flattened Porous-Core Honeycomb Lattice Terahertz Fiber for Ultra Low Loss Transmission. *IEEE J. Quantum Electron.* **2017**, *53*, 1–8. [CrossRef]

30. Islam, M.S.; Sultana, J.; Atai, J.; Abbott, D.; Rana, S.; Islam, M.R. Ultra low-loss hybrid core porous fiber for broadband applications. *Appl. Opt.* **2017**, *56*, 1232–1237. [CrossRef] [PubMed]

31. Islam, M.S.; Rana, S.; Islam, M.R.; Faisal, M.; Rahman, H.; Sultana, J. Porous core photonic crystal fibre for ultra-low material loss in THz regime. *IET Commun.* **2016**, *10*, 2179–2183. [CrossRef]

32. Islam, M.S.; Sultana, J.; Rifat, A.A.; Dinovitser, A.; Wai-Him Ng, B.; Abbott, D. Terahertz Sensing in a Hollow Core Photonic Crystal Fiber. *IEEE Sens. J.* **2018**, *18*, 4073–4080. [CrossRef]

33. Islam, M.S.; Sultana, J.; Dorraki, M.; Atai, J.; Islam, M.R.; Dinovitser, A.; Ng, B.W.-H.; Abbott, D. Low loss and low dispersion hybrid core photonic crystal fiber for terahertz propagation. *Photonic Netw. Commun.* **2018**, *35*, 364–373. [CrossRef]

34. Wang, K.; Mittleman, D.M. Guided propagation of terahertz pulses on metal wires. *J. Opt. Soc. Am. B* **2005**, *22*, 2001–2008. [CrossRef]

35. Wang, K.; Mittleman, D.M. Metal wires for terahertz wave guiding. *Nature* **2004**, *432*, 376. [CrossRef] [PubMed]

36. Argyros, A. Microstructures in Polymer Fibres for Optical Fibres, THz Waveguides, and Fibre-Based Metamaterials. *ISRN Opt.* **2012**, *2013*, 22. [CrossRef]

37. Ung, B.; Mazhorova, A.; Dupuis, A.; Rozé, M.; Skorobogatiy, M. Polymer microstructured optical fibers for terahertz wave guiding. *Opt. Express* **2011**, *19*, B848–B861. [CrossRef] [PubMed]

38. Cruz, A.L.S.; Migliano, A.C.C.; Franco, M.A.R. Polymer optical fibers for Terahertz: Low loss propagation and high evanescent field. In Proceedings of the IEEE International Microwave & Optoelectronics Conference (IMOC), Rio de Janeiro, Brazil, 4–7 August 2013; pp. 1–5.

39. Naftaly, M.; Miles, R.E. Terahertz Time-Domain Spectroscopy for Material Characterization. *Proc. IEEE* **2007**, *95*, 1658–1665. [CrossRef]

40. Lee, Y.-S. *Principles of Terahertz Science and Technology*; Springer: Boston, MA, USA, 2009; 340p, ISBN 978-0-387-09539-4.

41. Hassani, A.; Dupuis, A.; Skorobogatiy, M. Porous polymer fibers for low-loss Terahertz guiding. *Opt. Express* **2008**, *16*, 6340–6351. [CrossRef] [PubMed]

42. Atakaramians, S.; Afshar, S.V.; Ebendorff-Heidepriem, H.; Nagel, M.; Fischer, B.M.; Abbott, D.; Monro, T.M. THz porous fibers: Design, fabrication and experimental characterization. *Opt. Express* **2009**, *17*, 14053–14062. [CrossRef] [PubMed]

43. Bao, H.; Nielsen, K.; Rasmussen, H.K.; Jepsen, P.U.; Bang, O. Fabrication and characterization of porous-core honeycomb bandgap THz fibers. *Opt. Express* **2012**, *20*, 29507–29517. [CrossRef] [PubMed]

44. Cruz, A.L.S.; Migliano, A.C.C.; Hayashi, J.G.; Cordeiro, C.M.B.; Franco, M.A.R. Highly birefringent polymer terahertz fiber with microstructure of slots in the core. In Proceedings of the 22nd International Conference on Plastic Optical Fibers (POF), Rio de Janeiro, Brazil, 7 June 2013; pp. 290–294.

45. Bao, H.; Nielsen, K.; Bang, O.; Jepsen, P.U. Dielectric tube waveguides with absorptive cladding for broadband, low-dispersion and low loss THz guiding. *Sci. Rep.* **2015**, *5*, 7620. [CrossRef] [PubMed]

46. Xin, H.; Liang, M. 3-D-Printed Microwave and THz Devices Using Polymer Jetting Techniques. *Proc. IEEE* **2017**, *105*, 737–755. [CrossRef]

47. Zhang, B.; Guo, Y.X.; Zirath, H.; Zhang, Y.P. Investigation on 3-D-Printing Technologies for Millimeter-Wave and Terahertz Applications. *Proc. IEEE* **2017**, *105*, 723–736. [CrossRef]

48. Zhang, B.; Zirath, H. Metallic 3-D Printed Rectangular Waveguides for Millimeter-Wave Applications. *IEEE Trans. Compon. Packag. Manuf. Technol.* **2016**, *6*, 796–804. [CrossRef]

49. Chia, H.N.; Wu, B.M. Recent advances in 3D printing of biomaterials. *J. Biol. Eng.* **2015**, *9*, 4. [CrossRef] [PubMed]

50. Hermann, S.; Wolfgang, R.; Stephan, I.; Barbara, L.; Carsten, T. Three-dimensional printing of porous ceramic scaffolds for bone tissue engineering. *J. Biomed. Mater. Res. B Appl. Biomater.* **2005**, *74B*, 782–788. [CrossRef]

51. Godoi, F.C.; Prakash, S.; Bhandari, B.R. 3d printing technologies applied for food design: Status and prospects. *J. Food Eng.* **2016**, *179*, 44–54. [CrossRef]

52. Zadpoor, A.A.; Malda, J. Additive Manufacturing of Biomaterials, Tissues, and Organs. *Ann. Biomed. Eng.* **2017**, *45*, 1–11. [CrossRef] [PubMed]

53. Norman, J.; Madurawe, R.D.; Moore, C.M.V.; Khan, M.A.; Khairuzzaman, A. A new chapter in pharmaceutical manufacturing: 3D-printed drug products. *Adv. Drug Deliv. Rev.* **2016**, *108*, 39–50. [CrossRef] [PubMed]

54. Ngo, T.D.; Kashani, A.; Imbalzano, G.; Nguyen, K.T.Q.; Hui, D. Additive manufacturing (3D printing): A review of materials, methods, applications and challenges. *Compos. Part B* **2018**, *143*, 172–196. [CrossRef]

55. Wong, K.V.; Hernandez, A. A Review of Additive Manufacturing. *ISRN Mech. Eng.* **2012**, *2012*, 10. [CrossRef]

56. Phipps, A.R.; MacLachlan, A.J.; Zhang, L.; Robertson, C.W.; Konoplev, I.V.; Phelps, A.D.R.; Cross, A.W. Periodic structure towards the terahertz region manufactured using high resolution 3D printing. In Proceedings of the 2015 8th UK, Europe, China Millimeter Waves and THz Technology Workshop (UCMMT), Cardiff, UK, 14–15 Septembr 2015; pp. 1–4.

57. Weidenbach, M.; Jahn, D.; Rehn, A.; Busch, S.F.; Beltrán-Mejía, F.; Balzer, J.C.; Koch, M. 3D printed dielectric rectangular waveguides, splitters and couplers for 120 GHz. *Opt. Express* **2016**, *24*, 28968–28976. [CrossRef] [PubMed]

58. Furlan, W.D.; Ferrando, V.; Monsoriu, J.A.; Zagrajek, P.; Czerwińska, E.; Szustakowski, M. 3D printed diffractive terahertz lenses. *Opt. Lett.* **2016**, *41*, 1748–1751. [CrossRef] [PubMed]

59. Headland, D.; Withayachumnankul, W.; Webb, M.; Ebendorff-Heidepriem, H.; Luiten, A.; Abbott, D. Analysis of 3D-printed metal for rapid-prototyped reflective terahertz optics. *Opt. Express* **2016**, *24*, 17384–17396. [CrossRef] [PubMed]

60. Talataisong, W.; Ismaeel, R.; Marques, T.H.R.; Mousavi, S.A.; Beresna, M.; Gouveia, M.A.; Sandoghchi, S.R.; Lee, T.; Cordeiro, C.M.B.; Brambilla, G. Mid-IR Hollow-core microstructured fiber drawn from a 3D printed PETG preform. *Sci. Rep.* **2018**, *8*, 8113. [CrossRef] [PubMed]

61. Busch, S.F.; Weidenbach, M.; Balzer, J.C.; Koch, M. THz Optics 3D Printed with TOPAS. *J. Infrared Millim. Terahertz Waves* **2016**, *37*, 303–307. [CrossRef]

62. Sakellari, I.; Yin, X.; Nesterov, M.L.; Terzaki, K.; Xomalis, A.; Farsari, M. 3D Chiral Plasmonic Metamaterials Fabricated by Direct Laser Writing: The Twisted Omega Particle. *Adv. Opt. Mater.* **2017**, *5*, 1700200. [CrossRef]

63. Dupuis, A.; Stoeffler, K.; Ung, B.; Dubois, C.; Skorobogatiy, M. Transmission measurements of hollow-core THz Bragg fibers. *J. Opt. Soc. Am. B* **2011**, *28*, 896–907. [CrossRef]

64. Wu, Z.; Ng, W.-R.; Gehm, M.E.; Xin, H. Terahertz electromagnetic crystal waveguide fabricated by polymer jetting rapid prototyping. *Opt. Express* **2011**, *19*, 3962–3972. [CrossRef] [PubMed]

65. Yang, J.; Zhao, J.; Gong, C.; Tian, H.; Sun, L.; Chen, P.; Lin, L.; Liu, W. 3D printed low-loss THz waveguide based on Kagome photonic crystal structure. *Opt. Express* **2016**, *24*, 22454–22460. [CrossRef] [PubMed]

66. Wei, C.; Joseph Weiblen, R.; Menyuk, C.R.; Hu, J. Negative curvature fibers. *Adv. Opt. Photonics* **2017**, *9*, 504–561. [CrossRef]

67. Cruz, A.L.S.; Serrão, V.A.; Barbosa, C.L.; Franco, M.A.R. 3D Printed Hollow Core Fiber with Negative Curvature for Terahertz Applications. *J. Microw. Optoelectron. Electromagn. Appl.* **2015**, *14*, SI-45–SI-53.

68. van Putten, L.; Gorecki, J.; Fokoua, E.R.N.; Apostolopoulos, A.; Poletti, F. 3D-Printed Polymer Antiresonant Waveguides for Short Reach Terahertz Applications. *Appl. Opt.* **2018**, *57*, 3953–3958. [CrossRef] [PubMed]

69. Cruz, A.L.S.; Argyros, A.; Tang, X.; Cordeiro, C.M.B.; Franco, M.A.R. 3D-printed terahertz Bragg fiber. In Proceedings of the 2015 40th International Conference on Infrared, Millimeter, and Terahertz waves (IRMMW-THz), Hong Kong, China, 23–28 August 2015; pp. 1–2.

70. Li, J.; Nallappan, K.; Guerboukha, H.; Skorobogatiy, M. 3D printed hollow core terahertz Bragg waveguides with defect layers for surface sensing applications. *Opt. Express* **2017**, *25*, 4126–4144. [CrossRef] [PubMed]

71. Hong, B.; Swithenbank, M.; Greenall, N.; Clarke, R.G.; Chudpooti, N.P.; Akkaraekthalin, N.; Somjit, J.E.; Cunningham, I.D. Robertson Low-Loss Asymptotically Single-Mode THz Bragg Fiber Fabricated by Digital Light Processing Rapid Prototyping. *IEEE Trans. Terahertz Sci. Technol.* **2018**, *8*, 90–99. [CrossRef]

72. Jin, Y.-S.; Kim, G.-J.; Jeon, S.-G. Terahertz dielectric properties of polymers. *J. Korean Phys. Soc.* **2006**, *49*, 513–517.

73. Cruz, A.L.S.; Franco, M.A.R.; Cordeiro, C.M.B.; Rodrigues, G.S.; Osório, J.H.; da Silva, L.E. Exploring THz hollow-core fiber designs manufactured by 3D printing. In Proceedings of the 2017 SBMO/IEEE MTT-S International Microwave and Optoelectronics Conference (IMOC), Aguas de Lindoia, Brazil, 27–30 August 2017; pp. 1–5.

74. Younus, A.; Desbarats, P.; Bosio, S.; Abraham, E.; Delagnes, J.C.; Mounaix, P. Terahertz dielectric characterisation of photopolymer resin used for fabrication of 3D THz imaging phantoms. *Electron. Lett.* **2009**, *45*, 702–703. [CrossRef]

75. Yu, F.; Wadsworth, W.J.; Knight, J.C. Low loss silica hollow core fibers for 3–4 μm spectral region. *Opt. Express* **2012**, *20*, 11153–11158. [CrossRef] [PubMed]

fibers

MDPI

Article

Effect of Nested Elements on Avoided Crossing between the Higher-Order Core Modes and the Air-Capillary Modes in Hollow-Core Antiresonant Optical Fibers

Laurent Provino

PERFOS, Research Technology Organization of Photonics Bretagne, 4 rue Louis de Broglie,
22300 Lannion, France; lprovino@photonics-bretagne.com

Received: 14 May 2018; Accepted: 13 June 2018; Published: 18 June 2018

Abstract: Optimal suppression of higher-order modes (HOMs) in hollow-core antiresonant fibers comprising a single ring of thin-walled capillaries was previously studied, and can be achieved when the condition on the capillary-to-core diameter ratio is satisfied ($d/D \approx 0.68$). Here we report on the conditions for maximizing the leakage losses of HOMs in hollow-core nested antiresonant node-less fibers, while preserving low confinement loss for the fundamental mode. Using an analytical model based on coupled capillary waveguides, as well as full-vector finite element modeling, we show that optimal d/D value leading to high leakage losses of HOMs, is strongly correlated to the size of nested capillaries. We also show that extremely high value of degree of HOM suppression ($\sim$1200) at the resonant coupling is almost unchanged on a wide range of nested capillary diameter d_{Nested} values. These results therefore suggest the possibility of designing antiresonant fibers with nested elements, which show optimal guiding performances in terms of the HOM loss compared to that of the fundamental mode, for clearly defined paired values of the ratios d_{Nested}/d and d/D. These can also tend towards a single-mode behavior only when the dimensionless parameter d_{Nested}/d is less than 0.30, with identical wall thicknesses for all of the capillaries.

Keywords: hollow-core antiresonant fiber; numerical modeling; modal fiber properties

1. Introduction

A new form of silica hollow-core fiber consisting of a single ring of touching or non-touching antiresonant elements (ARE) surrounding a central hollow-core has emerged in recent years. This antiresonant fiber (ARF) has been investigated as a result of the discovery of the importance of core wall shape in the attenuation reduction in Kagome-structured hollow-core fiber in 2010 [1]. Accordingly, it was possible to reduce the Kagome cladding to just one single glass layer of ARE without significantly increasing fiber attenuation [2]. Thereafter, different hollow core fibers comprising a single ring of touching capillaries in the cladding void have been proposed and studied. It was proven that ARFs, with an inverted optical core boundary, possess large transmission bandwidth and low attenuations in the mid-infrared spectral region due to both low leakage losses and weak coupling of air-core modes with the cladding structure [3–5]. This type of design was extended to shorter wavelength transmission in the near-infrared and visible spectrum [6,7]. A modified form of the basic design with contactless capillaries has also been proposed and fabricated in order to remove the additional optical resonances in the transmission bands related to nodes between cladding elements [8]. In this way, the loss level can be further decreased in the mid-infrared wavelength range [8,9]. Recently, greatly reduced transmission loss at 750 nm in similar fiber was reported by Debord et al. [10]. By adding one or more nested capillaries within the node-less cladding structure, numerical simulations predicted leakage losses reduction by roughly two orders of magnitude in the middle and near infrared spectral regions [11,12]. Up to now, the fabrication of at least two fibers with nested antiresonant node-less elements (NANFs) has been reported for low loss operation. The fibers have been manufactured both with different [13] and closely identical [14] wall thicknesses between the inner and outer cladding capillaries. The wall thickness of the large capillaries was greater than a micrometer. According to performances of theses fabricated fibers, the positive effect of the nested capillaries on the loss is limited by their small hole diameter and the distinction between the wall thicknesses of the large and small capillaries. However, a minimum optical loss of 74 dB/km at 1.8 µm was obtained in a NANF with a 25 µm core diameter and a 2.3 µm average wall thickness of all the capillaries [14]. The leakage loss in the ARFs, as in the NANFs, is inversely proportional to the fourth power of core diameter [15], so lower losses are much easier to achieve at larger core diameters. For large core diameters (>25 µm), the ARFs and NANFs are however multimoded [11] and therefore not ideal for applications where a high modal purity is desirable, for example in high-power pulse delivery or in gas cells. To suppress HOMs while preserving low confinement loss for the fundamental core mode, a technique has already been proposed for different designs of ARFs with touching and non-touching capillaries [16–18]. It exploits resonant coupling between the higher-order core modes and the air-capillary modes. The approach is analogous to using defect modes in HC-PCFs [19,20]; however, there is no need to create defects in this case, since the capillaries that create cladding structure can also provide the resonant coupling. Specifically Uebel et al. [17] showed the importance of the dimensionless parameter d/D, with the inner capillary diameter d and the inner core diameter D, in order to achieve optimal suppression of HOMs over all wavelength bands where the ARF guides with low loss. For 6- and 7-capillary designs [17,18], an avoided crossing between TM_{01}, TE_{01}, and HE_{21} core modes and fundamental air-capillary mode is observed at $d/D \approx 0.68$ which leads to high leakage losses of those HOMs and hence, provides robust single-mode guidance at all wavelengths within the main transmission window, independent of the absolute core size of the structure. This condition on the capillary-to-core diameter can also be verified for any ARF structures composed of a smaller number of capillaries. However, a number of capillaries less than six would lead to a pronounced gap between capillaries that can affect the fundamental mode loss.

The purpose of this work is to study the impact that the nested elements can have on the avoided crossing between the higher-order core modes and the air-capillary modes, in order to verify that the condition of HOM suppression established for the ARFs could be applied or not to NANFs. We show computationally that it is also possible to suppress the HOMs in NANFs; however, the optimal d/D value, for which the leakage losses of HOMs are maximum, is strongly dependent on the geometric dimension of nested capillaries. The numerical results are also interpreted and verified by extending an already proposed analytical model based on coupled capillaries.

2. Design and Numerical Analysis

In order to analyze the influence of nested elements on the modal properties of ARFs, we propose in this work to maintain the six node-less capillary lattice structure described in [17] and add nested capillaries of inner diameter d_{Nested} with the same wall thickness t_{Nested} as the outer ones and attached to the cladding at the same azimuthal position, as shown in the left panel of Figure 1. Considering the fact that both the transmission band positions [11,12,16,17] and the leakage losses [13] are dependent on the wall thickness of the capillaries, the choice of setting $t_{Nested} = t$ allows for avoiding additional effects on the present study.

All numerical simulations reported here are based on a commercial full-vector finite-element based modal solver (Comsol Multiphysics). A circular Perfectly Matched Layer (PML) surrounding the simulated area is used to calculate the mode leakage losses. Only a quarter of the geometry is used in modeling fibers because of the symmetry of the modes [21]. As in [17], we adopted a core diameter $D = 30$ µm, a silica wall thickness $t = 0.30$ µm and a wavelength $\lambda = 1.50$ µm, in such a way that the ratios t/D and D/λ are equal to 0.01 and 20, respectively. The glass refractive index was set a constant value of 1.45 and the hollow regions were taken to be vacuum. The material loss is neglected since the material absorption is quite low at this wavelength [22].

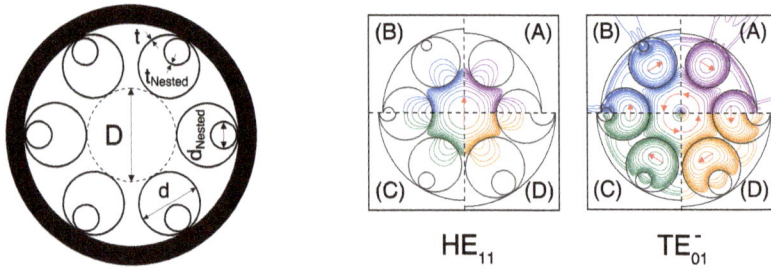

Figure 1. Left: Sketch of the Nested Antiresonant Node-less Fiber (NANF) cross-section, with its key parameters. Glass is marked in black and the hollow regions in white. Right: 3-dB contour plots of the HE_{11} fundamental modes and hybrid TE_{01}^{-} modes at the avoided crossing point, for different values of d_{Nested}/d equal to 0.00 (A), 0.20 (B), 0.30 (C) and 0.40 (D) respectively. The color code used is identical to that in Figure 2. The contour plots represent the normalized electric field intensity and the red arrows indicate the polarization direction of the transverse electric field.

In order to model NANF accurately, great care was taken to optimize both mesh and PML parameters. Typically, a maximum element size of $\lambda/4$ was used in the air regions while a rather dense mesh with a maximum element size of $\lambda/6$ in the thin glass regions was found to be essential to obtain reliable results. To ensure convergence of the numerical results, we first checked our model by reproducing the results of the [17] ($d_{Nested}/d = 0$), and afterward the modal properties for the first guided modes were simulated for d_{Nested}/d ratio respectively equal to 0.20, 0.30 and 0.40.

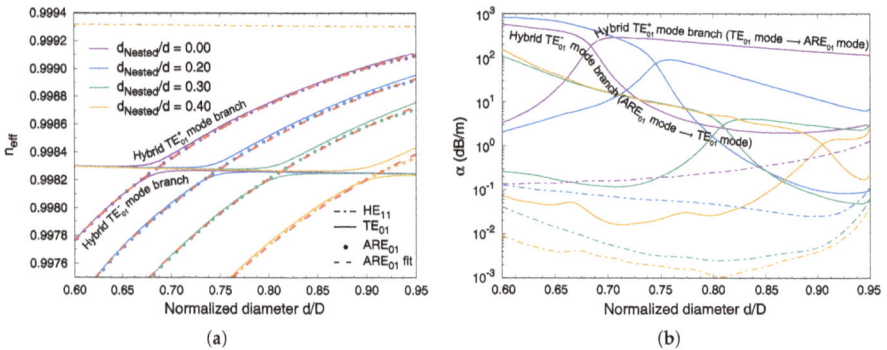

Figure 2. Numerically computed modal properties d/D dependence for different values of d_{Nested}/d with $t/D = 0.01$ and $D/\lambda = 20$. (a) Effective indices of the fundamental HE_{11} core mode (dot dashed curves), hybrid TE_{01}^{+} and TE_{01}^{-} modes (solid curves) and ARE_{01} air-capillary mode (dotted curves). The red dashed curve for each d_{Nested}/d value represents the effective index of the ARE_{01} mode of an isolated capillary, using the analytic model. (b) Corresponding mode confinement loss.

Figure 2a shows the effective index for the fundamental HE_{11} core mode, the TE_{01} core mode, and the ARE_{01} air-capillary mode as a function of the radio d/D. The TE_{01} mode is the higher-order core mode that has the lowest loss, and also the lowest effective index difference with the fundamental HE_{11} mode. Note that, according to the polarization profile of the guided modes, the core modes are indicated using the designation of vector modes HE_{nm}, EH_{nm}, TE_{0m} and TM_{0m} with n and m integers and the air-capillary mode of the node-less cladding structure is labeled ARE_{lm} with l and m integers, based on the notation of the linearly polarized modes. Figure 2b shows the corresponding confinement loss of the core modes.

For $d_{Nested}/d = 0$, we find again in Figure 2a the discontinuity of the TE_{01} core mode effective index for $0.62 < d/D < 0.75$ as in [17], typical of an avoided crossing between the core localized mode (herein the TE_{01} mode) and a cladding leaky mode (herein the ARE_{01} mode) [23]. In this region of d/D values, the TE_{01} core mode splits into two hybrid modes labelled TE_{01}^+ and TE_{01}^-, which evolve asymptotically in toward uncoupled ARE_{01} and TE_{01} modes. Indeed for largest values of effective index, the TE_{01} core mode gradually changes its nature to become an ARE_{01} air-capillary mode and conversely for smaller values of effective index, the ARE_{01} air-capillary mode progressively converges toward a TE_{01} core mode. For non-zero values of the ratio d_{Nested}/d, the evolution of the TE_{01} core mode effective index as a function of d/D is similar. However, we observe an increasing shift in the position of the discontinuity in relation to d/D when the ratio d_{Nested}/d rises. As with the d/D parameter, these numerical results show the importance of the nested capillaries's diameter on avoided crossing between the TE_{01} core mode and the ARE_{01} air-capillary mode, and the scalability of this phenomenon through the dimensionless parameter d_{Nested}/d. On the other hand, the effective index of the HE_{11} core mode remains independent of d/D, regardless of the d_{Nested}/d value.

To quantify the avoided crossing properties, we introduce the quantity $\mathfrak{D} = \partial^2 \Re e(n_{eff})/\partial(d/D)^2$. By analogy with chromatic dispersion, we can associate this quantity with the effective index dispersion of a mode in relation to d/D. In Figure 3a, we plotted the obtained curves for the four values of the parameter d_{Nested}/d. For each transition region, we observe that the effective index dispersion curve exhibits a concave profile and large positive coefficient around a specific $(d/D)_{max}$ value corresponding at the phase-matching to resonance point ($n_{eff}^{TE_{01}} = n_{eff}^{ARE_{01}}$). These values are summarized in Table 1. The numerical result presented for $d_{Nested}/d = 0$ is in good agreement with that of the [17], and thus ensures the validity of the optimal values for the ratio $d_{Nested}/d > 0$ for which an avoided crossing between the TE_{01} core mode and the ARE_{01} air-capillary mode exists. Similarly to what was already described in [24], the strength of the interaction between the core mode and the air-capillary mode or, in other words, the degree of overlap between the fields of the two modes is proportional to the magnitude of the d/D range over which the transformation takes place, and is conversely in proportion to the peak value that is smoothed. As a result, we can deduce that an increase of the nested capillary's diameter induces a stronger coupling between these modes because the full width at half maximum Δ_{FWHM} of the curves increases slightly with the d_{Nested}/d parameter and the peak value at $(d/D)_{max}$ decreases. This increased interaction is related to the nested capillaries which increasingly squeeze the air-capillary modes towards the fiber core when their sizes increase (see Figure 1 at the right).

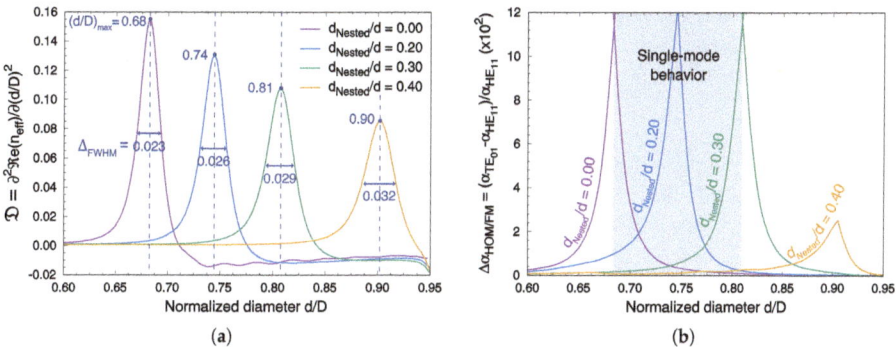

Figure 3. (a) Effective index dispersion of the TE_{01}^+ hybrid mode as a function of normalized diameter d/D, for four different values of parameter d_{Nested}/d; (b) Numerically calculated degree of HOM suppression, plotted against d/D using the simulated confinement losses above. The gray-shaded area shows the region where NANFs tend towards a single-mode behavior.

Table 1. Summary of main results and parameters.

d_{Nested}/d	$(d/D)_{max}$	Δ_{FWHM}	$(d/D)_{th}$	f_{co}	f_{cl}
0.00	0.68	0.023	0.68	1.07	0.98
0.20	0.74	0.026	0.75	1.07	0.90
0.30	0.81	0.029	0.81	1.07	0.83
0.40	0.90	0.032	0.91	1.07	0.73

In terms of confinement loss, we observe in each case that the loss of the two hybrid modes strongly increases up to the avoided crossing point and then slowly varies beyond the $(d/D)_{max}$ value. Conversely, the HE_{11} core mode has a weakly fluctuating low loss over the value range of d/D, independently of the set value of the d_{Nested}/d ratio. We thus find the same performances in the evolution of the losses in respect to d/D as described in [17] for the HE_{11} and hybrid $TE_{01}^{+,-}$ modes, which makes it possible to consider finding NANF designs with a strong degree of HOM suppression compared to the fundamental mode, defined as [17]

$$\Delta\alpha_{HOM/FM} = \frac{\alpha_{HOM} - \alpha_{FM}}{\alpha_{FM}} \tag{1}$$

where α_{FM} and α_{HOM} are the losses of the HE_{11} and hybrid $TE_{01}^{+,-}$ modes (in dB/m), respectively. This relation can give an indication of the single-modeness of the given fiber. However, the addition of nested capillaries decreases the overall losses for all guided modes between one to two orders of magnitude depending on the size of the nested AREs. For NANFs consisting of six nested elements with a ratio $d_{Nested}/d \geq 0.30$, the suppression of higher-order core modes while satisfying the condition on the parameter d/D can become problematic. This is due to the fact that the nested capillaries and the narrow inter-capillary distance effectively shield the electric field leaking into the silica outer cladding (see Figure 1 on the right hand side). A sufficiently large loss level cannot hence be reached for the higher-order TE_{01} mode. In Figure 3b the calculated degree of HOM suppression strongly increases at the anti-crossing position, peaking at the values around 1200 for the first three values of d_{Nested}/d. When $d_{Nested}/d = 0.40$, this value drops at ~250. The degree of HOM suppression for $d_{Nested}/d = 0.30$ is similar to that without nested capillaries, whereas the loss is almost two orders of magnitude smaller for the two core modes. Whatever the value of d_{Nested}/d, the single-mode or multimode behavior of NANF can only be defined with respect to the length of fiber needed to develop new applications, such as light-sources/lasers emitting in the deep-UV/UV or the mid-infrared. If only a few meters of fiber are sufficient so that the losses of the HE_{11} and TE_{01} core modes are practically negligible over this length (as for $d_{Nested}/d = 0.30$), in spite of the fact the TE_{01} mode presents 1000 times higher loss, it will not actually be effectively suppressed. Thus, in order to verify the single-mode behavior of NANFs over a specified wavelength interval, estimating the degree of HOM suppression is a necessary but not sufficient criteria and the lowest loss estimation of higher-order core modes is most important.

3. Analytical Model

To understand and predict the avoided crossings in ARFs with touching or non-touching capillaries, an analytical model in which the core and the cladding of AREs are treated as two coupled capillaries has been proposed in the [16,17]. We propose here to apply this analytical model to NANFs depicted in Figure 1. Generally, the effective index of guided modes in the inner region of capillary with infinite wall thickness can be estimated by the Marcatili-Schmeltzer formula [25], rewritten as

$$n_{eff}^{co,cl} = n_L - \frac{1}{2}\left(\frac{u_{pm}^{co,cl}}{\pi\sqrt{n_L}}\right)^2\left(\frac{\lambda}{d_{eff}^{co,cl}}\right)^2 \tag{2}$$

where the exponents *co* and *cl* indicate the capillary linked to the core and the cladding, respectively; n_L the refractive index of inner medium made of air ($n_L = n_{air} = 1$); the coefficient $u_{pm}^{co,cl}$ is the m-th zero of the Bessel function $J_p(u_{pm}) = 0$, with $p = n$ for vector modes and $p = l + 1$ for linearly polarized modes. The effective diameter $d_{eff}^{co} = f_{co} \times D$ or $d_{eff}^{cl} = f_{cl} \times d$ is the inner diameter of the capillary. The coefficients f_{co} and f_{cl} are used to adjust the value of diameters d and D in order to match the analytical values from Equation (2) to the numerical results of finite-element simulations. In Figure 2a, are plotted the fitted values for each ARE_{01} air-capillary mode (red dashed curves) calculated using Equation (2) with the fitting coefficients f_{cl} reported in Table 1. Analytical and numerical computed values are in excellent agreement. The dependence of the coefficient f_{cl} with respect to the dimensionless parameter d_{Nested}/d can be related to the space filling rate of the nested capillaries inside the outer ones. Indeed we observe in Figure 2a that the more the nested capillary diameter increases, the more the effective index of the ARE_{01} air-capillary mode decreases with both parameters D and d held constant. According to (2), the combination of two nested capillaries can then be defined by a single capillary with an effective diameter of smaller size. For the HE_{11} core mode, the effective index value showing a very low variation at constant core diameter ($D = 30$ μm), the application of Equation (2) allows by taking the average of the effective indices for each value of d_{Nested}/d to find a constant value of the fitting coefficient f_{co} equal to 1.07. We verified that this increase in core diameter by 7% can be applied to the first four higher-order core modes and found that the effective indices of the $HE_{21}^{even,\ odd}$, TM_{01} and TE_{01} core modes in NANFs match the effective indices of the modes in the corresponding capillary with a precision of order to 10^{-5}. At the avoided crossing point of the coupled modes, the equivalence of effective indices between the TE_{01} core mode and ARE_{01} air-capillary mode by using Equation (2) allows to derive a simple expression for the d/D parameter

$$n_{eff}^{co} = n_{eff}^{cl} \quad \Rightarrow \quad \left(\frac{d}{D}\right)_{th} = \frac{u_{11}^{cl}}{u_{01}^{co}} \cdot \frac{f_{co}}{f_{cl}(d_{Nested}/d)} \tag{3}$$

in which the value of coefficient $f_{cl}(d_{Nested}/d)$ depends on the value of the ratio d_{Nested}/d. The calculated theoretical values of $(d/D)_{th}$ from Equation (3) are summarized in Table 1 and are in good agreement with the $(d/D)_{max}$ values obtained from finite-element modeling. Thus, the analytical model based on coupled capillaries can be extended to NANFs and allows a simple prediction of the avoided crossing between the higher-order core modes and the air-capillary modes.

4. Conclusions

We investigated numerically and analytically the effect of nested thin-walled capillaries on the modal properties of a hollow-core fiber consisting of a single ring of six non-touching capillaries of the same wall thickness, mounted inside a thick-walled glass capillary. We observed a dependence with respect to the dimensionless parameter d_{Nested}/d on the $(d/D)_{max}$ value at which the resonant coupling between the higher-order core modes and an air-capillary mode, characterized by an avoided crossing and an extremely high degree of HOM suppression, is optimal. This leads to the possibility of designing a large range of NANFs with optimal transmission performances. However, we have noted that a high degree of mode suppression does not necessarily mean that the fibre is single-mode. The NANFs will perform effectively as a single-mode fiber, when the loss of the HOM are significantly higher (>30 dB/m) than that of the fundamental mode, for fiber lengths of a few meters. For values of $d_{Nested}/d < 0.30$, the NANFs can then tend towards a single-mode behavior, with performance in confinement loss superior to ARFs, for the fundamental mode. However, this type of NANFs seem much more difficult to fabricate than ARFs, compared to attempts made for at least three years. Indeed, for NANFs with a core diameter between 30 μm and 50 μm, the nested capillary diameters will not exceed 10 μm when $d_{Nested}/d < 0.30$, with the same wall-thickness as the outer capillaries of dimensions smaller than or of the order of micrometer. Therefore, there is a trade-off between transmission performance and technological difficulties related to fiber design.

Funding: This work is supported in part by the "Conseil Régional de Bretagne" and the "Fonds Européen de Développement Economique des Régions".

Acknowledgments: The author would like to thank Thierry Taunay for useful discussions throughout the work.

Conflicts of Interest: The authors declare no conflict of interest.

References

1. Wang, Y.Y.; Couny, F.; Roberts, P.J.; Benabid, F. Low loss broadband transmission in optimized core-shape Kagome hollow-core PCF. In Proceedings of the Lasers Electro-Optics, Quantum Electron, Laser Science Conference, San Jose, CA, USA, 16–21 May 2010; pp. 1–2.
2. Gérôme, F.; Jamier, R.; Auguste, J.L.; Humbert, G.; Blondy, J.M. Simplified hollow-core photonic crystal fiber. *Opt. Lett.* **2010**, *35*, 1157–1159. [CrossRef] [PubMed]
3. Pryamikov, A.D.; Biriukov, A.S.; Kosolapov, A.F.; Plotnichenko, V.G.; Semjonov, S.L.; Dianov, E.M. Demonstration of a waveguide regime for a silica hollow-core microstructured optical fiber with a negative curvature of the core boundary in the spectral region >3.5 μm. *Opt. Express* **2011**, *19*, 1441–1448. [CrossRef] [PubMed]
4. Yu, F.; Wadsworth, W.J.; Knight, J.C. Low loss silica hollow core fibers for 3–4 μm spectral region. *Opt. Express* **2012**, *20*, 11153–11158. [CrossRef] [PubMed]
5. Yu, F.; Knight, J.C. Spectral attenuation limits of silica hollow core negative curvature fiber. *Opt. Express* **2013**, *21*, 21466–21471. [CrossRef] [PubMed]
6. Jaworski, P.; Yu, F.; Maier, R.R.J.; Wadsworth, W.J.; Knight, J.C.; Shephard, J.D.; Hand, D.P. Picosecond and nanosecond pulse delivery through a hollow-core Negative Curvature Fiber for micro-machining applications. *Opt. Express* **2013**, *21*, 22742–22753. [CrossRef] [PubMed]
7. Jaworski, P.; Yu, F.; Carter, R.M.; Knight, J.C.; Shephard, J.D.; Hand, D.P. High energy green nanosecond and picosecond pulse delivery through a negative curvature fiber for precision micro-machining. *Opt. Express* **2015**, *23*, 8498–8506. [CrossRef] [PubMed]
8. Kolyadin, A.N.; Kosolapov, A.F.; Pryamikov, A.D.; Biriukov, A.S.; Plotnichenko, V.G.; Dianov, E.M. Light transmission in negative curvature hollow core fiber in extremely high material loss region. *Opt. Express* **2013**, *21*, 9514–9519. [CrossRef] [PubMed]
9. Belardi, W.; Knight, J.C. Hollow antiresonant fibers with low bending loss. *Opt. Express* **2014**, *22*, 10091–10096. [CrossRef] [PubMed]
10. Debord, B.; Amsanpally, A.; Chafer, M.; Baz, A.; Maurel, M.; Blondy, J.M.; Hugonnot, E.; Scol, F.; Vincetti, L.; Gérôme, F.; et al. Ultralow transmission loss in inhibited-coupling guiding hollow fibers. *Optica* **2017**, *4*, 209–217. [CrossRef]
11. Poletti, F. Nested antiresonant nodeless hollow core fiber. *Opt. Express* **2014**, *22*, 23807–23828. [CrossRef] [PubMed]
12. Habib, M.S.; Bang, O.; Bache, M. Low-loss hollow-core silica fibers with adjacent nested anti-resonant tubes. *Opt. Express* **2015**, *23*, 17394–17406. [CrossRef] [PubMed]
13. Belardi, W. Design and Properties of Hollow Antiresonant Fibers for the Visible and Near Infrared Spectral Range. *J. Lightwave Technol.* **2015**, *33*, 4497–4503. [CrossRef]
14. Kosolapov, A.F.; Alagashev, G.K.; Kolyadin, A.N. Hollow-core revolver fibre with a double-capillary reflective cladding. *Quantum Electron.* **2016**, *46*, 267–270. [CrossRef]
15. Vincetti, L. Empirical formulas for calculating loss in hollow core tube lattice fibers. *Opt. Express* **2016**, *24*, 10313–10325. [CrossRef] [PubMed]
16. Wei, C.; Kuis, R.A.; Chenard, F.; Menyuk, C.R.; Hu, J. Higher-order mode suppression in chalcogenide negative curvature fibers. *Opt. Express* **2015**, *23*, 15824–15833. [CrossRef] [PubMed]
17. Uebel, P.; Günendi, M.C.; Frosz, M.H.; Ahmed, G. Broadband robustly single-mode hollow-core PCF by resonant filtering of higher-order modes. *Opt. Lett.* **2016**, *41*, 1961–1964. [CrossRef] [PubMed]
18. Michieletto, M.; Lyngsø, J.K.; Jakobsen, C.; Lægsgaard, J.; Bang, O.; Alkeskjold, T.T. Hollow-core fibers for high power pulse delivery. *Opt. Express* **2016**, *24*, 7103–7120. [CrossRef] [PubMed]
19. Fini, J.M.; Nicholson, J.W.; Windeler, R.S.; Monberg, E.M.; Meng, L.; Mangan, B.; DeSantolo, A.; DiMarcello, F.V. Low-loss hollow-core fibers with improved single-modedness. *Opt. Express* **2013**, *21*, 6233–6242. [CrossRef] [PubMed]

20. Saitoh, K.; Florous, N.J.; Murao, T.; Koshiba, M. Design of photonic band gap fibers with suppressed higher-order modes: Towards the development of effectively single mode large hollow-core fiber platforms. *Opt. Express* **2006**, *14*, 7342–7352. [CrossRef] [PubMed]
21. McIsaac, P. Symmetry-Induced Modal Characteristics of Uniform Waveguides—I: Summary of Results. *IEEE Trans. Microw. Theory Tech.* **1975**, *23*, 421–429. [CrossRef]
22. Pilon, L.; Jonasz, M.; Kitamura, R. Optical constants of silica glass from extreme ultraviolet to far infrared at near room temperature. *Appl. Opt.* **2007**, *46*, 8118–8133.
23. Renversez, G.; Boyer, P.; Sagrini, A. Antiresonant reflecting optical waveguide microstructured fibers revisited: A new analysis based on leaky mode coupling. *Opt. Express* **2006**, *14*, 5682–5687. [CrossRef] [PubMed]
24. Engeness, T.D.; Ibanescu, M.; Johnson, S.G.; Weisberg, O.; Skorobogatiy, M.; Jacobs, S.; Fink, Y. Dispersion tailoring and compensation by modal interactions in OmniGuide fibers. *Opt. Express* **2003**, *11*, 1176–1196. [CrossRef]
25. Marcatili, E.A.J.; Schmeltzer, R.A. Hollow metallic and dielectric waveguides for long distance optical transmission and lasers. *Bell Syst. Tech. J.* **1964**, *43*, 1783–1809. [CrossRef]

![fibers logo]

fibers

MDPI

Article

Understanding Dispersion of Revolver-Type Anti-Resonant Hollow Core Fibers

Matthias Zeisberger [1], Alexander Hartung [1] and Markus A. Schmidt [1,2,3,]*

[1]   Leibniz Institute of Photonic Technology, Albert-Einstein-Str. 9, 07745 Jena, Germany;
      matthias.zeisberger@leibniz-ipht.de (M.Z.); alexander.hartung@leibniz-ipht.de (A.H.)
[2]   Otto Schott Institute of Materials Research (OSIM), Friedrich Schiller University of Jena, Fraunhoferstr. 6,
      07743 Jena, Germany
[3]   Abbe Center of Photonics and Faculty of Physics, Friedrich-Schiller-University Jena, Max-Wien-Platz 1,
      07743 Jena, Germany
*   Correspondence: markus.schmidt@leibniz-ipht.de; Tel.: +49-3641-206-140

Received: 24 August 2018; Accepted: 14 September 2018; Published: 20 September 2018

Abstract: Here, we analyze the dispersion behavior of revolver-type anti-resonant hollow core fibers, revealing that the chromatic dispersion of this type of fiber geometry is dominated by the resonances of the glass annuluses, whereas the actual arrangement of the anti-resonant microstructure has a minor impact. Based on these findings, we show that the dispersion behavior of the fundamental core mode can be approximated by that of a tube-type fiber, allowing us to derive analytic expressions for phase index, group-velocity dispersion and zero-dispersion wavelength. The resulting equations and simulations reveal that the emergence of zero group velocity dispersion in anti-resonant fibers is fundamentally associated with the adjacent annulus resonance which can be adjusted mainly via the glass thickness of the anti-resonant elements. Due to their generality and the straightforward applicability, our findings will find application in all fields addressing controlling and engineering of pulse dispersion in anti-resonant hollow core fibers.

Keywords: fiber optics; fiber design and fabrication; microstructured fibers; anti-resonant fibers

1. Introduction

Hollow core fibers (HCFs) allow for efficiently guiding light and are intensively investigated since they allow for accessing previously inaccessible fields for fiber optics or to substantially improve device performance within areas such as mid-IR gas lasers [1,2], broadband light sources [3,4], nonlinear optical effects [5], high-power pulse delivery [6], gas and liquid analytics [7,8], and pulse compression [9]. A comprehensive overview on HCFs can be found in Ref. [10]. Particular anti-resonant HCFs (ARHCFs) have recently gained substantial attraction by the Fiber Optics community since they uniquely combine low optical loss and cladding microstructures that demand only moderate fabrication efforts compared to more sophisticated fiber geometries such as photonic band gap HCFs [11–19]. The ARHCF geometry that is mostly addressed during recent times is the single-ring anti-resonant or revolver-type fiber (RTF) geometry [20–25], consisting of a finite number of thin-walled non-touching glass tubes arranged in a circle at constant azimuthal distances (Figure 1). This arrangement is mechanically stabilized by joining the individual tubes to the inner wall of a supporting capillary. As the guided field is concentrated in the core region, the supporting capillary plays a minor role for the optical properties in particular for the dispersion, and therefore we neglect it in our model. Each tube supports a well-defined number of modes (so-called annulus resonances), allowing for efficiently guiding light in the central fiber section in case these modes are not phase-matched to the core mode, i.e., core and annulus modes are anti-resonant. As a result, these glass annuluses are typically refereed to as anti-resonant elements (ARE) [24]. The anti-resonant effect relies on the interference of the two waves reflected at the inner and outer annulus interface, leading to low-loss transmission bands that are spectrally limited by annulus resonances [26]. It is important to note that already the most generic type of ARHCF geometry—the tube-type fiber (TTF, Figure 2) geometry—qualitatively shows all key features of an ARHCF, namely strong resonances imposed by the ARE-modes and a characteristic loss evolution within the transmission bands. This approximation was utilized by several authors [27–30] to simulate complex ARHCF structures on the basis of the properties of the TTF geometry. Recent experiments involving improved designs of RTFs indicate that these types of fibers show off-resonance losses as low as 7.7 dB/km within various spectral domains [31], making it highly attractive for numerous applications.

One particular striking application that has recently attracted substantial attention is nonlinear light generation in gas-filled HCFs due to low damage thresholds, high output energy densities and the possibility for spectrally tuning the output light via pressure modification [3,4,32]. In addition to sufficiently high nonlinearities, precise control on chromatic dispersion is essential to efficiently generate supercontinua from ultrashort optical pulses [33]. Within soliton-based supercontinuum generation, the key parameter to be controlled is the group velocity dispersion (GVD), which needs to be designed such that the used ultrashort pulse does not significantly disperse over the nonlinear length and that solitons of higher orders are supported. A straightforward-to-access design parameter which is widely used throughout the ultrafast nonlinear community is the wavelength at which the GVD vanishes—the so-called zero-dispersion wavelength (ZDW)—which needs to be carefully adjusted with respect to the pump laser wavelength to allow for efficient supercontinuum generation.

In this work, we present a detailed numerical and analytical study on the dependence of important dispersion parameters on all relevant structural parameters of the RTF geometry. We found that the behaviors of phase index, GVD and ZDW are dominated by the annulus resonances (i.e., ARE wall thickness) while the actual shape of the ARE-based cladding plays only a minor role. These findings allowed us to approximate the mentioned dispersion parameters by those of the most generic type of ARHCF, the TTF geometry, which was numerically confirmed for all practically relevant situations. Using an interface reflection model to approximate the phase index of the TTF geometry, analytic expressions for the mentioned dispersion parameters were obtained, allowing us to gain fundamental insights into the dispersion behavior of the ARHCF geometry.

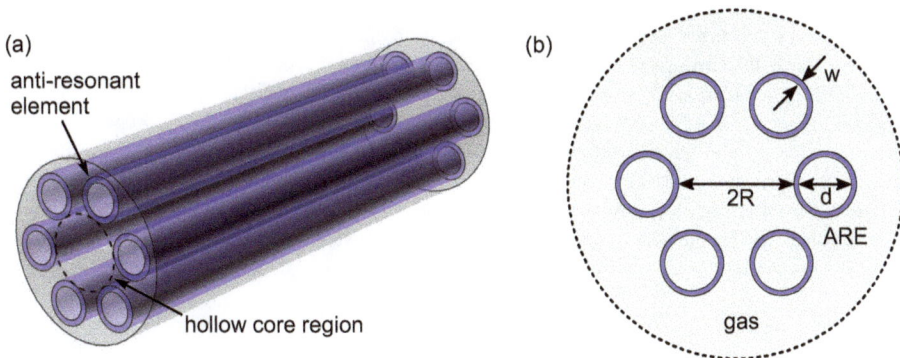

Figure 1. (**a**) illustration of the revolver-type anti-resonant hollow core fiber (RTF) geometry considered here: the fiber consists of six annulus-type anti-resonant elements (blue area) arranged around the central core region (indicted by the black dashed line); (**b**) cross section of that structure with all relevant geometric parameters. The annuluses are made from a dielectric material (typically silica glass), whereas a medium with a lower refractive index (typically gas) is located elsewhere.

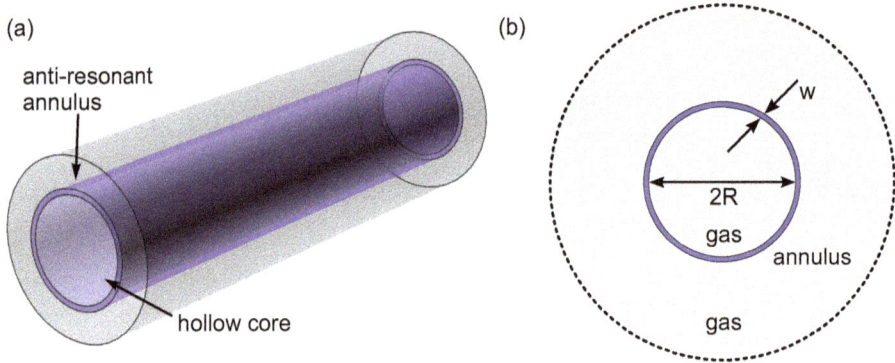

Figure 2. (**a**) sketch of the tube-type anti-resonant hollow core fiber (TTF) geometry that is used to approximate the dispersion properties of RTFs (a cross section of this geometry including the relevant parameters in shown in (**b**)). This structure is composed of a thin annulus of dielectric material containing a low index medium in the core and the most outer region.

2. Analytical Equations for GVD and ZDW

Before comparing the modal properties of RTF and TTF geometries, we would like to introduce the analytical model that allows approximating the chromatic dispersion of TTFs by an analytic expression. As shown by Marcatili et al. [34], the real part of the effective index n_{eff} (i.e., phase index) of a capillary can be expressed as $n_{eff} = 1 - a\lambda^2$ in case the capillary bore diameter is much larger than the operation wavelength λ (the parameter a depends on core radius, and mode order). The related GVD $D_\lambda = -(\lambda/c)\, d^2 n_{eff}/d\lambda^2$ (c speed of light in vacuum) only weakly depends on wavelength and, in case material dispersion is neglected, yielding an entirely analytic form $D_\lambda = 2a\lambda/c$. In contrast, the phase index of the ARHCF geometry includes an additional resonance term resulting from the interference of the waves reflected at the two glass/air interfaces, causing the GVD to strongly differ from that of a capillary particular close to the resonances [26]. For the TTF geometry (Figure 2), the spectral positions of the annulus resonances mostly depend on wall thickness w and on the refractive indices of ring and inner and outer media n_s and n_a, respectively: $\lambda_R = (2w/l)(n_s^2 - n_a^2)^{1/2}$ with $l = 1, 2, ...$ being the annulus resonance order. The same relation was reported for the RTF geometry by Uebel [24], which already indicates similarities between TTF and RTF geometries. In [26], we reported an approximate analytical model for the chromatic dispersion of the TTF geometry in case the core radius R is much larger than the operation wavelength λ (i.e., $\lambda/R \ll 1$). This model is based on a series expansion of the complex effective index with respect to λ/R up to the fourth power. For the discussion presented in this work, we are not addressing losses but chromatic dispersion only, allowing us to neglect the fourth power term that is solely related to modal attenuation and resulting in a real-valued analytic expression of the phase index (see Zeisberger for details [26]) as follows:

$$n_{eff}(\lambda) = n_a(\lambda) - AF(\lambda)\lambda^2 - BE(\lambda)C(\lambda)\lambda^3. \tag{1}$$

The parameters indicated by the capital letters are given as follows:

$$F(\lambda) = \frac{1}{n_a(\lambda)}, \quad E(\lambda) = \frac{n_s(\lambda)^2 + n_a(\lambda)^2}{n_a(\lambda)^3 \sqrt{n_s(\lambda)^2 - n_a(\lambda)^2}}, \tag{2}$$

$$C(\lambda) = \cot\phi, \quad \phi = \frac{W}{\lambda}, \quad W = 2\pi w \sqrt{n_s(\lambda)^2 - n_a(\lambda)^2}, \tag{3}$$

whereas particular emphasis should be placed on the parameter C that includes a cot-function. The parameter ϕ represents the accumulated phase of the waves propagating from the outer to the inner (or vice versa) interface of the ARE. For the fundamental mode (HE_{11} mode), the constants are

$$A = \frac{j_{01}^2}{8\pi^2 R^2}, \quad B = \frac{j_{01}^2}{16\pi^3 R^3}, \quad j_{01} = 2.40. \tag{4}$$

The refractive index of the medium in the core and the outer region n_a can either be that of air or, in the context of supercontinuum generation, argon or another low index medium. It is important to note that the strand resonances are included in the factor $C(\lambda)$ and are located at the phases with the values $\phi = l\pi$ ($l = 1, 2, ...$). The corresponding GVD (i.e., dispersion parameter) is obtained by taking the second derivative of the phase index with respect to wavelength $D_\lambda = -(\lambda/c)d^2 n_{eff}/d\lambda^2$. As mentioned in the introduction section, key parameters for controlling nonlinear light generation in waveguides are GVD and ZDW, which need to be adjusted appropriately. In the following, we use Equation (1) to derive analytic expressions for both parameters. Compared to the factors λ^2, λ^3, and $C(\lambda)$, the factors $F(\lambda)$ and $E(\lambda)$ show a negligible dependence on λ, allowing us to treat them as wavelength-independent constants throughout the remaining part of the manuscript. We have checked this approximation numerically for the wavelength range 0.4–2.0 μm using the material data of argon and silica and found a deviation of less than 3% between the approximation and the exact values. The first term of Equation (1) describes the material dispersion of the core medium, which we approximate in the following by a Cauchy model that fits very well to the properties of argon for wavelengths around 1 μm (see Appendix A for details):

$$n_a(\lambda) \approx a + \frac{b}{\lambda^2}. \tag{5}$$

Here, the term b/λ^2 accounts for the material dispersion of the low index medium. In combination with Equations (2)–(5), Equation (1) yields an analytic expression for the GVD of the TTF geometry:

$$D_\lambda = -\frac{6b}{c\lambda^3} + \frac{2AF\lambda}{c} + \frac{2BE}{c}\left(\frac{W^2\cos\phi(\lambda)}{\sin^3\phi(\lambda)} + \frac{2W\lambda}{\sin^2\phi(\lambda)} + \frac{3\lambda^2\cos\phi(\lambda)}{\sin\phi(\lambda)}\right). \tag{6}$$

The first term in Equation (6) represents the contribution of the core medium (e.g., argon), the second term is the same as for a capillary, and the last term is associated with the annulus resonances. Equation (6) clearly shows that, in contrast to step-index fibers and capillaries, the TTF geometry shows a sophisticated dispersion behavior and provides more degrees of freedom for tuning GVD and ZDW. In fact, this is a result of the annulus resonances causing strong variations of $d^2 n_{eff}/d\lambda^2$ upon λ including positive and negative values particularly close to the resonances. As the main features of the fiber discussed here are related to the resonances, we derive an approximation of Equation (6) for wavelengths close to the resonances. The first step is to use the phase ϕ rather than λ as variable, i.e., λ is substituted by ϕ using Equation (3):

$$D_\lambda = -\frac{6b\phi^3}{cW^3} + \frac{2AFW}{c\phi} + \frac{2BEW^2}{c}\left(\frac{\cos\phi}{\sin^3\phi} + \frac{2}{\phi\sin^2\phi} + \frac{3\cos\phi}{\phi^2\sin\phi}\right). \tag{7}$$

Values of the phase ϕ being close to the resonances can be expressed as $\phi = l\pi - \Delta\phi$ with the off-resonance parameter $|\Delta\phi| \ll 1$. With this assumption, the following approximations can be applied

$$\frac{\cos\phi}{\sin^3\phi} \approx -\frac{1}{\Delta\phi^3}, \quad \phi \approx l\pi, \quad \phi^4 \approx l^4\pi^4, \tag{8}$$

resulting in the following approximation for the GVD of the TTF geometry:

$$D_\lambda = -\frac{6bl^3\pi^3}{cW^3} + \frac{2AFW}{cl\pi} - \frac{2BEW^2}{c}\left(\frac{1}{\Delta\phi^3} - \frac{2}{l\pi\Delta\phi^2} + \frac{3}{l^2\pi^2\Delta\phi}\right). \tag{9}$$

With this approximation, the $\Delta\phi$ value related to the ZDW can be straightforwardly obtained from the condition $D_\lambda = 0$ resulting in the following criterion for zero dispersion:

$$\Delta\phi_{ZD} = \left(\frac{BEWl\pi}{AF}\right)^{1/3}\left(1 - \frac{3bl^4\pi^4}{AFW^4}\right)^{-1/3}. \tag{10}$$

With Equations (2)–(4), we obtain the following relation for the off-resonance parameter value corresponding to the ZDW:

$$\Delta\phi_{ZD} = \left(l\pi\frac{n_s^2 + n_a^2}{n_a^2}\frac{w}{R}\right)^{1/3}\left(1 - \frac{3l^4\pi^2}{2j_{01}^2}\frac{n_a}{(n_s^2 - n_a^2)^2}\frac{b}{w^2}\frac{R^2}{w^2}\right). \tag{11}$$

The related wavelength, i.e., the ZDW can be obtained using Equation (3), leading to

$$\lambda_{ZD} = \frac{2\pi w\sqrt{n_s^2 - n_a^2}}{l\pi - \Delta\phi_{ZD}}. \tag{12}$$

It is interesting to note that the calculation above results in a positive value of $\Delta\phi$. With Equation (12), this results in values of the ZDW being larger than the related resonance wavelength, which is given by

$$\lambda_R = \frac{2w}{l}\sqrt{n_s^2 - n_a^2}. \tag{13}$$

The related anti-resonance wavelength, which is defined by the criterion $\phi = (\pi/2)(2l-1)$ and $l = 1, 2, ...$, is given by

$$\lambda_A = \frac{4w}{2l-1}\sqrt{n_s^2 - n_a^2}. \tag{14}$$

Equations (11) and (12) clearly suggest that the spectral position of the ZDW is associated with that of the corresponding annulus resonance, which is evident by taking the ratio between ZDW and annulus resonance wavelength of the same order $\lambda_R/\lambda_{ZD} = 1 - \Delta\phi_{ZD}/l\pi$. The relation between resonances and ZDW can also be regarded from a more general point of view. Besides the resonance regions, the fibers regarded here show approximately the same dispersion $n_{eff}(\lambda)$ as an empty capillary with $d^2n_{eff}/d\lambda^2 < 0$. Therefore, zero GVD, which corresponds to an inflection point $d^2n_{eff}/d\lambda^2 = 0$ of the dispersion, requires an additional contribution with a sufficiently large positive value of $d^2n_{eff}/d\lambda^2$. According to the Kramers–Kronig relation, every loss peak at a certain wavelength is related to a Lorentzian shaped dispersion with $d^2n_{eff}/d\lambda^2 > 0$ above the resonance wavelength.

3. Dependence of Dispersion on Number and Diameter of AREs

Compared to TTFs, the RTF geometry has two additional geometric parameters, namely the number N and the diameter d of the AREs, which might have a strong impact on chromatic dispersion. Using a finite element mode solver (FEM, COMSOL, simulation details are presented in Ref. [35]), we simulated the spectral dependence of the phase index of the fundamental HE_{11}-mode for a range of N and d values of the RTF geometry (Figure 3, $N = 5, 6, 7$; annulus 5 μm $< d <$ 25 μm) assuming a constant strand width of $w = 0.5$ μm, and material data of silica [36] and argon [37]. Remarkably, both $n_{eff}(\lambda)$ and $D_\lambda(\lambda)$ only show a weak dependence on both geometric parameters within the investigated parameter range. In particular, the spectral evolutions of the GVDs strongly overlap across the entire bandwidth of the transmission band. Only in close proximity to the annulus resonances, differences are observed for small ARE diameters, showing that varying N and d mostly imposes a constant offset to the phase index, which is irrelevant for the GVD as it is correlated to the second derivative of n_{eff} with respect to wavelength. These results also show that the ZDW is mostly independent of N and d within practically relevant parameters' ranges, which has obvious implication on design issue related to ultrashort pulse propagation. Together with the results presented later, these findings suggest that chromatic dispersion in RTFs is dominated by the impact of the strand resonances and hardly depends on the specific properties of the actual AREs used, suggesting that the dispersion behavior of the RTF geometry is identical to that of the most generic type of anti-resonant fiber geometry, the TTF geometry. Since the core diameters of ARHCFs are substantially larger than the wavelengths considered, the above-derived analytic expressions for n_{eff} and $D_\lambda(\lambda)$ can be applied as confirmed in [26]. The resulting spectral evolutions calculated using Equations (1) and (6) (lines in Figure 3) overlap with the numerically obtained RTF results almost across the entire bandwidth of the transmission bands, whereas the match is particularly good for the GVD. An even better agreement is achieved when including a core diameter correction factor f that slightly increases the core radius R (as proposed in Ref. [24]) to account for the penetration of the core mode into the gaps between the AREs (here $f = 1.08$). It is important to note that such kind of agreement and independence on N and d is not achieved for the imaginary part of the complex effective index, which strongly depends on the microstructure used as shown in numerous works [20,22,31]. However, here we are only interested in the dispersion properties, i.e., in the real part of the effective index.

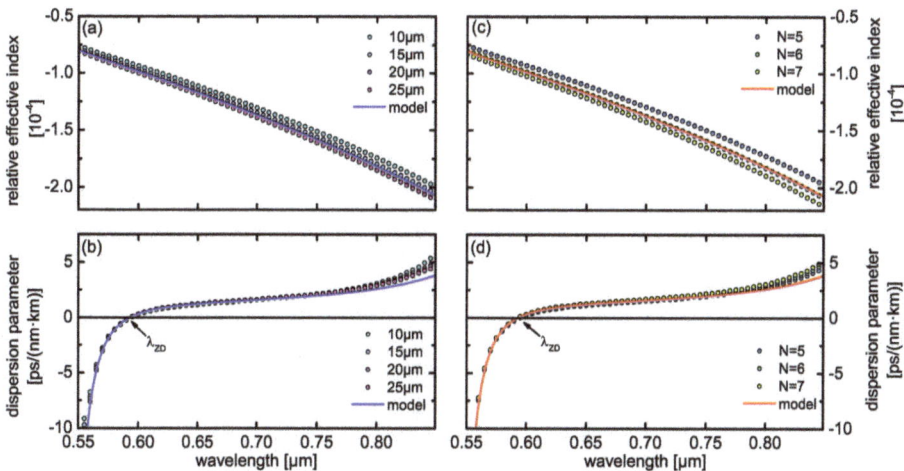

Figure 3. Spectral distributions of the real part of the effective index (top row) and of the group velocity dispersion (bottom row) for (**a,b**) different ARE diameters d ($N = 6$) as well as (**c,d**) different numbers of AREs N ($d = 20\,\mu m$). The symbols represent the results from the numerical FEM simulations considering an RTF, whereas the lines are calculated by the analytic TTF model (Equations (1) and (6)) taking into account an effective core radius parameter of $f = 1.08$. For all data sets, we used $2R = 30\,\mu m$, $w = 0.5\,\mu m$, and material data of silica [36] and argon [37].

4. Dependence on the Core Radius

As a next step, we numerically investigate the dependence of phase index and GVD of the fundamental mode of the RTF geometry on a central core radius (Figure 4) while keeping the annulus parameters fixed ($w = 0.5\,\mu m$, $d = 20\,\mu m$, $n = 6$, refractive index distribution as for Figure 3).

As expected, the spectral distributions of both $n_{eff}(\lambda)$ and $D_\lambda(\lambda)$ strongly change in case the core dimension is modified, with the evolutions provided by our TTF model (solid lines in Figure 4) matching the numerical results particularly good within the transmission bands (dots in Figure 4). Deviations between model and numerics are only visible towards the long-wavelength side of the transmission band. The GVD-evolutions that do not include the core diameter correction factor ($f = 1$) already show excellent overlaps between TTF model and numerics (Figure 4b,d), whereas an improved match is achieved for the phase index when using $f = 1.08$ (Figure 4c,d). It is important to note that in case the strand resonances are neglected in the TTF calculations (i.e., neglecting the λ^3-term in Equation (1)), no match between model and numerics is achieved. This is, for instance, highly visible in the spectral distributions of the GVD (insets in Figure 4b,d), which never crosses the zero within the spectral domain considered here.

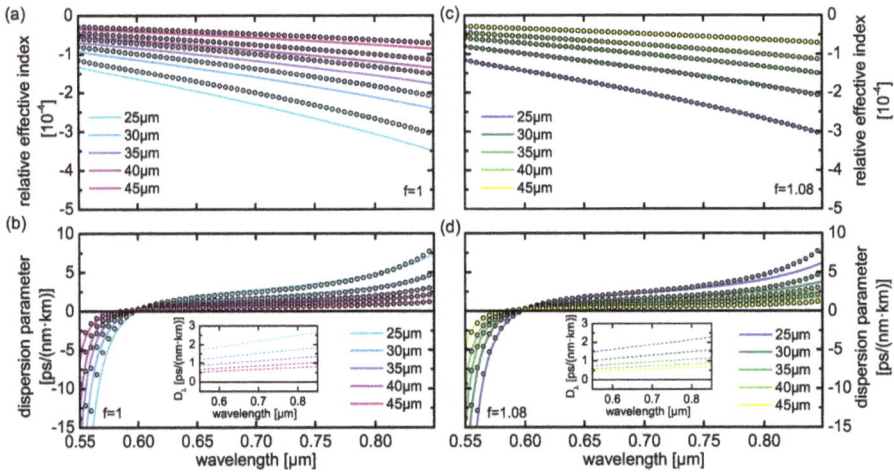

Figure 4. Spectral distributions of real part of the effective index (top row) and group velocity dispersion (bottom row) for different central core diameters of $2R = 25 - 45$ μm (**a,b**) excluding and (**c,d**) including the core diameter correction factor. The symbols represent data from the FEM simulations of the RTF geometry ($d = 20$ μm, $w = 0.5$ μm, $N = 6$), and the lines are calculated from the analytical equations for TTF (Equations (1) and (6), refractive indices defined in the caption of Figure 3). The two insets in (**b,d**) show the corresponding GVD distributions in case the resonance term in Equation (1) is ignored.

Due to its importance in nonlinear photonics and ultrashort pulse propagation, we take a closer look into the behavior of the ZDW in the following. Assuming the geometry considered above ($w = 0.5$ μm, $n = 6$), we have calculated the dependence of the ZDW on central core radius (Figure 5a) numerically from the FEM data of the RTF geometry, from the analytic expression of the GVD of the TTF model (Equation (7)), and by using the approximate equation for the ZDW (Equation (12)). Here, the TTF model (blue symbols in Figure 5a) accurately follows the evolution of the FEM simulations (dark blue symbols in Figure 5a), with the ZDWs matching within a spectral interval <2 nm. The analytic expression (Equation (12)) yields the correct trend, with a deviation of the ZDWs of about 8 nm. Considering that the transmission band has a spectral bandwidth of about 300 nm (see, e.g., Figure 4), this 8 nm difference yields a relative spectral deviation of about 2.5%, which yields an accuracy that is better than can be achieved in fabrication. In addition, we have also calculated the ZDW using Equations (1) and (11) within different bands, i.e., for different annulus resonance orders (Figure 5b) and plotted them together with the wavelengths of the related resonance (assuming $\Delta\phi = 0$ in Equation (12)) in Figure 5b. It is interesting to see that the ZDWs are always located at slightly longer wavelength compared to the corresponding resonance, which is a result of the small positive values of $\Delta\phi$, i.e., $\Delta\phi > 0$ and $|\Delta\phi| \ll 1$. From the practical perspective, the most important parameter that allows for adjusting the spectral positions of annulus resonances is the respective ring width, since $\lambda_R \propto w$ (see Equation (12) in case $\Delta\phi = 0$). As a result, increasing the ring width for a constant core diameter imposes the ZDW to increase accordingly (inset of Figure 5b), again emphasizing that the emergence of a ZDW in one transmission band is a result of the presence of the annulus resonance.

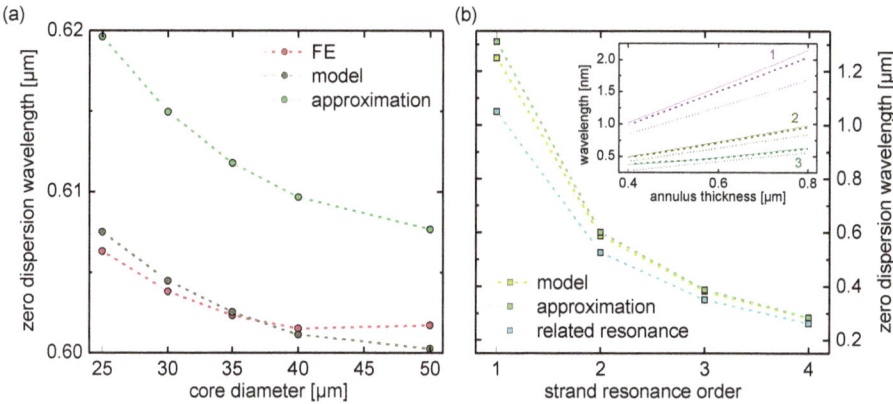

Figure 5. (**a**) dependence of the zero-dispersion wavelength on central core diameter ($w = 0.5$ µm). The three curves show the evolutions from the numerical FEM calculations (red) for $n = 6$ and $d = 20$ µm, the TTF model (Equation (6), gray) and the expression that approximate the ZDW (Equation (12), green); (**b**) ZDW vs. annulus resonance order (yellow: TTF model; green: approximate expression (Equation (12))). The cyan symbols show the related annulus resonance (according to Equation (13)). The connecting lines are only guides to the eye. The inset shows the dependence of the ZDW in case the ring thickness is changed ($2R = 50$ µm) for the three lowest resonance orders (solid lines: analytic equation (Equation (6)); dashed lines: model (Equation (12))); dotted lines: corresponding resonance (Equation (13)). The colors refer to the different resonance orders (indicated by the respective numbers). Both plots assume argon and silica as gas and glass, respectively, and do not include the core diameter correction factor (i.e., $f = 1.0$).

5. Discussion

The key finding of this work is that $\text{Re}(n_{eff})$ and $D_\lambda(\lambda)$ of the TTF geometry fit extremely well to the related quantities of the RTF geometry, suggesting that chromatic dispersion in ARHCF has a generic origin and is not related to a specific cladding microstructure. Here we believe that this coincidence, which is not obvious from the first view, is associated with the following reasoning: as shown in our previous work [26], the dispersion properties of the TTF geometry can by approximated to a very high degree by considering the reflection of a wave on a planar three-layer-system (e.g., here Ar/SiO$_2$/Ar) under nearly grazing incidence in the situation in which the core diameter is much larger than the operation wavelength. Here, we believe that the local fields at the surface of the individual ARE can be treated in a similar way, i.e., that the reflection process of an ARE is principally identical to that happening at a planar interface. Varying w/λ modifies the phase of the reflected wave accordingly, imposing a corresponding variation in the phase of the guided wave, i.e., in $\text{Re}(n_{eff})$. This basic principle is in fact independent on the shape of the core-cladding boundary, suggesting that the dispersion behavior of the RTF geometry is dominated by the impact of the annulus resonances. The main difference between the TTF and the RTF geometry lies in local fields around the core circumference, which mostly impacts losses but not dispersion. In Ref. [29], an empirical formula is presented that also provides an extension of the capillary dispersion [34] with a resonance term that includes empirical parameters. Equation (2) from [29] and Equation (1) from this paper can be transformed into the same mathematical form using approximations for a large core radius ($R \gg \lambda$) and a small off-resonance parameter $|\Delta\phi| \ll 1$. In this context, our results provide absolute values with analytical expressions for the strength of the resonance terms that are empirical parameters in Ref. [29]. From the practical perspective, it is important to know the modal losses in the spectral vicinity of the ZDWs. Our results show that the ZDW present in one transmission band is located in-between the anti-resonance and the related strand resonance wavelengths ($\lambda_R < \lambda_{ZD} < \lambda_A$). This is consistent with the results presented in Ref. [38] where Figure 3 shows that the ZDWs are located slightly above the resonance wavelengths, which corresponds to $0 < \Delta\phi_{ZD} \ll 1$ in terms of the work presented here. We checked all data from our FEM simulations performed in the context of this paper and found very low losses when operating close to the ZDW, whereas more sophisticated microstructures such as nested AREs yield even lower loss [23]. In the context of practical applications, the effect of geometric non-uniformity can be relevant. According to our model, the fiber dispersion can be interpreted as a combination of two effects, the dispersion of a hollow waveguide with a perfectly reflecting boundary (the first two terms in Equation (1)), and the impact of the resonances (the third term in Equation (1)). According to our model, we expect for an RTF with non-uniform wall thickness a corresponding modification of the resonances. Depending on the nature of the geometric non-uniformity, e.g., a discrete or a continuous distribution of the thickness w, we expect several discrete resonances or a broadening of the resonance, changing the overall dispersion. A detailed investigation of this effect ought to be verified by simulations, which is beyond the scope of this paper. The observed behaviors of phase index, GVD, and ZDW being dominated by the annulus resonances (i.e., ARE wall thickness), were also qualitatively found in reported experiments [39]. Even though a direct comparison to the experimental results is unfeasible due to the strand thickness variations across one fiber cross section and the additional struts in the experimentally investigated fiber, the key qualitative features (observation of one ZDW in one transmission band and the bending of the spectral distribution of the GVD in close proximity to the strand resonance) are also found experimentally, which clearly shows that the dispersion of the measured fiber is dominated by the strand resonances. Please note that the TTF-model presented here applies only to fibers with well-defined resonances that are imposed by the strands of the AREs. Photonic band gap HCFs, for instance, only exhibit a small number of transmission bands (in most cases, only one), which are separated by a comparably large spectral interval. The spectral positions of the high loss intervals cannot be described by the resonance of Equation (1) given in this work, i.e., by single strand resonances, with the consequence that the application of our model to photonic band gap HCFs is not possible.

6. Conclusions

Precise knowledge on modal dispersion is relevant for all applications involving the propagation of short pulses in optical waveguides and is particularly important within areas such as nonlinear photonics and ultrashort pulse delivery. Here, we show a detailed study of the dispersion behavior of RTFs, revealing that the resonances provided by the thin-walled annulus of the AREs surrounding the central core dominate the chromatic dispersion. Specifically, we found that the spectral distributions of phase index and GVD strongly depend on central core diameter and annulus width, whereas the modal behavior is almost independent of diameter and number of AREs used. One of the key findings of this work is that the dispersion properties (i.e., phase index and GVD) of the RTF geometry match those provided by a corresponding TTF. We derived analytic expressions for phase index and GVD that both match the corresponding features of the RTF geometry for all practically relevant situations, whereas a particular good match is obtained for the GVD. We also derived an analytic expression for the ZDW showing that each transmission band owns one ZDW, which is spectrally located in between the annulus resonance and the anti-resonance wavelengths. Moreover, the derived equations suggest that the emergence of one ZDW in one band is associated with the presence of one resonance, i.e., that the origin of the multiple ZDWs observed in the RTF geometry is solely related to annulus resonances that are formed by the interference within the thin glass membrane. As a result of these facts, we strongly believe that both the TTF model and the analytic expressions for GVD and ZDW yield straightforward-to-use design tools that are relevant for the development of future RTFs for applications in nonlinear photonics, ultrafast light transportation and any application that demands controlling pulse dispersion in ARHCFs.

Author Contributions: M.Z. performed the analytical calculations, and A.H. the FEM simulations; M.A.S. prepared the graphics; The manuscript was written by M.Z. and M.A.S.

Funding: This research was funded by German Research Foundation (Grant SCHM2655/6-1, SCHM2655/8-1); Thuringian State Projects (2015FGI0011, 2015-0021, 2016FGR0051); European Regional Development Fund (ERDF); European Social Funds (ESF).

Conflicts of Interest: The authors declare no conflict of interest.

Appendix A

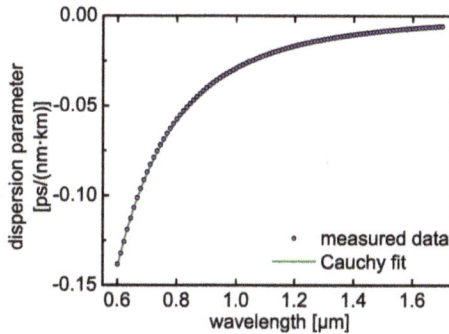

Figure A1. Spectral distribution of the GVD of argon calculated from empirical data (symbols) [37] and from Equation (A1) (line).

As shown in [37] the material dispersion of argon can be approximated by a Sellmeier expression. In case the operation wavelength is of the order of 1 μm this expression can further be approximated by first order series expansion with respect to λ^{-2} resulting in the Cauchy expression presented in Equation (5) with $a = 1.00028$ and $b = 1.507 \cdot 10^{-6}$ from which the following expression for the GVD of argon is obtained.

$$D_\lambda = -\frac{\lambda}{c}\frac{d^2 n_a}{d\lambda^2} = -\frac{6b}{c\lambda^3} \tag{A1}$$

References

1. Hassan, M.R.A.; Yu, F.; Wadsworth, W.J.; Knight, J.C. Cavity-based mid-IR fiber gas laser pumped by a diode laser. *Optica* **2016**, *3*, 218–221. [CrossRef]
2. Wang, Z.; Belardi, W.; Yu, F.; Wadsworth, W.J.; Knight, J.C. Efficient diode-pumped mid-infrared emission from acetylene-filled hollow-core fiber. *Opt. Express* **2014**, *22*, 21872–21878. [CrossRef] [PubMed]
3. Sollapur, R.; Kartashov, D.; Zürch, M.; Hoffmann, A.; Grigorova, T.; Sauer, G.; Hartung, A.; Schwuchow, A.; Bierlich, J.; Kobelke, J.; et al. Resonance-enhanced multi-octave supercontinuum generation in antiresonant hollow-core fibers. *Light Sci. Appl.* **2017**, *6*, e17124. [CrossRef] [PubMed]
4. Russell, P.S.J.; Hölzer, P.; Chang, W.; Abdolvand, A.; Travers, J.C. Hollow-core photonic crystal fibres for gas-based nonlinear optics. *Nat. Photonics* **2014**, *8*, 278–286. [CrossRef]

5. Ouzounov, D.G.; Ahmad, F.R.; Müller, D.; Venkataraman, N.; Gallagher, M.T.; Thomas, M.G.; Silcox, J.; Koch, K.W.; Gaeta, A.L. Generation of megawatt optical solitons in hollow-core photonic band-gap fibers. *Science* **2003**, *301*, 1702–1704. [CrossRef] [PubMed]

6. Jaworski, P.; Yu, F.; Carter, R.M.; Knight, J.C.; Shephard, J.D.; Hand, D.P. High energy green nanosecond and picosecond pulse delivery through a negative curvature fiber for precision micro-machining. *Opt. Express* **2015**, *23*, 8498–8506. [CrossRef] [PubMed]

7. Jin, W.; Cao, Y.; Yang, F.; Ho, H.L. Ultra-sensitive all-fibre photothermal spectroscopy with large dynamic range. *Nat. Commun.* **2015**, *6*, 6767. [CrossRef] [PubMed]

8. Nissen, M.; Doherty, B.; Hamperl, J.; Kobelke, J.; Weber, K.; Henkel, T.; Schmidt, M.A. UV Absorption Spectroscopy in Water-Filled Antiresonant Hollow Core Fibers for Pharmaceutical Detection. *Sensors* **2018**, *18*, 478. [CrossRef] [PubMed]

9. Heckl, O.H.; Saraceno, C.J.; Baer, C.R.; Südmezer, T.; Wang, Y.Y.; Cheng, Y.; Benabid, F.; Keller, U. Temporal pulse compression in a xenon-filled kagome-type hollow-core photonic crystal fiber at high average power. *Opt. Express* **2011**, *19*, 19142–19149. [CrossRef] [PubMed]

10. Harrington, J.A. *Infrared Fibers and Their Applications*; SPIE Press: Bellingham, WA, USA, 2004.

11. Cregan, R.F.; Mangan, B.J.; Knight, J.C.; Birks, T.A.; Russell, P.S.; Roberts, P.J.; Allan, D.C. Single-Mode Photonic Band Gap Guidance of Light in Air. *Science* **1999**, *285*, 1537–1539. [CrossRef] [PubMed]

12. Knight, J.C.; Broeng, J.; Birks, T.A.; Russell, P.S.J. Photonic band gap guidance in optical fibers. *Science* **1998**, *282*, 1476–1478. [CrossRef] [PubMed]

13. Benabid, F.; Roberts, P.J. Linear and nonlinear optical properties of hollow core photonic crystal fiber. *J. Mod. Opt.* **2011**, *58*, 87–124. [CrossRef]

14. Broeng, J. Photonic Crystal Fibers: A New Class of Optical Waveguides. *Opt. Fiber Technol.* **1999**, *5*, 305–330. [CrossRef]

15. Gebert, F.; Frosz, M.H.; Weiss, T.; Wan, Y.; Ermolov, A.; Joly, N.Y.; Schmidt, P.O.; Russell, P.S.J. Damage-free single-mode transmission of deep-UV light in hollow-core PCF. *Opt. Expr.* **2014**, *22*, 15388–15396. [CrossRef] [PubMed]

16. Russell, P. Photonic Crystal Fibers. *Science* **2003**, *299*, 358–362. [CrossRef] [PubMed]

17. Knight, J.C. Photonic crystal fibres. *Nature* **2003**, *424*, 847–851. [CrossRef] [PubMed]

18. Smith, C.M.; Venkataraman, N.; Gallagher, M.T.; Müller, D.; West, J.A.; Borrelli, N.F.; Allan, D.C.; Koch, K.W. Low-loss hollow-core silica/air photonic bandgap fibre. *Nature* **2003**, *424*, 657–659. [CrossRef] [PubMed]

19. Frosz, M.H.; Nold, J.; Weiss, T.; Stefani, A.; Babic, F.; Rammler, S.; Russell, P.S.J. Five-ring hollow-core photonic crystalfiber with 1.8 dB/km loss. *Opt. Lett.* **2013**, *38*, 2215–2217. [CrossRef] [PubMed]

20. Pryamikov, A.D.; Biriukov, A.S.; Kosolapov, A.F.; Plotnichenko, V.G.; Semjonov, S.L.; Dianov, E.M. Demonstration of a waveguide regime for a silica hollow-core microstructured optical fiber with a negative curvature. *Opt. Express* **2011**, *19*, 1441–1448. [CrossRef] [PubMed]

21. Kolyadin, A.N.; Kosolapov, A.F.; Pryamikov, A.D.; Biriukov, A.S.; Plotnichenko, V.G.; Dianov, E.M. Light transmission in negative curvature hollowcore fiber in extremely high material loss region. *Opt. Express* **2013**, *21*, 9514–9519. [CrossRef] [PubMed]

22. Belardi, W.; Knight, J.C. Hollow antiresonant fibers with low bending loss. *Opt. Express* **2014**, *22*, 10091–10096. [CrossRef] [PubMed]

23. Belardi, W.; Knight, J.C. Hollow antiresonant fibers with reduced attenuation. *Opt. Lett.* **2014**, *39*, 1853–1856. [CrossRef] [PubMed]

24. Uebel, P.; Günendi, M.C.; Frosz, M.H.; Ahmed, G.; Edavalath, N.N.; Menard, J.M.; Russell, P.S.J. Broadband robustly single-mode hollow-core PCF by resonant filtering of higher-order modes. *Opt. Lett.* **2016**, *41*, 1961–1964. [CrossRef] [PubMed]

25. Frosz, M.H.; Roth, P.; Günendi, M.C.; Russell, P.S.J. Analytical formulation for the bend-loss in singel-ring hollow-core photonic crystal fibers. *Photonics Res.* **2017**, *5*, 88–91. [CrossRef]

26. Zeisberger, M.; Schmidt, M.A. Analytic model for the complex effective index of leaky modes of anti-resonant single ring hollow core fibers. *Sci. Rep.* **2017**, *7*, 11761. [CrossRef] [PubMed]

27. Hayes, J.R.; Poletti, F.; Abokhamis, M.S.; Wheeler, N.V.; Baddila, N.K.; Richardson, D.J. Anti-resonant hexagram hollow core fibers. *Opt. Express* **2015**, *23*, 1289–1299. [CrossRef] [PubMed]

28. Wang, Y.; Ding, W. Confinement loss in hollow-core negative curvature fiber: A multi-layered model. *Opt. Express* **2017**, *25*, 33122–33133. [CrossRef]

29. Hasan, M.I.; Akhmediev, N.; Chang, W. Empirical formulae for the hollow-core antiresonant fibers: Dispersion and effective mode area. *J. Lightw. Technol.* **2018**, *36*, 4060–4065. [CrossRef]

30. Stawska, H.I.; Popenda, M.A.; Beres-Pawlik, E. Anti-resonant Hollow Core Fibers with Modified Shape of the Core for the Better Optical Performance in the Visible Spectral Region—A Numerical Study. *Polymers* **2018**, *10*, 899. [CrossRef]

31. Debord, B.; Amsanpally, A.; Chafer, M.; Baz, A.; Maurel, M.; Blondy, J.M.; Hugonnot, E.; Scol, F.; Vincetti, L.; Gerome, F.; et al. Ultralow transmission loss in inhibited-coupling guiding hollow fibers. *Optica* **2017**, *4*, 209–217. [CrossRef]

32. Travers, J.C.; Chang, W.; Nold, J.; Joly, N.Y.; Russell, P.S.J. Ultrafast nonlinear optics in gas-filled hollow-core photonic crystal fibers. *J. Opt. Soc. Am. B* **2011**, *28*, A11–A26. [CrossRef]

33. Dudley, J.M.; Genty, G.; Coen, S. Supercontinuum generation in photonic crystal fiber. *Rev. Mod. Phys.* **2006**, *78*, 1135–1184. [CrossRef]

34. Marcatili, E.A.J.; Schmeltzer, R.A. Hollow Metallic and Dielectric Waveguides for Long Distance Optical Transmission and Lasers. *Bell Syst. Tech. J.* **1964**, *43*, 1783–1809. [CrossRef]

35. Hartung, A.; Kobelke, J.; Schwuchow, A.; Wondraczek, K.; Bierlich, J.; Popp, J.; Frosch, T.; Schmidt, M.A. Double antiresonant hollow core fiber—Guidance in the deep ultraviolet by modified tunneling leaky modes. *Opt. Express* **2014**, *22*, 19131–19140. [CrossRef] [PubMed]

36. Palik, E.D. *Handbook of Optical Constants of Solids*; Academic Press: San Diego, CA, USA, 1998.

37. Bideau-Mehu, A.; Guern, Y.; Abjean, R.; Johannin-Gilles, A. Measurement of refractive inidces of neon, argon, krypton and xenon in the 253.7–140.4 nm wavelength range. Dispersion relations and estimated oscillator strengths of the resonance lines. *J. Quant. Spectrosc. Radiat. Transf.* **1981**, *25*, 395–402. [CrossRef]

38. Alagashev, G.K.; Pryamikov, A.D.; Kosolapov, A.F.; Kolyadin, A.N.; Lukovkin, A.Y.; Biriukov, A.S. Impact of the geometrical parameters on the optical properties of negative curvature hollow-core fibers. *Laser Phys.* **2015**, *25*, 055101. [CrossRef]

39. Grigorova, T.; Sollapur, R.; Hoffmann, A.; Schwuchow, A.; Bierlich, J.; Kobelke, J.; Schmidt, M.A.; Spielmann, C. Measurement of the dispersion of an antiresonant hollow core fiber. *IEEE Photonics J.* **2018**, *10*, 7104406. [CrossRef]

![fibers logo] *fibers*

MDPI

Article

Geometry of Chalcogenide Negative Curvature Fibers for CO_2 Laser Transmission

Chengli Wei [1], Curtis R. Menyuk [2] and Jonathan Hu [1,*]

[1] Department of Electrical and Computer Engineering, Baylor University, Waco, TX 76798, USA; cwei@umhb.edu

[2] Department of Computer Science and Electrical Engineering, University of Maryland Baltimore County, Baltimore, MD 21227, USA; menyuk@umbc.edu

* Correspondence: jonathan_hu@baylor.edu; Tel.: +1-254-710-1853

Received: 12 July 2018; Accepted: 27 September 2018; Published: 30 September 2018

Abstract: We study the impact of geometry on leakage loss in negative curvature fibers made with As_2Se_3 chalcogenide and As_2S_3 chalcogenide glasses for carbon dioxide (CO_2) laser transmission. The minimum leakage loss decreases when the core diameter increases both for fibers with six and for fibers with eight cladding tubes. The optimum gap corresponding to the minimum loss increases when the core diameter increases for negative curvature fibers with six cladding tubes. For negative curvature fibers with eight cladding tubes, the optimum gap is always less than 20 μm when the core diameter ranges from 300 μm to 500 μm. The influence of material loss on fiber loss is also studied. When material loss exceeds 10^2 dB/m, it dominates the fiber leakage loss for negative curvature fiber at a wavelength of 10.6 μm.

Keywords: CO_2 lasers; negative curvature fibers; chalcogenide glass; fiber loss; mid-IR

1. Introduction

Carbon dioxide (CO_2) lasers have been widely used in surgery, medicine, and material processing [1–3]. Step index fibers are commonly used to transmit CO_2 laser light. The material loss of silica glass in the mid-infrared limits the transmission of mid-infrared light using silica step-index fibers. However, it is possible in principle to obtain a lower loss in hollow-core fiber than in step-index fiber because air does not contribute to material loss [4,5]. In addition, the nonlinearity in the glass sets a limit to the transmitted power. Hollow-core fibers have low nonlinearity, because the light is mostly transmitted in air, which does not contribute to the nonlinearity. Recently, hollow-core negative curvature fibers have drawn a large amount of interest due to their attractive properties including low loss, broad bandwidth, and a high damage threshold [6–12]. The delivery of mid-infrared radiation has also been demonstrated using chalcogenide negative curvature fibers for a CO_2 laser at a wavelength of 10.6 μm [13–15]. Previous study shows that chalcogenide glass should be used for wavelength larger than 4.5 μm [16]. The relative simplicity of the negative curvature structure could enable the fabrication of fiber devices for mid-IR applications using non-silica glasses, such as chalcogenide [13–15].

The guiding mechanism in negative curvature fibers is inhibited coupling [10,17,18]. A large amount of research [10,19] has been carried out to determine the impact of fiber parameters on leakage loss [20] in negative curvature fibers and then optimize these parameters to minimize the loss. These parameters include the curvature of the core boundary, the thickness of the tubes, the number of cladding tubes, and the nested cladding tubes [17,18,21–24]. By introducing a gap between cladding tubes, the loss can be decreased in negative curvature fibers [24,25]. When the tubes touch, modes exist in the localized node area. A gap between the cladding tubes removes the additional resonances due to the localized node. Fibers with a gap between tubes are also expected to be easier to fabricate, since surface tension would assist to maintain the circular shape of the tubes [22]. On the other hand, when the gap is too big, the core mode can leak through the gaps, which increases the loss in negative curvature fibers [26]. Therefore, an optimum gap exists. The optimal gap corresponding to the minimum loss in a fiber with six cladding tubes is three times as large as the optimal gap in fibers with eight or ten cladding tubes [26]. In a fiber with six cladding tubes, a larger gap is needed to remove the weak coupling between the core mode and tube modes [26].

In previous studies, the optimum gap was found in negative curvature fibers with a fixed core diameter [26]. Chalcogenide negative curvature fibers with different core diameters of 170 μm to 380 μm have been fabricated [13–15]. In this paper, we find optimal structures of chalcogenide negative curvature fibers for CO_2 laser transmission, in which we minimize the loss in the two-dimensional parameter space that consists of the core diameter and the gap size. In previous studies, the optimum gap was found in negative curvature fibers with a fixed core diameter [26]. We find that the minimum leakage loss decreases when the core diameter increases both for fibers with six and for fibers with eight cladding tubes. The optimum gap increases when the core diameter increases for negative curvature fibers with six cladding tubes. The optimum gap is always less than 20 μm when the core diameter increases for negative curvature fibers with eight cladding tubes when the core diameter ranges from 300 to 500 μm. We find optimal structures of chalcogenide negative curvature fibers for CO_2 laser transmission, in which we minimize the loss in the two-dimensional parameter space that consists of the core diameter and the gap size.

2. Geometry

Negative curvature fibers with six and eight cladding tubes have been fabricated by several research groups [17,25,27,28]. Figure 1 shows schematic illustrations of negative curvature fibers with six and eight cladding tubes. The white regions represent air, and the gray regions represent glass. The inner tube diameter, d_{tube}, the core diameter, D_{core}, the tube wall thickness, t, the minimum gap between the cladding tubes, g, and the number of tubes, p, are related by the expression: $D_{core} = (d_{tube} + 2t + g)/\sin(\pi/p) - (d_{tube} + 2t)$ [29]. We calculate the leakage loss for negative curvature fibers using Comsol Multiphysics, a commercial full-vector mode solver based on the finite-element method. Perfectly matched layers are added outside the cladding region in order to reduce the size of the simulation window [30]. The wavelength of 10.6 μm for a CO_2 laser is used in our simulation.

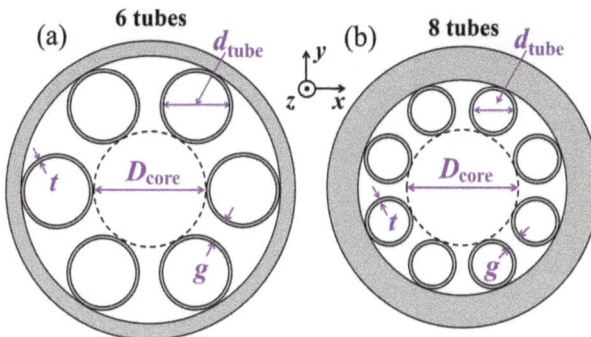

Figure 1. Schematic illustration of negative curvature fibers with (**a**) six and (**b**) eight cladding tubes.

3. As$_2$Se$_3$ Chalcogenide Glass

In this section, we study the loss in negative curvature fibers made with As_2Se_3 chalcogenide glass. We use a refractive index of 2.8 and a material loss of 10.6 dB/m for As_2Se_3 chalcogenide glass in our simulations [31]. The tube thickness, t, is fixed at 5.2 μm corresponding to the third antiresonance. A glass thickness corresponding to the third antiresonance has been drawn in the past [15]. A thicker tube wall with a higher-order antiresonance makes fabrication easier. Geometries that use tube thicknesses corresponding to the first, second, or third antiresonance have similar minimum losses in the transmission band [16,26]. We first study negative curvature fibers with six cladding tubes. We define d_{6max} as the maximum possible tube diameter for the fiber with 6 cladding tubes, which equals $D_{core} - 2t$. Figure 2a shows the contour plot of loss as a function of core diameter, D_{core}, and normalized tube diameter, d_{tube}/d_{6max}. For a fixed D_{core}, the loss decreases and then increases when d_{tube}/d_{6max} increases from 0.2 to 1.0. The minimum loss occurs when $d_{tube}/d_{6max} = 0.62$, and it does not change when D_{core} increases from 300 to 500 μm. The loss decreases when D_{core} increases. In addition, we show the loss as a function of the core diameter, D_{core}, and the gap, g, in Figure 2b. The loss first decreases and then increases as the gap, g, increases. When there is no gap, a mode exists in the node that is created by the two touching tubes [25]. When the gap is too large, core mode leaks through the gap [17,26]. Previous study shows that the electric field intensity in the middle of the gap between cladding tubes can increase by a factor of 15 when the gap increases from 5 to 10 μm in a silica negative curvature fiber with a glass index of 1.45 and a core diameter of 30 μm at a wavelength of 1 μm [10]. Here, we study chalcogenide negative curvature fibers with a glass index of 2.8 at a wavelength of 10.6 μm. The electric field intensity in the middle of the gap between cladding tubes increases by a factor of 15 when the gap increases from 50 to 100 μm in a negative curvature fiber with a core diameter of 300 μm. We also plot the loss as a function of gap, g, for different core diameters in Figure 3a. In order to quantify the minimum loss and the corresponding optimum gap for different core diameters, we also plot the minimum loss and the corresponding optimum gap, g, using blue solid curve and red dashed curves, respectively, in Figure 3b. When the core diameter increases from 300 to 500 μm, the minimum loss decreases by more than one order of magnitude and the corresponding optimum gap, g, increases from 60 to 90 μm. Hence, a larger gap is needed for a fiber with a larger core diameter to decrease the loss in negative curvature fibers with six cladding tubes.

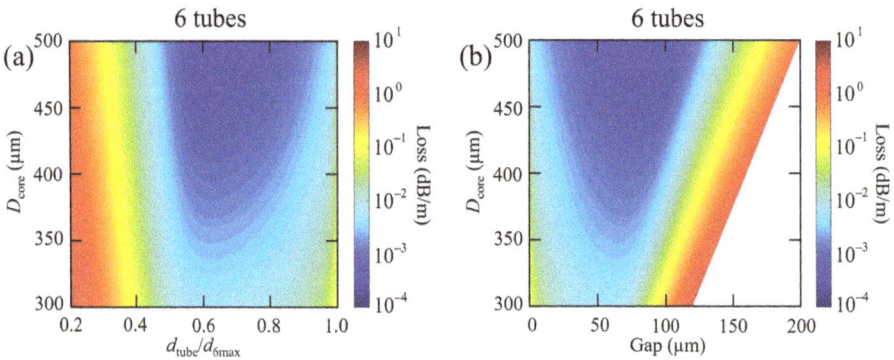

Figure 2. (a) Contour plot of loss as a function of core diameter and normalized tube diameter. (b) Contour plot of loss as a function of core diameter and gap. The number of cladding tubes is six.

We next carry out the same loss analysis on negative curvature fibers with eight cladding tubes. Figure 4a shows the contour plot of loss as a function of core diameter, D_{core}, and normalized tube diameter, d_{tube}/d_{8max}, where d_{8max} is defined as the maximum possible tube diameter for the fiber with 8 cladding tubes, which is $D_{core} \sin(\pi/8)/[1 - \sin(\pi/8)] - 2t$ [32]. Figure 4b shows the contour plot of loss as a function of core diameter, D_{core}, and gap, g. The minimum loss occurs at a larger value of d_{tube}/d_{8max}, or a smaller value of g, than is the case for negative curvature fibers with six cladding tubes. In Figure 5a, we show the loss as a function of the gap, g, for different core diameters. The optimum gap corresponding to the minimum loss is less than 20 μm for fibers with different core diameters and the loss increases slowly when gap further increases. The minimum loss and the corresponding gap, g, are plotted using blue solid curve and red dashed curves, respectively, in Figure 5b. The minimum loss decreases by around one order of magnitude when the core diameter increases from 300 to 500 μm. Different from fibers with six cladding tubes, the corresponding optimum gap, *g*, is much smaller and is always less than 20 μm when the core diameter increases from 300 to 500 μm in fibers with eight cladding tubes. There is a wide range of gaps that realize low loss in the fibers with eight cladding tubes, as shown in Figure 5a. The loss is less sensitive to the gap in the region between 10 and 50 μm. Since the tube diameter is much smaller than the diameter of core, the coupling between the core mode and tube modes is weak. It has been shown that the power ratio in the air region of cladding tubes is always less than 0.1% in the negative curvature fiber with eight cladding tubes, while the power ratio in tube air could be more than 0.8% for fibers with six cladding tubes [26]. In negative curvature fibers with six cladding tubes, a larger gap is needed to remove the weak coupling between the core and cladding tube modes.

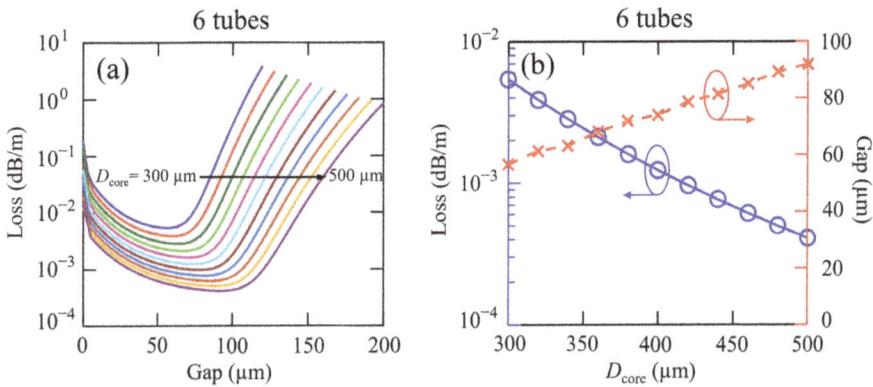

Figure 3. (a) Loss as a function of gap in fibers with different core diameters. (b) Minimum loss and the corresponding optimum gap in fibers with different core diameters. The number of cladding tubes is six.

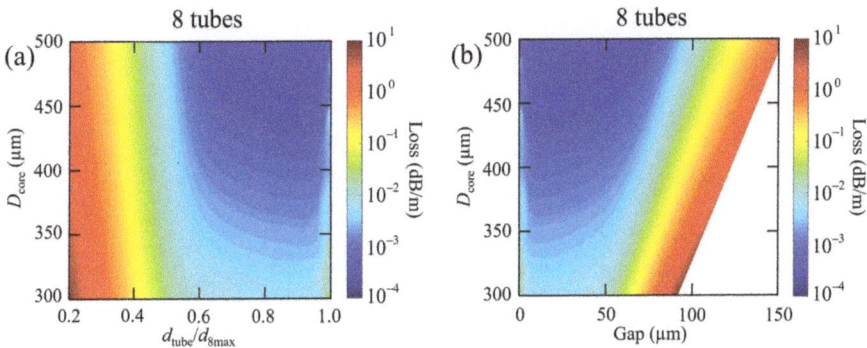

Figure 4. (a) Contour plot of loss as a function of core diameter and normalized tube diameter. (b) Contour plot of loss as a function of the core diameter and gap. The number of cladding tubes is eight.

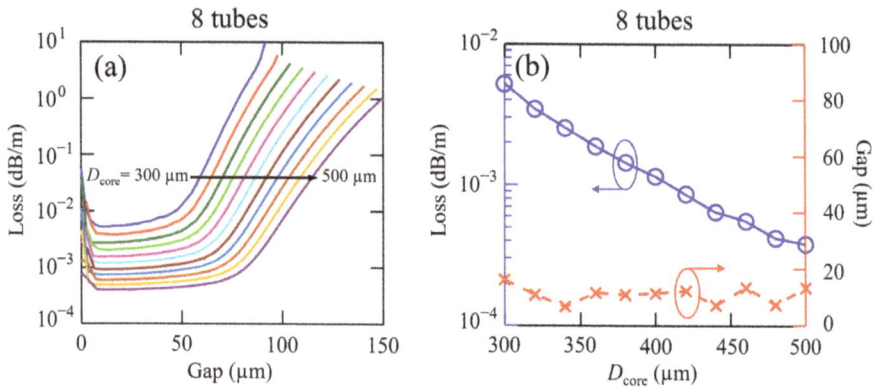

Figure 5. (**a**) Loss as a function of the gap in fibers with different core diameters. (**b**) Minimum loss and the corresponding gap in fibers with different core diameters. The number of cladding tubes is eight.

4. As$_2$S$_3$ Chalcogenide Glass

In this section, we carried out the same loss analysis in negative curvature fibers made with As$_2$S$_3$ chalcogenide glass. We use a refractive index of 2.4 and a material loss of 500 dB/m for As$_2$S$_3$ chalcogenide glass in our simulations [15,16]. The tube thickness, t, is fixed at 6.1 μm corresponding to the third antiresonance. Figure 6a shows the loss as a function of gap, g, when the core diameter increases from 300 to 500 μm in As$_2$S$_3$ chalcogenide fiber with six cladding tubes. Compared with the loss in Figure 3a, the losses in the fiber using As$_2$S$_3$ chalcogenide glass, shown in Figure 6a, are higher and have a flatter minimum. In Figure 6b, we show the minimum loss and the corresponding gap, g, as blue solid curve and red dashed curve, respectively. We also study the fiber leakage loss with and without material loss in an As$_2$S$_3$ chalcogenide fiber with six cladding tubes. In Figure 7a, we show the results in order to explain the broad, low-loss region in Figure 6a. The core diameter is fixed at 300 μm. The solid curve shows the fiber loss with material loss of 500 dB/m for As$_2$S$_3$ chalcogenide glass, which is the same as the blue solid curve in Figure 6a. The dashed curve shows the fiber loss without material loss, which is similar to the curve in Figure 3a. The high material loss of As$_2$S$_3$ chalcogenide glass dominates and leads to a flat minimum in the fiber loss curve, as shown by the blue solid curve in Figure 7a.

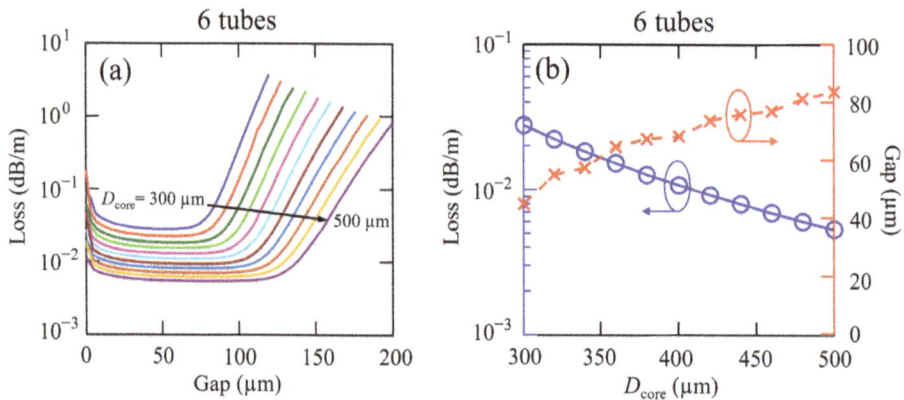

Figure 6. (**a**) Loss as a function of gap in fibers with different core diameters. (**b**) Minimum loss and corresponding optimum gap in fibers with different core diameters. There are six cladding tubes.

In order to better illustrate the influence of the material loss on the total fiber loss, we study the fiber loss as a function of material loss both for As$_2$S$_3$ chalcogenide glass and As$_2$Se$_3$ chalcogenide glass, shown in Figure 7b as the red dashed and blue solid curves, respectively. The core diameter is 300 µm and the gap is 60 µm. The fiber loss changes little when the material loss increases from 0.1 to 10 dB/m, and the fiber loss is dominated by the confinement loss in the blue region for both curves. The loss of fiber that is made with As$_2$Se$_3$ chalcogenide glass is located in the blue region, which is marked with the blue circle on the blue solid curve. The fiber loss begins to increase when the material loss increases from 10 to 10^2 dB/m, and the influence of the material loss becomes visible. When the material loss further increases, the fiber loss increases sharply, and the fiber loss is dominated by the material loss in the red region for both curves, when the material loss is higher than 10^2 dB/m. The loss of fiber made with As$_2$S$_3$ chalcogenide glass is located in the red region, which is marked with the red triangle on the red dashed curve. Due to the inhibited coupling between the core mode and glass modes, the power ratios in the glass of negative curvature fibers for the two points marked by circle and triangle in Figure 7b are 0.0016% and 0.002%, respectively. With this low power ratio in glass [33], the fiber leakage loss in negative curvature fibers is more than three orders of magnitude lower than the material loss of glass, as shown in Figure 7b.

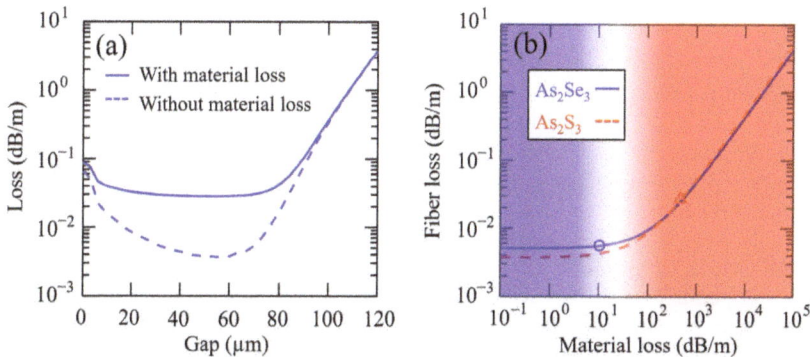

Figure 7. (**a**) Loss as a function of gap in fibers with and without material loss. (**b**) Fiber loss as a function of material loss in As$_2$Se$_3$ chalcogenide glass fiber and As$_2$S$_3$ chalcogenide glass fiber with six cladding tubes, a core diameter of 300 µm, and a gap of 60 µm.

Figure 8a shows the loss as a function of gap, *g*, in As$_2$S$_3$ chalcogenide fiber with eight cladding tubes. In Figure 8b, we show the minimum loss and the corresponding gap, *g*, using a blue solid curve and a red dashed curve, respectively. The minimum loss decreases by less than one order of magnitude and the corresponding optimum gap, *g*, is always less than 20 µm, which agrees with the results in the As$_2$Se$_3$ chalcogenide fiber with 8 cladding tubes. Small loss variation near zero gap occurs due to the glass modes existed near the node area between two tubes in Figure 8a.

Chalcogenide negative curvature fibers with eight cladding tubes have been successfully fabricated. The fiber loss was measured to be 2.1 dB/m at 10 µm for a fiber with a core diameter of 172 µm and a gap of 9 µm. Due to the structure distortion during fabrication, the losses of fabricated fibers are two orders of magnitude higher than the losses in simulation, indicating there are room to improve the fabrication [34]. The distortion of the negative curvature fiber structure has an evident impact on the transmission window and the leakage loss [34]. We also observed higher-order modes in the negative curvature fibers [35].

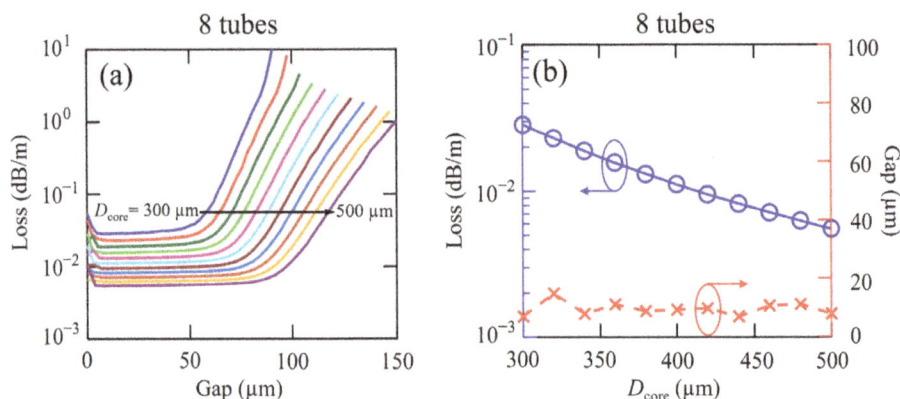

Figure 8. (**a**) Loss as a function of gap in fibers with different core diameters. (**b**) Minimum loss and corresponding gap in fibers with different core diameters. The number of cladding tube is eight.

5. Conclusions

In this paper, we optimize the structure of negative curvature fibers for CO_2 laser transmission. We investigate the impact of the size of the gap between cladding tubes on the loss of negative curvature fibers made with As_2Se_3 and As_2S_3 chalcogenide glasses. For As_2Se_3 chalcogenide fibers with six cladding tubes, the minimum loss decreases by an order of magnitude and the corresponding optimum gap, g, increases from 60 to 90 µm when the core diameter increases from 300 to 500 µm. A greater gap is needed for a fiber with greater core diameter to reduce the coupling between the core mode and tube mode. For a fiber with eight cladding tubes, the optimum gap, g, that corresponds to the minimum loss is always less than 20 µm when the core diameter ranges from 300 to 500 µm. We also study As_2S_3 chalcogenide fibers, which has a higher material loss at a wavelength of 10.6 µm. It is found that material loss dominates the fiber leakage loss. The fiber loss is dominated by the material loss, when the material absorption loss is higher than 10^2 dB/m.

Author Contributions: Supervision, C.R.M. and J.H.; Validation, C.R.M. and J.H.; Writing: original draft, C.W.; Writing: review and editing, C.W., C.R.M. and J.H.

Funding: Work at Baylor was supported by the National Science Foundation (ECCS-1809622). Work at UMBC was supported by the Naval Research Laboratory.

Conflicts of Interest: The authors declare no conflict of interest.

References

1. Snakenborg, D.; Klank, H.; Kutter, J.P. Microstructure fabrication with a CO_2 laser system. *J. Micromech. Microeng.* **2004**, *14*, 182–189. [CrossRef]
2. Hædersdal, M.; Sakamoto, F.H.; Farinelli, W.A.; Doukas, A.G.; Tam, J.; Anderson, R.R. Fractional CO_2 laser-assisted drug delivery. *Lasers Surg. Med.* **2010**, *42*, 113–122. [CrossRef] [PubMed]
3. Witteman, W.J. *The CO₂ Laser*; Enoch, J.F., Macadam, D.L., Schawlow, A.L., Shimoda, K., Tamir, T., Eds.; Springer: Berlin, Germany, 1987; pp. 1–4, ISBN 978-3-540-47744-0.
4. Poletti, F.; Petrovich, M.N.; Richardson, D.J. Hollow-core photonic bandgap fibers: Technology and applications. *Nanophotonics* **2013**, *2*, 315–340. [CrossRef]
5. Roberts, P.J.; Couny, F.; Sabert, H.; Mangan, B.J.; Williams, D.P.; Farr, L.; Mason, M.W.; Tomlinson, A.; Birks, T.A.; Knight, J.C.; et al. Ultimate low loss of hollow-core photonic crystal fibres. *Opt. Express* **2005**, *13*, 236–244. [CrossRef] [PubMed]

6. Wang, Y.Y.; Couny, F.; Roberts, P.J.; Benabid, F. Low loss broadband transmission in optimized core-shaped Kagome hollow-core PCF. In Proceedings of the Lasers Electro-Optics, Quantum Electron, Laser Science Conference, San Jose, CA, USA, 16–21 May 2010.

7. Wang, Y.Y.; Wheeler, N.V.; Couny, F.; Roberts, P.J.; Benabid, F. Low loss broadband transmission in hypocycloid-core Kagome hollow-core photonic crystal fiber. *Opt. Lett.* **2011**, *36*, 669–671. [CrossRef] [PubMed]

8. Pryamikov, A.D.; Biriukov, A.S.; Kosolapov, A.F.; Plotnichenko, V.G.; Semjonov, S.L.; Dianov, E.M. Demonstration of a waveguide regime for a silica hollow-core microstructured optical fiber with a negative curvature of the core boundary in the spectral region >3.5 μm. *Opt. Express* **2011**, *19*, 1441–1448 . [CrossRef] [PubMed]

9. Yu, F.; Wadsworth, W.J.; Knight. J.C. Low loss silica hollow core fibers for 3–4 μm spectral region. *Opt. Express* **2012**, *20*, 11153–11158. [CrossRef] [PubMed]

10. Wei, C.; Weiblen, R.J.; Menyuk, C.R.; Hu, J. Negative curvature fibers. *Adv. Opt. Photon.* **2017**, *9*, 504–561. [CrossRef]

11. Michieletto, M.; Lyngs, J.K.; Jakobsen, C.; Lgsgaard, J.; Bang, O.; Alkeskjold, T.T. Hollow-core fibers for high power pulse delivery. *Opt. Express* **2016**, *24*, 7103–7119. [CrossRef] [PubMed]

12. Wei, C.; Menyuk, C.R.; Hu, J. Polarization-filtering and polarization-maintaining low-loss negative curvature fibers. *Opt. Express* **2018**, *26*, 9528–9540. [CrossRef] [PubMed]

13. Kosolapov, A.F.; Pryamikov, A.D.; Biriukov, A.S.; Shiryaev, V.S.; Astapovich, M.S.; Snopatin, G.E.; Plotnichenko, V.G.; Churbanov, M.F.; Dianov, E.M. Demonstration of CO_2-laser power delivery through chalcogenide glass fiber with negative-curvature hollow core. *Opt. Express* **2011**, *19*, 2572–25728. [CrossRef] [PubMed]

14. Shiryaev, V.S. Chalcogenide glass hollow-core microstructured optical fibers. *Front. Mater.* **2015**, *2*, 24. [CrossRef]

15. Gattass, R.R.; Rhonehouse, D.; Gibson, D.; McClain, C.C.; Thapa, R.; Nguyen, V.Q.; Bayya, S.S.; Weiblen, R.J.; Menyuk, C.R.; Shaw, L.B.; et al. Infrared glass-based negative-curvature anti-resonant fibers fabricated through extrusion. *Opt. Express* **2016**, *14*, 25697–25703 . [CrossRef] [PubMed]

16. Wei, C.; Hu, J.; Menyuk, C.R. Comparison of loss in silica and chalcogenide negative curvature fibers as the wavelength varies. *Front. Phys.* **2016**, *4*, 30. [CrossRef]

17. Debord, B.; Amsanpally, A.; Chafer, M.; Baz, A.; Maurel, M.; Blondy, J.M.; Hugonnot, E.; Scol, F.; Vincetti, L.; Gérôme, F.; et al. Ultralow transmission loss in inhibited-coupling guiding hollow fibers. *Optica* **2017**, *4*, 209–217 . [CrossRef]

18. Debord, B.; Alharbi, M.; Bradley, T.; Fourcade-Dutin, C.; Wang, Y.Y.; Vincetti, L.; Gérôm, F.; Benabid, F. Hypocycloid-shaped hollow-core photonic crystal fiber Part I: Arc curvature effect on confinement loss. *Opt. Express* **2013**, *21*, 28597–28608. [CrossRef] [PubMed]

19. Yu, F.; Knight, J.C. Negative curvature hollow-core optical fiber. *IEEE J. Sel. Top. Quantum Electron.* **2016**, *22*, 4400610. [CrossRef]

20. Hu, J.; Menyuk, C.R. Understanding leaky modes: Slab waveguide revisited. *Adv. Opt. Photonics* **2009**, *1*, 58–106. [CrossRef]

21. Alagashev, G.K.; Pryamikov, A.D.; Kosolapov, A.F.; Kolyadin, A.N.; Lukovkin, A.Y.; Biriukov, A.S. Impact of geometrical parameters on the optical properties of negative curvature hollow core fibers. *Laser Phys.* **2015**, *25*, 055101. [CrossRef]

22. Poletti, F. Nested antiresonant nodeless hollow core fiber. *Opt. Express* **2014**, *22*, 23807–23828. [CrossRef] [PubMed]

23. Habib, M.S.; Bang, O.; Bache, M. Low-loss hollow-core silica fibers with adjacent nested anti-resonant tubes. *Opt. Express* **2015**, *23*, 17394–17406. [CrossRef] [PubMed]

24. Belardi, W.; Knight, J.C. Hollow antiresonant fibers with reduced attenuation. *Opt. Lett.* **2014**, *39*, 1853–1856. [CrossRef] [PubMed]

25. Kolyadin, A.N.; Kosolapov, A.F.; Pryamikov, A.D.; Biriukov, A.S.; Plotnichenko, V.G.; Dianov, E.M. Light transmission in negative curvature hollow core fiber in extremely high material loss region. *Opt. Express* **2013**, *21*, 9514–9519. [CrossRef] [PubMed]

26. Wei, C.; Menyuk, C.R.; Hu, J. Impact of cladding tubes in chalcogenide negative curvature fibers. *IEEE Photonics J.* **2016**, *8*, 2200509. [CrossRef]

27. Uebel, P.; Günendi, M.C.; Frosz, M.H.; Ahmed, G.; Edavalath, N.N.; Ménard, J.-M.; Russell, P.S.J. Broadband robustly single-mode hollow-core PCF by resonant filtering of higher-order modes. *Opt. Lett.* **2016**, *41*, 1961–1964. [CrossRef] [PubMed]

28. Liu, X.; Ding, W.; Wang, Y.Y.; Gao, S.; Cao, L.; Feng, X.; Wang, P. Characterization of a liquid-filled nodeless anti-resonant fiber for biochemical sensing. *Opt. Lett.* **2017**, *42*, 863–866. [CrossRef] [PubMed]

29. Wei, C; Menyuk, C.R.; Hu, J. Bending-induced mode non-degeneracy and coupling in chalcogenide negative curvature fibers. *Opt. Express* **2016**, *24*, 12228–12239. [CrossRef] [PubMed]

30. Saitoh, K.; Koshiba, M. Leakage loss and group velocity dispersion in air-core photonic bandgap fibers. *Opt. Express* **2003**, *11*, 3100–3109. [CrossRef] [PubMed]

31. Caillaud, C.; Renversez, G.; Brilland, L.; Mechin, D.; Calvez, L.; Adam, J.-L.; Troles, J. Photonic Bandgap Propagation in All-Solid Chalcogenide Microstructured Optical Fibers. *Materials* **2014**, *7*, 6120–6129. [CrossRef] [PubMed]

32. Wei, C.; Kuis, R.A.; Chenard, F.; Menyuk, C.R.; Hu, J. Higher-order mode suppression in chalcogenide negative curvature fibers. *Opt. Express* **2015**, *23*, 15824–15832. [CrossRef] [PubMed]

33. Belardi, W.; Knight, J.C. Negative curvature fibers with reduced leakage loss. In Proceedings of the Optical Fiber Communication Conference, San Francisco, CA, USA, 9–13 March 2014.

34. Weiblen, R.J.; Menyuk, C.R.; Gattass, R.R.; Shaw, L.B.; Sanghera, J.S. Fabrication tolerances in As_2S_3 negative-curvature antiresonant fibers. *Opt. Lett.* **2016**, *41*, 2624–2627. [CrossRef] [PubMed]

35. Hayes, J.R.; Sandoghchi, S.R.; Bradley, T.D.; Liu, Z.; Slavik, R.; Gouveia, M.A.; Wheeler, N.V.; Jasion, G.; Chen, Y.; Fokoua, E.N.; et al. Antiresonant hollow core fiber with an octave spanning bandwidth for short haul data communications. *J. Lightw. Technol.* **2017**, *35*, 437–442. [CrossRef]

![fibers logo] *fibers* MDPI

Article

Fabrication of Shatter-Proof Metal Hollow-Core Optical Fibers for Endoscopic Mid-Infrared Laser Applications

Katsumasa Iwai [1], Hiroyuki Takaku [1], Mitsunobu Miyagi [2], Yi-Wei Shi [3] and Yuji Matsuura [4,*]

[1] National Institute of Technology, Sendai College, Sendai 989-3128, Japan; iwai@sendai-nct.ac.jp (K.I.); info@do-ko.sakura.ne.jp (H.T.)
[2] Headquarters, Miyagi Gakuin, Sendai 981-8557, Japan; mmiyagi@mgu.ac.jp
[3] School of Information Science and Engineering, Fudan University, Shanghai 200433, China; ywshi@fudan.edu.cn
[4] Graduate School of Biomedical Engineering, Tohoku University, Sendai 980-8579, Japan
* Correspondence: yuji@ecei.tohoku.ac.jp; Tel.: +81-22-795-7108

Received: 27 March 2018; Accepted: 14 April 2018; Published: 18 April 2018

Abstract: A method for fabricating robust and thin hollow-core optical fibers that carry mid-infrared light is proposed for use in endoscopic laser applications. The fiber is made of stainless steel tubing, eliminating the risk of scattering small glass fragments inside the body if the fiber breaks. To reduce the inner surface roughness of the tubing, a polymer base layer is formed prior to depositing silver and optical-polymer layers that confine light inside the hollow core. The surface roughness is greatly decreased by re-coating thin polymer base layers. Because of this smooth base layer surface, a uniform optical-polymer film can be formed around the core. As a result, clear interference peaks are observed in both the visible and mid-infrared regions. Transmission losses were also low for the carbon dioxide laser used for medical treatments as well as the visible laser diode used for an aiming beam. Measurements of bending losses for these lasers demonstrate the feasibility of the designed fiber for endoscopic applications.

Keywords: hollow optical fiber; carbon dioxide laser; endoscopic laser applications

1. Introduction

Mid-infrared lasers are increasingly being used in medical applications because mid-infrared light is strongly absorbed by the water, proteins, and lipids in human tissue. Mid-infrared lasers can be used in combination with a flexible endoscope or catheter for minimally invasive treatments of tumors and other diseased tissue. Such applications require an optical fiber that is sufficiently thin and flexible for insertion into the working channel of endoscopes or thin catheters. The fiber must also be able to deliver laser light to the target tissue with minimal transmission loss.

Common silica-glass fibers cannot transmit mid-infrared light with wavelengths longer than 2 μm because of absorption in the silica-glass material [1], so many types of infrared optical fibers have been developed, such as chalcogenide-glass fibers [2,3], metal-halide polycrystalline fibers [4,5], and hollow-core optical fibers [6,7]. Hollow-core optical fibers that confine light in an air core have some advantages over solid-core fibers for high-power laser delivery. Two types of hollow-core optical fibers have been developed so far. One is composed of glass capillary tubing with a silver film on the inside and a dielectric thin film on top of the silver [7–9] and this type of fiber is already used in medical applications [10]. The other is made entirely of glass and utilizes photonic crystal structures to confine light in the central air core [11,12].

Recently, an endoscopic submucosal dissection (ESD) process using a carbon dioxide (CO_2) laser has been developed for treating early-stage gastric cancer [13], and this process uses a hollow-glass optical fiber to deliver the CO_2 laser light. ESD is performed using a fiber (inner diameter 700 µm, length 2.65 m) that transmits laser light with power up to 12 W. When using such a common hollow-core optical fiber for endoscopic applications, a surgeon must be careful not to break the fiber by bending it too sharply because the breakage may scatter small glass fragments inside the body. This damage could be fatal. Glass shards can be avoided by using plastic tubing for the hollow-core optical fiber [14–16], but the power capacity of plastic-based hollow fibers is usually lower than that of glass-based fibers, especially when the fiber is bent.

Temelkuran et al. have developed a hollow waveguide with polymer dielectric multilayers that confines light by Bragg reflection [17]. Although this type of fiber transmits CO_2 laser light of relatively high power with low losses [18], the chalcogenide material used for cladding may cause trouble in endoscopic applications.

Another option for fabricating a robust and flexible hollow-core optical fiber is to use metal tubing as the base material [19,20]. However, the inner surface of metal tubing is usually much rougher than that of glass tubing, even after chemical etching. This roughness causes scattering loss of the transmitted light. In clinical applications of infrared lasers, delivery of a visible targeting laser beam along with mid-infrared laser is necessary to make the irradiated spot visible to the surgeon. When the inner surface of the hollow core is relatively rough, the visible light is strongly scattered and does not form a targeting dot.

In this paper, we propose a fabrication method for a stainless steel hollow-core optical fiber for mid-infrared light. To reduce the roughness of the tubing inner surface, a polymer base layer is formed before depositing a silver film. This base coating smooths the surface, and as a result, the transmission loss is reduced for both CO_2 laser light and visible light.

2. Material and Methods

Figure 1 shows a schematic of the proposed hollow-core optical fiber based on stainless-steel (SUS) tubing. The fiber begins as commercially available SUS tubing with an electrochemically polished inner surface. Although the root-mean-square (RMS) roughness of the surface is as low as 0.3 to 0.9 µm, it still causes scattering losses, especially for visible light. To smooth the inner surface of the SUS tubing, we form a relatively thick base coat of polymer on the surface. Then, a silver (Ag) film is coated on top of the base layer using the silver mirror reaction. Then, another polymer layer is formed on the top of silver layer and functions as an optical interference film. This polymer needs to be transparent to mid-infrared light since this layer optically functions as a dielectric film, so we use cyclic olefin polymer (COP) [21].

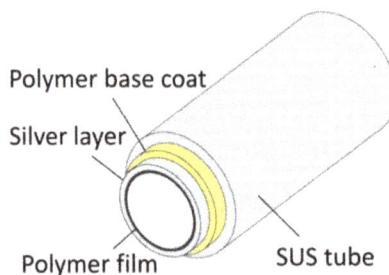

Figure 1. Schematic structure of the stainless-steel (SUS)-based hollow optical fiber.

As the base-coating material, a commonly used two-liquid-reaction acrylic-silicone resin (AlcoSP, NATOCO, Aichi, Japan) was chosen. We chose this material partly because we found that it gives excellent adhesion to the silver film formed on top of the base coating [22]. The resin solution was made to flow through the SUS tubes (inner/outer diameter 550/780 µm) at 8 cm/min. Then nitrogen gas was flushed through the tubes at the rate of 50 mL/min for 1 h at room temperature to dry the resin coating. By this coating process, the resin film with a thickness of around 0.5 μm is formed. When forming the acrylic-silicone resin film, we found that re-coating several thin films yields a smoother surface than the thick film formed with a single coating. Based on the results of preliminary experiments, we chose a resin concentration of 45.5 wt % for the resin base coating.

Before the silver coating process, the surface of the resin layer was sensitized with $SnCl_2$ solution to increase the adhesion of Ag film. A silver film was formed on the base coat by passing silver nitrate solution and reducing solution through the tube for 3 min. In the next step, we formed a COP thin film on the top of silver layer. The silver layer is usually only around 0.2 μm thick, so the cyclohexane solvent used to dissolve the COP may penetrate to the base layer through the thin silver film. In a preliminary test of whether the cyclohexane solvent damages the base coat, we removed the COP layer using cyclohexane and formed the COP again, comparing the loss spectra of the first and second COP-coated fibers.

In the measurement of the loss spectra in the visible to near-infrared region from 0.4 to 1.6 μm wavelength, of the fabricated fibers (550 μm inner diameter and 1 m length), an optical spectrum analyzer was used (AQ6315A, Yokogawa, Tokyo, Japan). Light from a halogen lamp was coupled to the fiber through a multimode silica-glass fiber with a core diameter of 400 μm. The output light from the tested fiber was then delivered to the spectrum analyzer using a silica-glass fiber with a core diameter of 600 μm. For measurement of the loss spectra in the mid-infrared region (2–12 μm), we used a Fourier transform infrared spectroscope (FT-IR) (FT/IR-350, JASCO, Tokyo, Japan). In this measurement setup, light from the FT-IR was coupled to the measured fiber by focusing the light with an off-axis mirror of focal length 50 mm. The light output from the fiber was detected using a liquid N_2-cooled HgCdTe detector (MCT, JASCO, Tokyo, Japan).

Bending losses of the fabricated fibers were measured with a CO_2 laser (LezawinCHS, J. MORITA, Kyoto, Japan) of wavelength 10.6 μm and a green laser diode with wavelength of 532 nm. In the experiment, the laser light was firstly injected into a short hollow-core optical fiber (length 15 cm, inner diameter 530 μm) that functions as a mode filter. The tested fibers were butt-coupled to this short fiber. In the bending test, the first 25 cm of the fiber was kept straight, and the middle part was bent to different bending angles with a bending radius of 20.25 cm. The output power was measured with a laser power meter.

3. Results and Discussion

Figure 2 shows the surface roughness of the films as observed by an atomic force microscope (AFM) (AFM5100N, Hitachi, Tokyo, Japan) with two and four base coats applied. For this observation, we made samples by cutting base-coated SUS tubes into small pieces. The roughness of different points were measured 5–12 times for each sample, and the mean values, excluding the maximum and minimum values, are plotted in the figure. The error bars show the uncertainty in the measurements. The RMS roughness of the inner surface of the original SUS tube was as large as 0.6–0.9 μm, and it was made drastically smoother by the base coat. One can see that the roughness decreased with each thin film coat, and we found that the minimum mean roughness of around 70 nm was obtained with four coats. Figure 3 shows typical AFM images of the inner surfaces of SUS tubing after applying two and four re-coatings. One can see that the inner surface of the SUS tubing was smoothed by applying the resin base coating. As mentioned above, the thickness of the resin coating formed by a single process is around 0.5 μm and thus, the total thickness of the film formed by four re-coatings is around 2 μm.

Figure 4 shows changes in the loss spectra in the visible to near-infrared regions of fibers fabricated with the COP film. In this experiment, light from white light source was coupled to the hollow optical fibers via a graded-index, silica-glass fiber. When such an incoherent light is injected into a hollow optical fiber, many high-order modes are excited in the fiber and as a result, the attenuation losses become high such as shown in Figure 5a. Firstly a COP film was formed on the Ag layer and the loss spectrum was measured. Then the COP film was removed using cyclohexane solvent, and another COP film was formed on the Ag after the solvent was evaporated. We observed clear interference peaks resulting from the high uniformity of the COP film in both spectra, which shows that the acrylic-silicone base coating is not affected by the cyclohexane solvent. These clear interference fringes allow low transmission loss of the visible-wavelength target laser. In this figure, the loss spectrum of a fiber without COP film is also shown for comparison. Please note that losses of the fiber without COP film are lower than those with COP film in visible and near-infrared region. This is because silver itself provides a high reflectance in visible region and the COP coating on the top of silver somewhat reduce the reflectance of silver. However, in mid-infrared region, this relationship is inverted because silver does not show a high reflectance in the mid infrared and therefore, the COP film is essential for the fibers to obtain a low transmission loss for mid-infrared lasers.

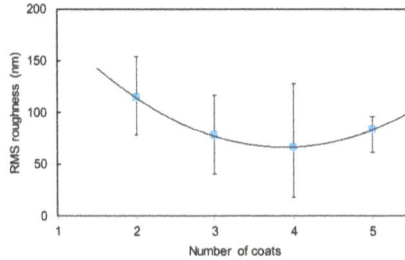

Figure 2. Surface roughness of inner surface of SUS tubing as a function of number of coats of the acrylic resin.

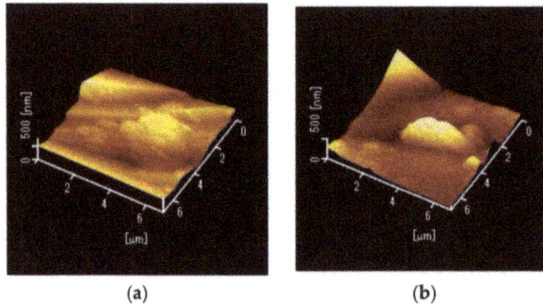

Figure 3. Inner surface of SUS tubing observed by atomic force microscopy (AFM) after applying two and four layers of base coating. The RMS roughness are 119 nm for two coats and 83 nm for four coats. (**a**) two times re-coating; (**b**) four times re-coating.

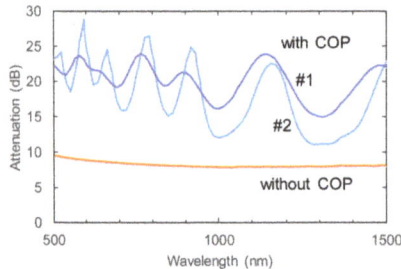

Figure 4. Loss spectra of cyclic olefin polymer (COP)-coated hollow fibers in the visible to near-infrared region before and after removal and re-coating of the COP film.

Figure 5a shows an example attenuation-loss spectrum of the COP-coated hollow-core optical fiber based on SUS tubing, measured in the visible region. The thickness of the COP film was finely tuned to match the interference fringe with the target wavelength of 532 nm that is used as the visible aiming beam of the medical CO_2 laser system. The thickness of COP film as estimated from the spectrum was 0.97 μm. Figure 5b is the loss spectrum of the COP-coated fiber measured in the mid-infrared region. The loss spectrum of a fiber without COP film is also shown for comparison. We confirmed that, with a COP film of the above thickness, two interference peaks appear around 2.2 μm and 4.4 μm and that a low-loss region was obtained owing to the interference effect of the COP film in the wavelength region of 9–11 μm. Although some sharp absorption peaks appear for the COP in the mid-infrared region, we confirm that no peak appears at the CO_2 laser wavelength of 10.6 μm.

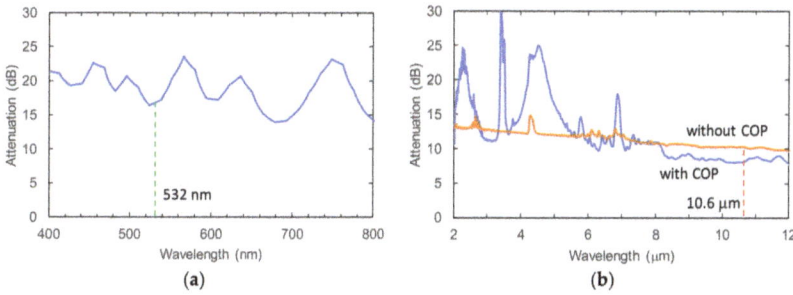

Figure 5. Loss spectra of the COP-coated SUS fibers in (**a**) visible region and (**b**) mid-infrared region.

Figure 6 shows the bending losses of the COP-coated fibers based on SUS tubing measured with (a) a 10.6-μm CO_2 laser and (b) a 532-nm laser diode. In this test, the laser light was coupled to the measured fiber via a short (15-cm long) hollow optical fiber. As mentioned above, this short fiber tip was used as a mode coupler to eliminate high-order modes. As a result, the low order modes are efficiently excited in the measured fiber and therefore, the measured losses become much lower than those in the loss spectra (Figures 4 and 5) measured with an incoherent light.

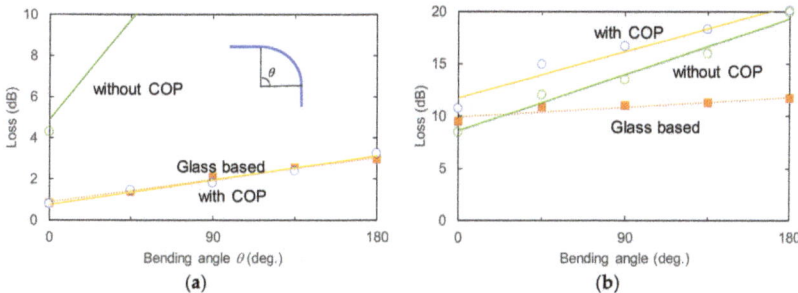

Figure 6. Bending losses of the SUS COP-coated fibers measured with (**a**) 10.6-μm CO_2 laser and (**b**) 532-nm laser. The inner diameter of the fiber is 550 μm and the length is 1 m.

In Figure 6a,b, the measured bending losses of a COP-coated fiber based on silica-glass capillary tubing are also shown for comparison. For mid-infrared CO_2 laser light (Figure 6a), the straight loss of the COP-coated fiber was 0.75 dB. This loss increases to 3.2 dB when the fiber is bent 180 degrees. The measured losses of the SUS fiber coincide with those of the silica-glass based fiber. This result confirms that the acrylic-silicone resin coating functions well as a base layer that smooths the inner surface of SUS tubing. The losses of the fiber without a COP layer are much higher than the COP-coated fiber, and therefore, we find that a highly uniform COP layer is formed inside the SUS tubing with the help of the resin base coating.

For the visible laser, as shown in Figure 6b, the transmission loss is increased by the COP film formed on top of the silver layer because the reflection of silver is itself very high in the visible region.

However, we confirm that the COP film does not largely affect the bending loss when the thickness is optimized, as shown in Figure 5a. Although the loss in the 180-degree bent fiber is as high as 20 dB, the green light transmitted through the fiber was clearly visible, confirming the feasibility of a COP-coated fiber based on SUS tubing for endoscopic applications in which a visible beam is needed at the target end. We also performed mechanical bending tests for samples of the fabricated fibers. The results of bending tests showed that the minimum elastic bending radius is 85 mm and that the fiber starts to snap at a bending radius of 2.5 mm.

The measured losses shown in Figure 6 are summarized in Table 1. Comparing the losses of the SUS-based and the glass-based fibers with COP coating, the losses of the SUS-based fibers are still larger than those of the glass-based fibers for the wavelength of 532 nm because of the larger inner-surface roughness of the SUS-based fibers (70 nm in RMS for the SUS-based fibers and 10 nm in RMS for the glass-based fibers). However, for the mid-infrared laser with a longer wavelength of 10.6 μm, the effect of roughness becomes smaller and the SUS-based fibers show low losses that are comparable to those of the glass-based fibers.

Table 1. Measured losses of SUS-based and glass-based hollow optical fibers.

Fiber Type		Straight Loss (dB/m)		Additonal Loss by 90-Deg Bending (dB)	
		λ = 532 nm	10.6 μm	532 nm	10.6 μm
SUS based	With COP	10.6	0.75	5.9	1.0
	Without COP	8.4	4.3	5.0	8.9
Glass based	With COP	9.4	0.8	1.5	1.2

4. Conclusions

We have proposed a fabrication method for a robust and thin hollow-core optical fibers based on SUS tubing. This fiber is sufficiently flexible and optically robust for endoscopic mid-infrared laser applications. To reduce the inner surface roughness of the SUS tubing, we formed a polymer base layer before depositing silver and optical-polymer layers. By re-coating the resin base layer, the inner surface of the SUS tubing was made smooth enough for an effective optical layer to be formed. Clear interference peaks were observed in both the visible and the mid-infrared regions, so the hollow-core fibers promise low transmission losses for both the CO_2 laser and the visible laser diode used in endoscopic procedures.

So far, we have succeeded to fabricate the SUS-based hollow optical fibers with inner/outer diameters 550/780 μm only. However, it is technically possible to make fibers with larger diameters up to 1000 μm although the flexibility will be limited low for the fibers with such a large diameter. For smaller diameters, we have not succeeded to fabricate fibers with inner diameter smaller than 550 μm because of relatively high viscosity of the acrylic-silicone resin solution, which makes it difficult to form a uniform base coat for the fibers with small diameters. The maximum length of the fibers fabricated by the proposed method is also limited to 1 m for the same reason at this stage.

Though the SUS hollow-core optical fibers showed low bending losses for both mid-infrared and visible laser lights, the minimum elastic bending radius was limited to 85 mm because the diameter of the SUS tube is still relatively large at 780 μm. Although this diameter is small enough to be inserted into the working channel of endoscopes used for laser ESD, a bending radius of around 15 mm is sometimes necessary for these applications. To allow this degree of bending, we are working toward the fabrication of SUS fibers with inner/outer diameters 300/450 μm by optimizing the coating conditions of base coating.

Acknowledgments: This research is partly supported by JSPS KAKENHI Grant Numbers JP15K06045 and JP16K06329.

Author Contributions: Katsumasa Iwai and Hiroyuki Takaku performed the experiments and analyzed the data; Mitsunobu Miyagi and Yi-Wei Shi conceived and designed the experiments; Yuji Matsuura wrote the paper.

Conflicts of Interest: The authors declare no conflict of interest.

References

1. Harrington, J.A. *Infrared Fibers and Their Applications*; SPIE: Bellingham, WA, USA, 2004; ISBN 9780819452184.
2. Tang, Z.; Shiryaev, V.S.; Furniss, D.; Sojka, L.; Sujecki, S.; Benson, T.M.; Seddon, A.B.; Churbanov, M.F. Low loss Ge-As-Se chalcogenide glass fiber, fabricated using extruded preform, for mid-infrared photonics. *Opt. Mater. Express* **2015**, *5*, 1722–1737. [CrossRef]
3. Kanamori, T.; Terunuma, Y.; Takahashi, S.; Miyashita, T. Chalcogenide glass fibers for mid-infrared transmission. *J. Lightw. Technol.* **1984**, *2*, 607–613. [CrossRef]
4. Israeli, S.; Katzir, A. Attenuation, absorption, and scattering in silver halide crystals and fibers in the mid-infrared. *J. Appl. Phys.* **2014**, *115*, 023104. [CrossRef]
5. Artyushenko, V.; Bocharnikov, A.; Colquhoun, G.; Leach, C.; Lobachev, V.; Sakharova, T.; Savitsky, D. Mid-IR fibre optics spectroscopy in the 3300–600 cm^{-1} range. *Vib. Spectrosc.* **2008**, *48*, 168–171. [CrossRef]
6. Croitoru, N.; Dror, J.; Gannot, I. Characterization of hollow fibers for the transmission of infrared radiation. *Appl. Opt.* **1990**, *9*, 1805–1809. [CrossRef] [PubMed]
7. Matsuura, Y.; Abel, T.; Harrington, J. Optical properties of small-bore hollow glass waveguides. *Appl. Opt.* **1995**, *34*, 6842–6847. [CrossRef] [PubMed]
8. Radii, C.D.; Harrington, J.A. Mechanical properties of hollow glass waveguides. *Opt. Eng.* **1999**, *38*, 1490–1499.
9. Iwai, K.; Miyagi, M.; Shi, Y.; Zhu, X.; Matsuura, Y. Infrared hollow fiber with a vitreous film as the dielectric inner coating layer. *Opt. Lett.* **2008**, *32*, 3420–3422. [CrossRef]
10. Darafsheh, A.; Melzer, J.; Harrington, J.; Kassaee, A.; Finlay, J. Radiotherapy fiber dosimeter probes based on silver-only coated hollow glass waveguides. *J. Biomed. Opt.* **2018**, *23*, 015006. [CrossRef] [PubMed]
11. Kolyadin, A.N.; Kosolapov, A.F.; Pryamikov, A.D.; Biriukov, A.S.; Plotnichenko, V.G.; Dianov, E.M. Light transmission in negative curvature hollow core fiber in extremely high material loss region. *Opt. Express* **2013**, *21*, 9514–9519. [CrossRef] [PubMed]
12. Urich, A.; Maier, R.R.J.; Yu, F.; Knight, J.C.; Hand, D.P.; Shephard, J.D. Silica hollow core microstructured fibres for mid-infrared surgical applications. *J. Non-Cryst. Solids* **2013**, *377*, 236–239. [CrossRef]
13. Obata, D.; Morita, Y.; Kawaguchi, R.; Ishii, K.; Hazama, H.; Awazu, K.; Kutsumi, H.; Azuma, T. Endoscopic submucosal dissection using a carbon dioxide laser with submucosally injected laser absorber solution (porcine model). *Surg. Endosc.* **2013**, *27*, 4241–4249. [CrossRef] [PubMed]
14. Alaluf, M.; Dror, J.; Dahan, R.; Croitoru, N. Plastic hollow fibers as a selective infrared radiation transmitting medium. *J. Appl. Phys.* **1992**, *72*, 3878–3883. [CrossRef]
15. George, R.; Harrington, J.A. Infrared transmissive, hollow plastic waveguides with inner Ag–AgI coatings. *Appl. Opt.* **2005**, *44*, 6449–6455. [CrossRef] [PubMed]
16. Nakazawa, M.; Shi, Y.W.; Matsuura, Y.; Iwai, K.; Miyagi, M. Hollow polycarbonate fiber for Er:YAG laser light delivery. *Opt. Lett.* **2006**, *31*, 1373–1375. [CrossRef] [PubMed]
17. Temelkuran, B.; Hart, S.D.; Benoit, G.; Joannopoulos, J.D.; Fink, Y. Wavelength-scalable hollow optical fibres with large photonic bandgaps for CO$_2$ laser transmission. *Nature* **2002**, *420*, 650–653. [CrossRef] [PubMed]
18. Torres, D.; Weisberg, O.; Shapira, G.; Anastassiou, C.; Temelkuran, B.; Shurgalin, M.; Jacobs, S.A.; Ahmad, R.U.; Wang, T.; Kolodny, U.; et al. OmniGuide photonic bandgap fibers for flexible delivery of CO$_2$ laser energy for laryngeal and airway surgery. In Proceedings of the Photonic Therapeutics and Diagnostics, San Jose, CA, USA, 25 April 2005; Volume 5686, pp. 310–321.
19. Iwai, K.; Hongo, A.; Takaku, H.; Miyagi, M.; Ishiyama, J.; Wu, X.X.; Shi, Y.W.; Matsuura, Y. Fabrication and transmission characteristics of infrared hollow fiber based on silver-clad stainless steel pipes. *Appl. Opt.* **2009**, *48*, 6207–6212. [CrossRef] [PubMed]
20. Iwai, K.; Takaku, H.; Miyagi, M.; Shi, Y.W. Improvement of transmission properties of visible light for Ag hollow fiber based on stainless tube. *Rev. Laser Eng.* **2016**, *44*, 684–687.
21. Abe, Y.; Matsuura, Y.; Shi, Y.; Wang, Y.; Uyama, H.; Miyagi, M. Polymer-coated hollow fiber for CO$_2$ laser delivery. *Opt. Lett.* **1998**, *23*, 89–90. [CrossRef] [PubMed]

22. Iwai, K.; Takaku, H.; Miyagi, M.; Shi, Y.W.; Matsuura, Y. Silver hollow optical fibers with acrylic silicone resin coating as buffer layer for sturdy structure. In Proceedings of the Optical Fibers and Sensors for Medical Diagnostics and Treatment Applications XVI, San Francisco, CA, USA, 13–14 February 2016; Volume 9702, p. 97020Z.

fibers

MDPI

Article

Combining Hollow Core Photonic Crystal Fibers with Multimode, Solid Core Fiber Couplers through Arc Fusion Splicing for the Miniaturization of Nonlinear Spectroscopy Sensing Devices

Hanna Izabela Stawska *,†, Maciej Andrzej Popenda *,† and Elżbieta Bereś-Pawlik

Department of Telecommunications and Teleinformatics, Wroclaw University of Science and Technology, 50-370 Wroclaw, Poland; elzbieta.pawlik@pwr.edu.pl
* Correspondence: hanna.stawska@pwr.edu.pl (H.I.S.); maciej.popenda@pwr.edu.pl (M.A.P.); Tel.: +48-71-340-7642 (H.I.S. & M.A.P.)
† These authors contributed equally to this work.

Received: 15 September 2018; Accepted: 4 October 2018; Published: 11 October 2018

Abstract: The presence of fiber optic devices, such as couplers or wavelength division multiplexers, based on hollow-core fibers (HCFs) is still rather uncommon, while such devices can be imagined to greatly increase the potential of HCFs for different applications, such as sensing, nonlinear optics, etc. In this paper, we present a combination of a standard, multimode fiber (MMF) optic coupler with a hollow core photonic bandgap fiber through arc fusion splicing and its application for the purpose of multiphoton spectroscopy. The presented splicing method is of high affordability due to the low cost of arc fusion splicers, and the measured splicing loss (*SL*) of the HCF-MMF splice is as low as (0.32 ± 0.1) dB, while the splice itself is durable enough to withstand a bending radius (r_{bend}) of 1.8 cm. This resulted in a hybrid between the hollow core photonic bandgap fiber (HCPBF) and MMF coupler, delivering 20 mW of average power and 250-fs short laser pulses to the sample, which was good enough to test the proposed sensor setup in a simple, proof-of-concept multiphoton fluorescence excitation-detection experiment, allowing the successful measurement of the fluorescence emission spectrum of 10^{-5} M fluorescein solution. In our opinion, the presented results indicate the possibility of creating multi-purpose HCF setups, which would excel in various types of sensing applications.

Keywords: microstructured optical fiber splicing; optical fiber sensors; hollow core fibers; photonic crystal fibers; multiphoton fluorescence spectroscopy

1. Introduction

The appearance of hollow core fibers (HCFs) about two decades ago [1] revealed new and versatile opportunities for the investigation of light-matter interactions. Owing to their unique structures and optical properties, HCFs successfully found application in many fields of science and techniques, including biological, chemical and environmental sensing. HCFs have been used in different kinds of sensing and measurement applications, i.e., temperature measurement [2–5], gas [6–9], strain [10], magnetic field [11], hydrostatic pressure [12], flying particle [13,14] and plasmonic sensors [15].

In HCF, light is confined in the hollow core and, depending on the cladding structure of the fiber, two basic propagation mechanisms are used to explain their waveguide properties—photonic bandgap (PB) effect and the anti-resonant reflective optical waveguide (ARROW) mechanism. These propagation models are not the only ones used to describe optical parameters of HCFs [16]—indeed, in the literature one can find, e.g., two others, namely omnidirectional reflection [17–19] and effective medium reflection [20]—but fibers whose way of operation can be explained using such propagation models are less common. In case of hollow core photonic bandgap fibers (HCPBF) their cladding forms a photonic crystal, which is a periodically structured (lattice-like) medium with the lattice constant being on the order of the wavelength of the light. In these kind of fibers the light is trapped in the defect of the structure which constitutes the air core. However, photonic bandgaps only occur for a limited range of wavelengths, which means that the transmission bandwidth of HCPBF is also limited [21,22]. Unlike in the case of HCPBF, the waveguide properties of anti-resonant fibers (ARFs) result from constructive interference of the radiation that is reflected from the core–cladding interface [23]. Up to now, different types of ARFs have been proposed, e.g., Kagomé HCF with hypocycloid core contour [24], Kagomé HCF with hypocycloid core contour with modified shape of core [25], revolver fiber (RF) [23], double revolver fiber [26], square-core hollow fibers [27], HCF with anisotropic, anti-resonant elements [28] and HCF with lotus-shaped core [29]. The common feature of all these fibers is the negative-curvature core shape, hence their name—negative curvature HCFs (NCHCFs).

As a result of remarkable progress in the manufacturing technology of HCFs and continuous development of different structures, HCFs are capable of guiding light with remarkably low levels of optical loss and nonlinearity. Additionally, this type of structure offers remarkable diversity, resulting in endless possibilities for engineering their dispersion, birefringence, effective mode field diameter, dispersion, transmission bandwidth, etc. [20,30–35]. All these attributes enable HCFs to resolve several challenging issues, such as delivering ultrashort, high power signals over a wide range of wavelengths to the sample under investigation and/or collecting low-level response signals. In medical applications, HCFs are a promising candidate for developing new kinds of biomedical devices, suitable, for example, for the identification of different diseases. Generally, these considerations motivate research efforts to take advantage of the intrinsic properties of HCFs and transform them into multifunctional biosensors. One example of such a sensor is the "lab-on-a-fiber" (LOF), which combines on-chip nanophotonic biosensors with optical fibers [36,37]. One can distinguish three different classes of LOF technology, depending on the specific location where functional materials are integrated, and interaction between light and this material may take place. According to [36], these three classes are: "lab around fiber" (LAF) devices, where materials under investigation are placed onto the outer surface of the fiber (i.e., around the fiber's main axis); "lab on tip" (LOP) devices, where the material is integrated with the distal end of the fiber; and "lab in fiber" (LIF) devices, where e.g., fluidic or gaseous material is placed inside the HCF or holey structure of microstructured optical fiber. Another interesting group is optical fiber devices which employ additional optical elements (e.g., lenses) in order to test or image a sample at a given distance. We will refer to these as "remote fiber lab" (RFL). Considering the application of HCFs, the most straightforward choice of the form of the developed devices is LIF or RFL, which results from the structure of the HCF itself, and from the possibility of contamination of the examined fiber's core. For the purpose of LIF, NCHCF were recently used for the optical detection of chemical and biological analytes [38,39]. It was shown that after filling the fiber with aqueous solutions its waveguiding properties were preserved, namely the transmission bandwidth from 540 to 1700 nm, the confinement loss at the level of 0.1 dB/m, and the single mode guidance. Due to the moderate core diameter (32 μm), it was possible to obtain a large analyte–light overlap integral and a fast liquid flow rate. Thus, liquid filled NCHCF can be considered for creating all-fiber, multifunctional optofluidic devices allowing a wide range of applications, i.e., Raman spectroscopy, UV spectroscopy, non-invasive biochemical analysis and/or interferometric sensing. In the case of UV spectroscopy, LIF devices based on HCF can be useful in many applications, e.g., for monitoring water quality [40] and for the real-time monitoring of isoprene in breath [41].

In particular, the application of HCFs as Raman scattering probes has attracted the attention of researchers due to the fact that Raman spectroscopy is a non-destructive, nonlinear technique that provides information about the molecular structure of the sample [42–47]. A key limitation of the Raman effect is its weak signal. The simplest way to increase the response signal is to use high-power laser light and a longer acquisition time, however this exposes biological samples to the risk of damage. It has been shown that the application of HCF can effectively increase the Raman signal due to the larger analyte-light interaction area and relatively low optical losses [43,45]. It has additionally been shown that Raman signals can be enhanced with metal nanoparticles. This technique is known as Surface Enhanced Raman Scattering (SERS) [48] and enhances the Raman signal by a factor of 10^4–10^8, enabling the detection of molecules even at a single-molecule scale [49]. A combination of SERS with HCFs has been successfully implemented in biomedical applications, e.g., for monitoring leukemia cells [50], the detection of serological liver cancer biomarkers [51], human breath analysis [52] and real time monitoring of heparin concentration in serum [53]. An additional advantage of using HCF with SERS is the extremely low sample volume required, about 20 nl, which is desirable in clinical diagnostics [49].

Biosensing via Raman scattering can be also implemented by means of an RFL device. In this case, HCF usually serves to guide the excitation signal to the sample and/or response signal from the sample. The application of HCF is especially justified in the case of Raman endoscopy. HCF can guide ultrafast, high-power signals with negligible group velocity dispersion (GVD) and losses, which are their key advantages over solid-core fibers. Additionally, it was shown that no Coherent Anti-Stokes Raman Scattering (CARS) or Stimulated Raman Scattering (SRS) signals were generated within the HCF, leading to excellent image quality. In 2011, Brustlein et al. [54] demonstrated, for the first time, the deployment of double-clad HCF to perform CARS and SRS in an endoscopy-like scheme. Moreover, in 2018, Lombardini et al. [55] presented a flexible fiber optic scanning endoscope dedicated to high-resolution coherent Raman imaging. They used double-clad HCF with Kagomé lattice that enables guiding both excitation signals and response signal, and a combination of microsphere and miniature objective lens was used to focus the excitation beam onto the sample and collect the response signal.

Furthermore, in 2018, Yerolatsitis et al. [56] showed an interesting RF which has an additional outer ring comprising of eight solid multimode cores. Owing to the simplified structure of the RF, the background Raman emission from silica was over 1000 times lower than that of a conventional, solid-core fiber. The collection efficiency was similar to that received by means of the solid fiber, but without the need to use other fibers or distal optics. Furthermore, compared to the other setups with HCF, the generated silica background was an order of magnitude smaller [56–59].

HCFs are additionally useful also in other nonlinear imaging and spectroscopy techniques, such as multiphoton excited fluorescence (MPEF) and higher harmonic generation. These techniques require ultrashort pulses to be delivered to the sample with the lowest possible temporal distortions, and HCFs fulfil these requirements. For example, in 2004, Tai et al. [60] demonstrated a HCPBF-based, two-photon fluorescence microscope for acquiring fluorescence images of mesophyll tissue in the leaf of *Rhaphidophora aurea*. Moreover, in 2014, a chirped HCF was applied in multiphoton imaging system in order to obtain autofluorescence images of the yew leaf, mouse tendon and human skin [61].

In 2016, Sherlock et al. [62] presented an NCHCF as a good candidate for delivering excitation pulses for two–photon microscopy. The NCHCF was also applied in the multiphoton fluorescence setup, where it served as the excitation beam delivery fiber, while the MPEF was collected with four surrounding plastic optical fibers [63].

HCFs have also found application in other fields of medical diagnostics, such as multi-element HCF for infrared thermal imaging [64]. Infrared thermal imaging is a non-invasive, real-time method that is useful in, e.g., early detection of breast cancer and other malignant tumors [64,65].

To take full advantage of the potential of HCFs, they must be integrated with other components and devices for creating systems with the required functionality. Due to the very limited availability of most of the optical components (fiber couplers, microlenses, Bragg gratings, polarization controllers, etc.) there is still a problem with developing all-fiber, compact, portable medical equipment based on HCFs. Nevertheless, the fabrication and development of HCF-based optical elements is of great interest. For example, long-period Bragg gratings based on HCF have been fabricated using CO_2 laser [66–68] and using the pressure-assisted electrode arc discharge (EAD) technique [69]. Additionally, other HCF optical components have been demonstrated, e.g., polarization controllers and polarizers [70–74]. Although the idea of integrating microlenses with photonic crystal fibers (PCFs) is not new, having been demonstrated in [75–77], in the case of HCFs it is not so obvious due to the microstructure collapse. As an alternative approach, in our recent work we have proposed to combine a short segment of the standard fiber ended with microlens with HCF. It was shown that attaching such a fiber segment has negligible influence on the broadening of the ultrafast signal which propagates through the system designed in such a way [78]. It is also worth noting that there has been fruitful research in developing low-loss splicing methods between HCFs and standard fibers. In 2005, Xiao et al. [79] presented very valuable work on the topic of selective injection of microstructured optical fibers (MOFs), in which they worked with a classic, honeycomb cladding HCPBF and a conventional arc fusion splicer. Using the previous works of Tachikura [80] and Yablon [81] on the topic of arc current and energy distribution in the vicinity of the splicer's electrodes, as well as general observations and remarks regarding fusion splicing of MOFs [82–84], they studied the effect of microstructure collapse under different values of fusion arc current, fusion duration and fusion offset. This idea was further developed by Thapa et al. [85], who, according to our knowledge, were the first ones to present a reliable, low-loss arc fusion splicing of HCPBFs and solid core, single-mode fibers (SMFs). Since then, the technology of splicing different types of MOFs has been developing, and numerous approaches can be found [86,87]. These results suggest the possible appearance of HCPBF-based optical elements, such as fiber couplers. However, although some attempts have been made to design and fabricate such devices [88,89], they are not fully integrated with the fibers, i.e., the use of bulk optics is necessary to connect such a coupler with input/output HCFs.

To overcome these obstacles, we here demonstrate a combination of HCPBF and Y-type MMF coupler, and its application in a simple, two-photon excited fluorescence (TPEF) fiber sensor setup. Both elements are combined via a standard fusion splicing technique, allowing the creation of a hybrid, all-fiber-optic device for the purpose of TPEF excitation and detection. The sensing tip is reduced to the size of a single fiber, while the coupler itself allows for the division of excitation and emission signals without the use of a dichroic mirror. Additionally, the TPEF sensing tip is equipped with a microlens, providing the focusing of the excitation signal, required for the increased efficiency of the TPEF. The influence of the HCPBF-MMF coupler hybrid on the temporal shape of the transmitted ultrashort laser pulses is also determined. Finally, the proposed fiber-optic setup is used in a simple, multiphoton fluorescence spectroscopy experiment.

2. Materials and Methods

The main goal in the paper was to present a truly all-fiber-optic TPEF sensor setup, which requires a reliable connection between the HCPBF and MMF elements. The latter was provided by the fiber fusion splicing procedure, performed with a conventional arc fusion splicer (FSU 975, Ericsson, Stockholm, Sweden). The splicing procedure of HCPBF and MMF is a challenging task on its own, and is further discussed in Section 3.1.1 of this paper. The HCPBF (HC-800-02, NKT Photonics, Birkerød, Denmark) was spliced with one of the arms of a Y-type, multimode, 50:50 coupling ratio (CR, at 850 and 1300 nm) fiber optic coupler (Cellco, Kobylanka, Poland). It should be noted that the coupler's CR was different at 780 nm, and to determine its performance at the spectral windows of interest, namely 770–790 nm (excitation window, W_{ex}) and 500–650 nm (TPEF window, W_{fluo}), a white light source (AQ4305, Yokogawa, Tokyo, Japan) was coupled into the coupler's common port, and the output spectra of the remaining ports were recorded with a USB spectrometer (USB-2000, OceanOptics, Largo, FL, USA). The obtained values of CR were ~80:20 for both W_{ex} and W_{fluo}, and the arm with higher coupling efficiency was chosen to couple with the excitation signal, hence its given name—excitation arm (EA). The fusion splicer was also used to create a microlens at the tip of the fiber coupler's common arm, which helped in increasing the efficiency of both the excitation and collection of the TPEF signal. To characterize the optical performance of the fabricated microlensed fiber tip (MFT), its output beam profile was measured with a beam profiler (BP109-IR, Thorlabs, Newton, NJ, USA).

The presence of a solid-core MMF coupler was expected to introduce significant temporal broadening of the coupled, fs-duration laser pulses. Thus, prior to the fluorescence excitation detection experiments, it was necessary to determine the sensor's influence on the temporal profile of the transmitted ultrashort laser pulses via measurements of the autocorrelation function (ACF). The problem of dispersion in the MMF coupler was addressed in two ways: firstly, the lengths of the excitation and output arms of the coupler were each reduced from the initial ~1 m to about 30 cm; then, the HCPBF fiber was introduced into the setup, one of the characteristic features of which is positive value of the dispersion parameter D at 780 nm ($D_{HCPBF780} \approx 10$ fs/(nm*m)), according to [90]), which can be expected to pre-compensate the chromatic dispersion of the MMF coupler. The ACF measurements were performed with an autocorrelator (pulseCheck, A.P.E. GmbH, Berlin, Germany), for four different cases—laser output pulse, HCPBF output pulse, fiber coupler with reduced arm lengths output pulse, and HCPBF+fiber coupler output pulse. Fiber output beams were collimated with a 10× microscope objective, and then pointed onto the autocorrelator's aperture. Coupling between the HCPBF and MMF coupler's excitation arm was performed with the fusion splicer, and the pulse widths of the pre-spliced (butt-coupled) and spliced fibers were also measured. This was of great importance, as splicing photonic crystal fibers always requires lower temperatures, making them much more prone to core displacement, which, in turn, can cause different mode excitation in the MMF and influence the final output pulse width.

The two-photon fluorescence spectroscopy setup, based on almost exclusively optical fiber components, is presented in Figure 1a.

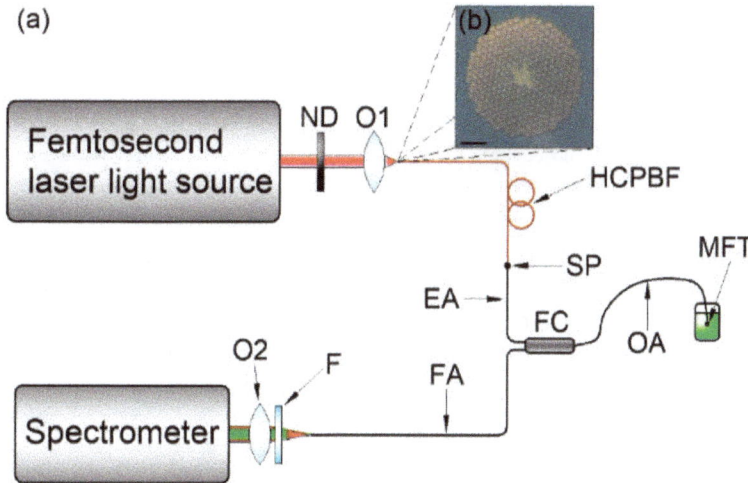

Figure 1. (**a**) Multiphoton fluorescence measurement setup schematic. ND: variable neutral density filter; O1, O2: 10×/0.24 microscope objectives; HCPBF: hollow-core photonic bandgap fiber; SP: splice point; FC: multimode fiber coupler; EA, OA, FA: fiber coupler's excitation, output and fluorescence arms, respectively; MFT: microlensed fiber tip; F: colored glass filter. The MFT output power is controlled with the ND, while the HCPBF allows the laser output pulse to be pre-compensated, reducing its temporal broadening due to the presence of FC. The FC had its EA and OA shortened to ~30 cm each. During the measurement, the MFT was immersed in a glass cuvette with a fluorescein solution, and multiphoton excited fluorescence was recollected and transmitted through the FA to the USB spectrometer. (**b**) Cross section of the HCPBF structure used in the experiment. Scale bar (lower left corner) is 10 μm.

A femtosecond laser light source (EFOA-SH, Atseva, Fort Collins, CO, USA), providing ultrashort light pulses of λ_{laser} = 780 nm and τ_{laser} = 130 fs, at f_{rep} = 78 MHz and P_{avg} = 120 mW, was coupled via a 10× microscope objective (NA_{MO} = 0.26) into the HCPBF ($NA_{HCPBF} \approx 0.2$). The total length of the HCPBF was ~9 m, and the the maximum power at its output was 60 mW (half the input), mainly due to the attenuation characteristic of the fiber, as well as the mismatch between the NA_{MO} and NA_{HCPBF}. The coupling was performed in free space, with a use of the 3D translation stage (MBT616D, Thorlabs, Newton, NJ, USA). The laser output power was additionally controlled with a variable neutral density (ND) filter (ND-100C-4M, Thorlabs, Newton, NJ, USA). The MFT was immersed in a 10^{-2} M NaOH solution of fluorescein (Chempur, Piekary Śląskie, Poland) with concentration C_{fluo} = 10^{-5} M. The fluorescence signal was recollected by the MFT, and the remaining arm of the fiber coupler (fluorescence arm–FA, see Figure 1a) was used to transmit it (fluorescence signal) to the USB spectrometer. The FA's output signal was additionally filtered (FGB37-A, Tholabs, Newton, NJ, USA) to reduce the amount of backscattered excitation light, and then focused onto the slit of the spectrometer with another 10× microscope objective. The spectrometer's integration time (t_{int}) during the fluorescence measurements was 2 s.

3. Results and Discussion

3.1. Combining the HCPBF and MM, Solid-Core Fiber with An Arc Fusion Splicer

3.1.1. Arc Fusion Splicing of HCPBF and Solid-Core MMF

When splicing dissimilar fibers, the loss figure of such splicing is mainly governed by the fibers' mode field mismatch, causing the splice to be non-symmetrical in terms of the transmission direction (i.e., the splice loss will be lower from the smaller MFD fiber to the larger MFD fiber than in the opposite direction). This loss can be reliably estimated by calculating the mode field overlap integral of the spliced fibers [91], however the complicated mode structure of the HCPBFs makes this method hard to implement [85]. Moreover, the mechanics of splicing MOFs is inherently different due to the presence of the photonic structure. The effect of its collapse during the splicing poses a major challenge, and many researchers have studied it extensively [79,84,86]. In general, the microstructured cladding collapses due to the silica's viscosity being greatly reduced in the high-temperature region of fusion arc discharge. Additionally, the presence of air holes in HCPBFs (and MOFs) severely distracts and reduces the rate of heat transfer from the fiber's outer surface to its core, causing a difference in its temperature and directly influencing both the optical losses and mechanical properties of the splice. This has resulted in a general preference for using sophisticated glass processing stations, such as filament fusion splicers or CO_2 laser-based fiber splicers, which provide a very stable and uniform heat distribution across the splice area, making them an extremely reliable device for the purpose of splicing MOFs. However, the price of such systems greatly reduces their affordability, whereas arc fusion splicers, due to their common presence and ease of use, make for a very interesting alternative, even though they are not as effective due to the electrical arc's susceptibility to the environment, wear of electrodes, etc. [81].

The initial splicing parameters used during this study were based on the previously mentioned work presented by Thapa et al. [85]. However, using the proposed parameters to splice the HCPBF and MMF used in our experiment resulted in extremely fragile splices, which broke almost immediately after removing the fibers from the splicers clamps. We attribute this behavior to the fact that the photonic structure diameter (d_{struct}) of our HCPBF is much smaller than in the case of HCPBF used by our predecessors [85] (45 and 70 µm, respectively), and in turn the total volume of glass is increased in our HCPBF. As a result, the latter requires higher temperatures (and higher splicing currents) to heat it properly and form a strong splice; this conclusion has been supported by the observations of Xiao et al. [86]. A detailed description of the splicing procedure, as well as the values of the fusion currents, times and offset, crucial for the quality of the splice, can be found in Appendix A. The measured values of splicing loss (SL) and bending radius (r_{bend}) of five consecutive splices, fabricated with the proposed program, are presented in Table 1. Each splice withstood bending, with $r_{bend} \approx 1.8$ cm, while the average SL (transmission direction 'HCPBF to MMF') was 0.32 dB. Bending the fabricated splices with $r_{bend} = 1.8$ cm had minimal influence on overall loss figure: splice bending loss (SBL) did not exceed the value of 0.07 dB. An example of a successful splice is presented in Figure 2a. The collapse of the photonic structure causes bright scattering and reflection of the transmitted signal at the HCPBF-MMF interface, observable on the splicers screen and with the naked eye (Figure 2b,c). Nevertheless, as mentioned previously, one can expect that when a small core fiber (HCPBF) is spliced with a large core fiber (MMF), the splice loss should be relatively small when signal transmission occurs in the 'HCPBF to MMF' direction.

Table 1. Splice loss, bending loss and bending radius of five consecutive HCPBF-MMF splices, fabricated with optimal splicing parameters presented in Table A1. The last row contains the average values of the presented results, together with their standard deviation.

Splice No.	Splice Loss SL (dB)	Bending Radius (r_{bend}) (cm)	Splice Bending Loss SBL (dB)
1	0.23		0.04
2	0.26		0.03
3	0.46	≤ 1.8	0.07
4	0.27		0.03
5	0.36		0.05
Average	(0.32 ± 0.1) dB	-	(0.04 ± 0.02) dB

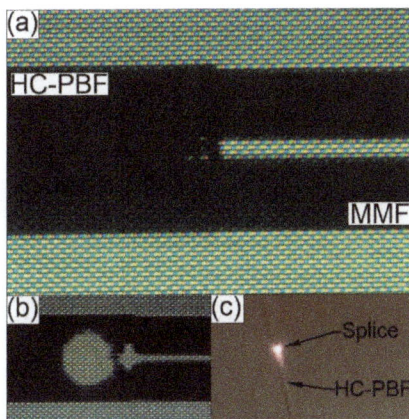

Figure 2. (a) Splice between the HCPBF and MMF. (b) Scattering and reflection of the transmitted $\lambda = 780$ nm laser signal, observed in the form of a large bright circle near the splice line between the HCPBF and MMF. This effect is caused by the collapse of HCPBF's microstructure, which impairs its ability to guide light and causes the transmitted signal to diverge quickly in the collapsed region. (c) Photograph of HCPBF-MMF splice during the transmission of laser signal, exhibiting the previously described microstructure collapse effect.

3.1.2. Microlensed Tip of MMF Coupler for the Enhancement of Two-Photon Excited Fluorescence

Fabrication and analysis of the optical performance of MFTs has previously been described by several researchers [75,77,78], while the fusion splicer program details and parameters used for this particular MFT are presented in Appendix A. The main goal was to obtain a small fiber end-face curvature, allowing the fiber output beam to be focused and the efficiency of TPEF to be enhanced. A picture of the fabricated MFT and its output beam profile are presented in Figure 3a,b, respectively. The beam profile was measured in the far field, showing the Gaussian-like intensity distribution in both the X and Y directions, and thus demonstrating good spatial beam quality. The working distance of the MFT (WD_{MFT}), measured according to the method presented in [78], was ~180 μm.

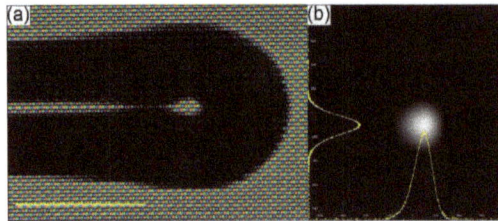

Figure 3. (**a**) Picture of the microlensed tip (MFT) of the MMF coupler output arm. (**b**) Far field beam profile of the MFT output beam. The profiles in the X and Y directions are both Gaussian-like. Scale bar in 3a is 125 μm.

3.2. ACF Traces of the Proposed HCPBF-MMF Coupler Setup and Its Application for the Purpose of Multiphoton Fluorescence Spectroscopy

Dispersion effects introduced by the solid-core fibers can have a destructive influence on the fundamental laser pulse width, and in turn impair the efficiency of nonlinear optical effects such as MPEF. The obtained ACF traces of the fiber-coupled fs laser beam are presented in Figure 4. Each of the ACFs was fitted with a Lorentzian-shaped curve, as it exhibited the lowest approximation errors (R^2 values for each fit were as good as 0.998). As mentioned before, HCPBF's dispersion parameter D at 780 nm is $D_{HCPBF780} \approx 10$ fs/(nm*m), suggesting that the total pulse broadening within ~9 m length of this fiber is about 90 fs. Figure 4a shows the ACF trace of HCPBF coupled pulse, whose calculated width (τ_{HCPBF}) was 190 fs.

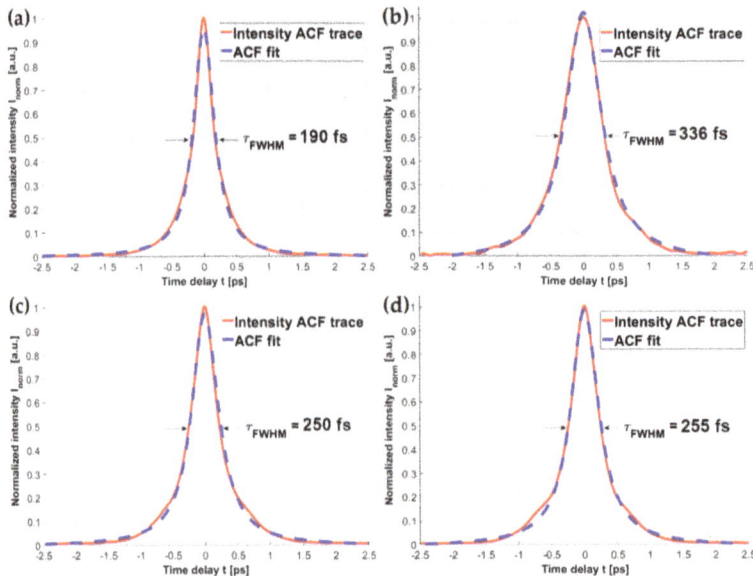

Figure 4. Autocorrelation function traces of femtosecond laser pulses transmitted through: (**a**) HCPBF; (**b**) MMF coupler; (**c**) HCPBF+MMF coupler, butt-coupled (prior to splicing); (**d**) HCPBF+MMF coupler, spliced. In 4b, c and d, the coupler output arm tip is microlensed. A noticeable reduction of the pulse width (τ_{FWHM}) can be observed in the case of the combined HCPBF and MMF coupler, showing the feasibility of compensating dispersion effects, introduced by the solid-core fiber elements, with HCPBF.

Comparing this with the τ of the laser beam, coupled into the MMF coupler (τ_{MMF} = 336 fs), one can observe that τ_{MMF} is nearly twice the τ_{HCPBF}. Combining these two types of fibers (Figure 4c,d), the effect of dispersion compensation can be observed, resulting in the reduction of the τ_{MMF} by about 86 fs ($\tau_{HCPBF+MMF} \approx 250$ fs), which corresponds to the total value of $D_{HCPBF780}$ after 9 m length of fiber, proving the anomalous dispersion of the HCPBF at λ = 780 nm. Additionally, due to the splicing procedure, which softened the fibers and could potentially result in the displacement of their cores, we measured the width of the spliced HCPBF and MMF coupler output pulse (Figure 4d). The value of τ_{FWHM} in this case was 255 fs, showing that, although there is a slight difference between the butt-coupled and spliced setups, this was not large enough to consider splicing as a procedure that significantly influences the pulse width. The experimental error in these measurements, apart from the resolution of the autocorrelator ($\Delta\tau = \pm2$ fs), is connected mainly with the problem of exciting different modes during the coupling of HCPBF and MMF. We did not study this effect, however it was not necessary due to the mechanical requirements of the splice. Mechanically strong splices should be geometrically symmetrical, which means that the MMF excitation conditions should be constant, since the core of MMF and HCPBF should run along the same optical axis after coupling prior to the splicing. In this case, the main source of the error should be the eccentricity of the cores of the spliced fibers; however, it does not influence the final pulse width significantly, which was proven by the pulse widths obtained for pre-spliced and spliced fibers. It should also be mentioned that the total pulse broadening was also increased by the two 10× objectives present in the setup; however, since it should be on the order of few fs, we considered it to be negligible.

To prove that the proposed HCPBF-MMF coupler hybrid can be considered a candidate for the purpose of MPEF spectroscopy, the TPEF emission spectra of fluorescein were measured, as depicted in Figure 5. The average power delivered to the sample was 20 mW. Two emission peaks can be observed, one in the ~760–800 nm range, originating from the excitation laser, and another one, spanning ~ 480 to 580 nm, with a maximum at 518 nm, which is connected with examined fluorescein solution. The obtained results are in good agreement with those presented in [92]. Although the collected spectrum is contaminated with the backscattered laser radiation (both by the sample and due to the weak isolation of the MMF coupler), the presented idea can be considered a promising step towards simple, fiber-based optical biopsy devices.

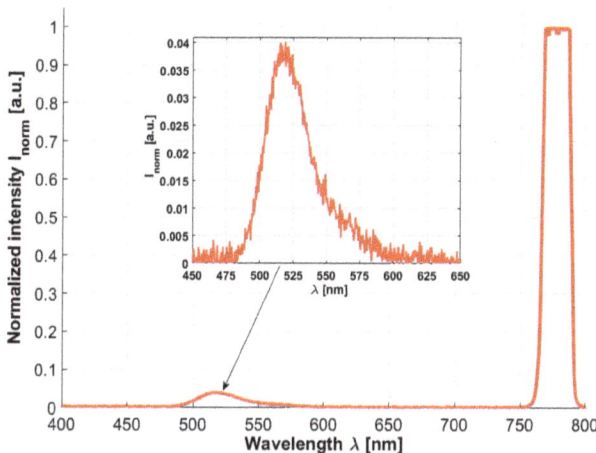

Figure 5. Two-photon excited fluorescence (TPEF) emission spectrum of 10^{-2} M NaOH fluorescein solution (concentration $C_{fluo} \approx 10^{-5}$ M). A fluorescence peak, visible on the enlarged inset graph, exhibits its maximum at ~518 nm. A backscattered excitation signal peak at 780 nm is also visible, and even though it exceeds the Y-axis maximum, it does not saturate the spectrometer's detector, and in turn has no influence on the recorded fluorescence spectrum.

4. Conclusions

This work presented a hybrid between an HCF and an MMF Y-type coupler, and its application to multiphoton fluorescence spectroscopy. Arc fusion splicing of HCF and MMF was presented, resulting in low splicing and bending loss (0.32 and 0.04 dB (at bending radius $\geq$ 1.8 cm), respectively) splices, which shows the opportunity of combining these two distinct types of fibers in a simple and affordable fashion. The presented fiber coupler hybrid was used in a proof-of-concept multiphoton spectroscopy experiment, with the coupler's common arm acting as a fluorescence excitation-detection arm. The tip of this arm was curved with the use of the same arc fusion splicer, creating a microlens with a working distance of ~180 μm and Gaussian-like output beam profile. This tip ensured focusing of the output beam, which is an important factor for the efficiency of exciting fluorescence due to the multiphoton absorption. Fluorescence was excited in a 10^{-2} NaOH fluorescein solution ($C_{fluo} \approx 10^{-5}$ M). The recollected fluorescence signal was recorded with a USB spectrometer. The presented fiber-optic setup has three features which are very interesting from the sensing point of view, namely size, price and simplicity. It should be noted that the MMF coupler used in this research is not an optimal solution for the spectral region of interest (400–800 nm), and thus one can suppose that the efficiency of the proposed setup could be increased by replacing the coupler with a one designed for the VIS spectral region. Nevertheless, the proposed HCPBF and MMF coupler combination allowed for the delivery of 130-fs short laser pulses with moderate dispersive broadening (255 fs at the HCPBF+MMF coupler output), making it a solution considerable for applications in the field of non-linear optics.

Author Contributions: H.I.S. and M.A.P. designed the experiment, conducted the measurements, discussed the results and prepared the data. All the authors edited the manuscript. E.B.-P. supervised the research.

Funding: This research was funded by the Wroclaw University of Science and Technology Grants No. 0401/0140/17 and 0402/0149/17.

Acknowledgments: We would like to thank prof. Sławomir Sujecki for his support during this research.

Conflicts of Interest: The authors declare no conflict of interest.

Appendix A. Arc Fusion Splicing Procedures–Technical Notes

Appendix A.1. Arc Fusion Splicing of HCPBF and MMF

The optimum splicing parameters were found by performing over 30 splices, and the splice pass/fail criteria were as follows: splice loss (SL) $\leq$ 0.5 dB and splice bending radius (r_{bend}) $\leq$ 1.8 cm. It could be possible to improve these parameters, however they satisfied the needs of our experiment, as only a single splice was required. Prior to splicing, about 2 cm of both HCPBF and MMF were stripped, cleaved (FC-6S, Sumitomo Electric, Raleigh, NC, USA) and cleaned with dry, lint-free tissue, resulting in good quality, clean fiber facets. Specific splicing parameters are presented in Table A1. As the splice itself is formed during splicing phases 2 and 3 (i.e., when the fusion currents 2 and 3 are applied), the changes were implemented mainly for these parameters. By connecting the MMF output to the optical power meter (S121C with PM100D, Thorlabs, Newton, NJ, USA) the evolution of the output power (P_{out_MM}) was observed (Figure A1) and it was noted that during splice phase 1, when a 10 mA current was applied for 0.2 s, P_{out_MM} fell quickly, which was a hint that the optimum current value lays between the 6.3 and 10 mA.

Table A1. Arc fusion splicer parameters adjusted for splicing hollow-core photonic crystal fiber (HC-800-02, NKT Photonics, Birkerød, Denmark) and solid-core, multimode, type OM2 fibers.

Splicing Parameter Name	Parameter Values
Prefuse time (s)	0.1
Prefuse current (mA)	10
Gap (μm)	10
Overlap (μm)	10
Fusion time 1 t_{F1} (s)	0.2
Fusion current 1 I_{F1} (mA)	10
Fusion time 2 t_{F2} (s)	6
Fusion current 2 I_{F2} (mA)	9.4
Fusion time 3 t_{F3} (s)	6
Fusion current 3 I_{F3} (mA)	8
Offset (-)	260

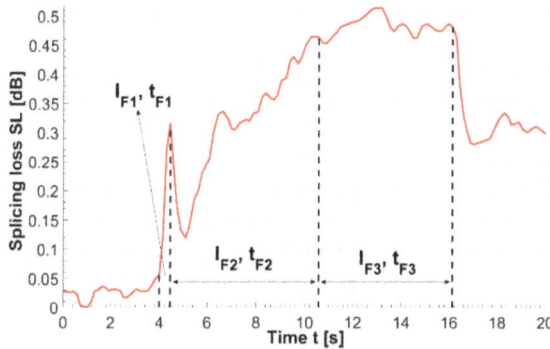

Figure A1. Evolution of splicing loss over time measured during the HCPBF-MMF splicing procedure. Dotted lines indicate moments of start and stop of three main splicing phases. *SL* rise is most intense during phase 1 of splicing (i.e., when the fiber tips are softened and prepared for forming a splice with a high fusion current). The *SL* value falls almost immediately after the splicing has stopped; the splicers program stops pushing the fibers towards each other, which probably reduces the displacement between the cores of both fibers and in turn positively influences coupling between them in the splice region.

Appendix A.2. Fabrication of the Fiber Microlens at the Tip of the MMF Coupler

MFT fabrication details are presented in Table A2. The same arc fusion splicer as in the case of splicing HCPBF and MMF was used. Moving the MMF tip from the center of the arc curvature by 50 μm (offset = 305) allowed for better control of the lens formation process.

Table A2. Arc fusion splicer parameters for the fabrication of microlensed fiber tip. The program was based on the splicers tapering procedure, which means that the fiber tip was constantly being pulled away from the splicers arc. A large offset (approximately 50 μm from the center arc) ensured that the lens curvature was small enough (about 150 μm), while large values of the arc currents made the process quicker than in the case of standard fiber tapering.

Parameter Name	Value
Prefuse time (s)	0.2
Prefuse current (mA)	10
Fusion time 1 t_{FL1} (s)	2
Fusion current 1 I_{FL1} (mA)	15
Fusion time 2 t_{F2} (s)	2
Fusion current 2 I_{F2} (mA)	10
Fusion time 3 t_{F3} (s)	1
Fusion current 3 I_{F3} (mA)	10
Offset (-)	305

References

1. Cregan, R.F.; Mangan, B.J.; Knight, J.C.; Birks, T.A.; Russell, P.S.J.; Roberts, P.J.; Allan, D.C. Single-mode photonic band gap guidance of light in air. *Science* **1999**, *285*, 1537–1539. [CrossRef] [PubMed]
2. Chen, M.-Q.; Zhao, Y.; Xia, F.; Peng, Y.; Tong, R.-J. High sensitivity temperature sensor based on fiber air-microbubble fabry-perot interferometer with pdms-filled hollow-core fiber. *Sens. Actuators A Phys.* **2018**, *275*, 60–66. [CrossRef]
3. Sun, H.; Luo, H.; Wu, X.; Liang, L.; Wang, Y.; Ma, X.; Zhang, J.; Hu, M.; Qiao, X. Spectrum ameliorative optical fiber temperature sensor based on hollow-core fiber and inner zinc oxide film. *Sens. Actuators B Chem.* **2017**, *245*, 423–427. [CrossRef]
4. Zhang, Z.; Liao, C.; Tang, J.; Wang, Y.; Bai, Z.; Li, Z.; Guo, K.; Deng, M.; Cao, S.; Wang, Y. Hollow-core-fiber-based interferometer for high-temperature measurements. *IEEE Photonics J.* **2017**, *9*, 1–9. [CrossRef]
5. Wei, C.; Young, J.T.; Menyuk, C.R.; Hu, J. Temperature sensor using fluid-filled negative curvature fibers. In Proceedings of the Conference on Lasers and Electro-Optics, San Jose, CA, USA, 13–18 May 2018.
6. Sardar, M.R.; Faisal, M. Gas sensor based on octagonal hollow core photonic crystal fiber. In Proceedings of the 2017 IEEE International Conference on Imaging, Vision & Pattern Recognition (icIVPR), Dhaka, Bangladesh, 13–14 February 2017.
7. Yang, F.; Jin, W.; Lin, Y.; Wang, C.; Lut, H.; Tan, Y. Hollow-core microstructured optical fiber gas sensors. *J. Lightwave Technol.* **2017**, *35*, 3413–3424. [CrossRef]
8. Lin, Y.; Liu, F.; He, X.; Jin, W.; Zhang, M.; Yang, F.; Ho, H.L.; Tan, Y.; Gu, L. Distributed gas sensing with optical fibre photothermal interferometry. *Opt. Express* **2017**, *25*, 31568–31585. [CrossRef] [PubMed]
9. Nikodem, M.; Krzempek, K.; Dudzik, G.; Abramski, K. Hollow core fiber-assisted absorption spectroscopy of methane at 3.4 μm. *Opt. Express* **2018**, *26*, 21843–21848. [CrossRef] [PubMed]
10. Shi, Q.; Lv, F.; Wang, Z.; Jin, L.; Hu, J.J.; Liu, Z.; Kai, G.; Dong, X. Environmentally stable fabry-p rot-type strain sensor based on hollow-core photonic bandgap fiber. *IEEE Photonics Technol. Lett.* **2008**, *20*, 237–239. [CrossRef]
11. Zhao, Y.; Lv, R.-Q.; Ying, Y.; Wang, Q. Hollow-core photonic crystal fiber fabry–perot sensor for magnetic field measurement based on magnetic fluid. *J. Light. Technol.* **2012**, *44*, 899–902. [CrossRef]
12. Jin, L.; Guan, B.-O.; Wei, H. Sensitivity characteristics of fabry-perot pressure sensors based on hollow-core microstructured fibers. *J. Light. Technol.* **2013**, *31*, 2526–2532.
13. Bykov, D.S.; Schmidt, O.A.; Euser, T.G.; Russell, P.S.J. Flying particle sensors in hollow-core photonic crystal fibre. *Nat. Photonics* **2015**, *9*, 461. [CrossRef]
14. Zeltner, R.; Bykov, D.S.; Xie, S.; Euser, T.G.; Russell, P.S.J. Fluorescence-based flying-particle sensor in liquid-filled hollow-core photonic crystal fiber. In Proceedings of the 2016 Conference on Lasers and Electro-Optics (CLEO), San Jose, CA, USA, 5–10 June 2016.
15. Momota, M.R.; Hasan, M.R. Hollow-core silver coated photonic crystal fiber plasmonic sensor. *Opt. Mater.* **2018**, *76*, 287–294. [CrossRef]
16. Zeisberger, M.; Schmidt, M.A. Analytic model for the complex effective index of the leaky modes of tube-type anti-resonant hollow core fibers. *Sci. Rep.* **2017**, *7*, 11761. [CrossRef] [PubMed]
17. Johnson, S.G.; Ibanescu, M.; Skorobogatiy, M.; Weisberg, O.; Engeness, T.D.; Soljačić, M.; Jacobs, S.A.; Joannopoulos, J.D.; Fink, Y. Low-loss asymptotically single-mode propagation in large-core omniguide fibers. *Opt. Express* **2001**, *9*, 748–779. [CrossRef] [PubMed]
18. Fink, Y.; Winn, J.N.; Fan, S.; Chen, C.; Michel, J.; Joannopoulos, J.D.; Thomas, E.L. A dielectric omnidirectional reflector. *Science* **1998**, *282*, 1679. [CrossRef] [PubMed]
19. Engeness, T.D.; Ibanescu, M.; Johnson, S.G.; Weisberg, O.; Skorobogatiy, M.; Jacobs, S.; Fink, Y. Dispersion tailoring and compensation by modal interactions in omniguide fibers. *Opt. Express* **2003**, *11*, 1175–1196. [CrossRef] [PubMed]
20. Zeisberger, M.; Tuniz, A.; Schmidt, M.A. Analytic model for the complex effective index dispersion of metamaterial-cladding large-area hollow core fibers. *Opt. Express* **2016**, *24*, 20515–20528. [CrossRef] [PubMed]
21. Poletti, F.; Petrovich Marco, N.; Richardson David, J. Hollow-core photonic bandgap fibers: Technology and applications. *Nanophotonics* **2013**, *2*, 315. [CrossRef]

22. Fokoua, E.N.; Richardson, D.J.; Poletti, F. Impact of structural distortions on the performance of hollow-core photonic bandgap fibers. *Opt. Express* **2014**, *22*, 2735–2744. [CrossRef] [PubMed]

23. Bufetov, I.; Kosolapov, A.; Pryamikov, A.; Gladyshev, A.; Kolyadin, A.; Krylov, A.; Yatsenko, Y.; Biriukov, A. Revolver hollow core optical fibers. *Fibers* **2018**, *6*, 39. [CrossRef]

24. Wang, Y.; Couny, F.; Roberts, P.J.; Benabid, F. Low loss broadband transmission in optimized core-shape kagome hollow-core PCF. In Proceedings of the CLEO/QELS: 2010 Laser Science to Photonic Applications, San Jose, CA, USA, 16–21 May 2010.

25. Stawska, H.; Popenda, M.; Bereś-Pawlik, E. Anti-resonant hollow core fibers with modified shape of the core for the better optical performance in the visible spectral region—A numerical study. *Polymers* **2018**, *10*, 899. [CrossRef]

26. Kosolapov, A.F.; Alagashev, G.K.; Kolyadin, A.N.; Pryamikov, A.D.; Biryukov, A.S.; Bufetov, I.A.; Dianov, E.M. Hollow-core revolver fibre with a double-capillary reflective cladding. *Quantum Electron.* **2016**, *46*, 267. [CrossRef]

27. Chen, Y.; Saleh, M.F.; Joly, N.Y.; Biancalana, F. Low-loss single-mode negatively curved square-core hollow fibers. *Opt. Lett.* **2017**, *42*, 1285–1288. [CrossRef] [PubMed]

28. Habib, M.S.; Bang, O.; Bache, M. Low-loss single-mode hollow-core fiber with anisotropic anti-resonant elements. *Opt. Express* **2016**, *24*, 8429–8436. [CrossRef] [PubMed]

29. Nawazuddin, M.B.S.; Wheeler, N.V.; Hayes, J.R.; Bradley, T.; Sandoghchi, S.R.; Gouveia, M.A.; Jasion, G.T.; Richardson, D.J.; Poletti, F. Lotus shaped negative curvature hollow core fibre with 10.5 db/km at 1550 nm wavelength. In Proceedings of the 2017 European Conference on Optical Communication (ECOC), Gothenburg, Sweden, 17–21 September 2017; pp. 1–3.

30. Roberts, P.J.; Couny, F.; Sabert, H.; Mangan, B.J.; Williams, D.P.; Farr, L.; Mason, M.W.; Tomlinson, A.; Birks, T.A.; Knight, J.C.; et al. Ultimate low loss of hollow-core photonic crystal fibres. *Opt. Express* **2005**, *13*, 236–244. [CrossRef] [PubMed]

31. Roberts, P.J.; Williams, D.P.; Sabert, H.; Mangan, B.J.; Bird, D.M.; Birks, T.A.; Knight, J.C.; Russell, P.S.J. Design of low-loss and highly birefringent hollow-core photonic crystal fiber. *Opt. Express* **2006**, *14*, 7329–7341. [CrossRef] [PubMed]

32. Amezcua-Correa, R.; Broderick, N.G.R.; Petrovich, M.N.; Poletti, F.; Richardson, D.J. Design of 7 and 19 cells core air-guiding photonic crystal fibers for low-loss, wide bandwidth and dispersion controlled operation. *Opt. Express* **2007**, *15*, 17577–17586. [CrossRef] [PubMed]

33. Jung, Y.; Sleiffer, V.A.J.M.; Baddela, N.; Petrovich, M.N.; Hayes, J.R.; Wheeler, N.V.; Gray, D.R.; Fokoua, E.N.; Wooler, J.P.; Wong, N.H.L.; et al. First demonstration of a broadband 37-cell hollow core photonic bandgap fiber and its application to high capacity mode division multiplexing. In Proceedings of the 2013 Optical Fiber Communication Conference and Exposition and the National Fiber Optic Engineers Conference (OFC/NFOEC), Anaheim, CA, USA, 17–21 March 2013.

34. Belardi, W.; Knight, J.C. Effect of core boundary curvature on the confinement losses of hollow antiresonant fibers. *Opt. Express* **2013**, *21*, 21912–21917. [CrossRef] [PubMed]

35. Belardi, W. Design and properties of hollow antiresonant fibers for the visible and near infrared spectral range. *J. Light. Technol.* **2015**, *33*, 4497–4503. [CrossRef]

36. Vaiano, P.; Carotenuto, B.; Pisco, M.; Ricciardi, A.; Quero, G.; Consales, M.; Crescitelli, A.; Esposito, E.; Cusano, A. Lab on fiber technology for biological sensing applications. *Laser Photonics Rev.* **2016**, *10*, 922–961. [CrossRef]

37. Ricciardi, A.; Crescitelli, A.; Vaiano, P.; Quero, G.; Consales, M.; Pisco, M.; Esposito, E.; Cusano, A. Lab-on-fiber technology: A new vision for chemical and biological sensing. *Analyst* **2015**, *140*, 8068–8079. [CrossRef] [PubMed]

38. Liu, X.-L.; Ding, W.; Wang, Y.-Y.; Gao, S.-F.; Cao, L.; Feng, X.; Wang, P. Characterization of a liquid-filled nodeless anti-resonant fiber for biochemical sensing. *Opt. Lett.* **2017**, *42*, 863–866. [CrossRef] [PubMed]

39. Liu, X.-L.; Wang, Y.-Y.; Ding, W.; Gao, S.-F.; Cao, L.; Feng, X.; Wang, P. Liquid-core nodeless anti-resonant fiber for biochemical sensing. In Proceedings of the Conference on Lasers and Electro-Optics, San Jose, CA, USA, 14–19 May 2017.

40. Nissen, M.; Doherty, B.; Hamperl, J.; Kobelke, J.; Weber, K.; Henkel, T.; Schmidt, A.M. Uv absorption spectroscopy in water-filled antiresonant hollow core fibers for pharmaceutical detection. *Sensors* **2018**, *18*, 478. [CrossRef] [PubMed]

41. Iwata, T.; Katagiri, T.; Matsuura, Y. Real-time analysis of isoprene in breath by using ultraviolet-absorption spectroscopy with a hollow optical fiber gas cell. *Sensors* **2016**, *16*, 2058. [CrossRef] [PubMed]

42. Yan, H.; Gu, C.; Yang, C.; Liu, J.; Jin, G.; Zhang, J.; Hou, L.; Yao, Y. Hollow core photonic crystal fiber surface-enhanced raman probe. *Appl. Physics Lett.* **2006**, *89*, 204101. [CrossRef]

43. Wang, C.; Zeng, L.; Li, Z.; Li, D. Review of optical fibre probes for enhanced raman sensing. *J. Raman Spectrosc.* **2017**, *48*, 1040–1055. [CrossRef]

44. Khetani, A.; Riordon, J.; Tiwari, V.; Momenpour, A.; Godin, M.; Anis, H. Hollow core photonic crystal fiber as a reusable raman biosensor. *Opt. Express* **2013**, *21*, 12340–12350. [CrossRef] [PubMed]

45. Stoddart, P.R.; White, D.J. Optical fibre sers sensors. *Anal. Bioanal. Chem.* **2009**, *394*, 1761–1774. [CrossRef] [PubMed]

46. Geng, Y.; Xu, Y.; Tan, X.; Wang, L.; Li, X.; Du, Y.; Hong, X. A simplified hollow-core photonic crystal fiber sers probe with a fully filled photoreduction silver nanoprism. *Sensors* **2018**, *18*, 1726. [CrossRef] [PubMed]

47. Yang, X.; Shi, C.; Wheeler, D.; Newhouse, R.; Chen, B.; Zhang, J.Z.; Gu, C. High-sensitivity molecular sensing using hollow-core photonic crystal fiber and surface-enhanced raman scattering. *J. Opt. Soc. Am. A* **2010**, *27*, 977–984. [CrossRef] [PubMed]

48. Albrecht, M.G.; Creighton, J.A. Anomalously intense raman spectra of pyridine at a silver electrode. *J. Am. Chem. Soc.* **1977**, *99*, 5215–5217. [CrossRef]

49. Laing, S.; Gracie, K.; Faulds, K. Multiplex in vitro detection using sers. *Chem. Soc. Rev.* **2016**, *45*, 1901–1918. [CrossRef] [PubMed]

50. Khetani, A.; Momenpour, A.; Alarcon, E.I.; Anis, H. Hollow core photonic crystal fiber for monitoring leukemia cells using surface enhanced raman scattering (sers). *Biomed. Opt. Express* **2015**, *6*, 4599–4609. [CrossRef] [PubMed]

51. Dinish, U.S.; Balasundaram, G.; Chang, Y.T.; Olivo, M. Sensitive multiplex detection of serological liver cancer biomarkers using sers-active photonic crystal fiber probe. *J. Biophotonics* **2013**, *7*, 956–965. [CrossRef] [PubMed]

52. Chow, K.K.; Short, M.; Lam, S.; McWilliams, A.; Zeng, H. A raman cell based on hollow core photonic crystal fiber for human breath analysis. *Med. Phys.* **2016**, *41*, 092701. [CrossRef] [PubMed]

53. Khetani, A.; Tiwari, V.S.; Harb, A.; Anis, H. Monitoring of heparin concentration in serum by raman spectroscopy within hollow core photonic crystal fiber. *Opt. Express* **2011**, *19*, 15244–15254. [CrossRef] [PubMed]

54. Brustlein, S.; Berto, P.; Hostein, R.; Ferrand, P.; Billaudeau, C.; Marguet, D.; Muir, A.; Knight, J.; Rigneault, H. Double-clad hollow core photonic crystal fiber for coherent raman endoscope. *Opt. Express* **2011**, *19*, 12562–12568. [CrossRef] [PubMed]

55. Lombardini, A.; Mytskaniuk, V.; Sivankutty, S.; Andresen, E.R.; Chen, X.; Wenger, J.; Fabert, M.; Joly, N.; Louradour, F.; Kudlinski, A.; et al. High-resolution multimodal flexible coherent raman endoscope. *Light Sci. Appl.* **2018**, *7*, 10. [CrossRef]

56. Yerolatsitis, S.; Yu, F.; McAughtrie, S.; Tanner, M.G.; Fleming, H.; Stone, J.M.; Campbell, C.J.; Birks, T.A.; Knight, J.C. Ultra-low background raman sensing using a negative-curvature fibre. In Proceedings of the OSA Advanced Photonics Congress 2018 (BGPP, IPR, NP, NOMA, Sensors, Networks, SPPCom, SOF), Zurich, Switzerland, 2–5 July 2018.

57. Konorov, S.O.; Addison, C.J.; Schulze, H.G.; Turner, R.F.B.; Blades, M.W. Hollow-core photonic crystal fiber-optic probes for raman spectroscopy. *Opt. Lett.* **2006**, *31*, 1911–1913. [CrossRef] [PubMed]

58. Couny, F.; Benabid, F.; Light, P.S. Large-pitch kagome-structured hollow-core photonic crystal fiber. *Opt. Lett.* **2006**, *31*, 3574–3576. [CrossRef] [PubMed]

59. Ghenuche, P.; Rammler, S.; Joly, N.Y.; Scharrer, M.; Frosz, M.; Wenger, J.; Russell, P.S.J.; Rigneault, H. Kagome hollow-core photonic crystal fiber probe for raman spectroscopy. *Opt. Lett.* **2012**, *37*, 4371–4373. [CrossRef] [PubMed]

60. Tai, S.-P.; Chan, M.-C.; Tsai, T.-H.; Guol, S.-H.; Chen, L.-J.; Sun, C.-K. Two-photon fluorescence microscope with a hollow-core photonic crystal fiber. *Opt. Express* **2004**, *12*, 6122–6128. [CrossRef] [PubMed]

61. Yu, J.; Zeng, H.; Lui, H.; Skibina, J.S.; Steinmeyer, G.; Tang, S. Characterization and application of chirped photonic crystal fiber in multiphoton imaging. *Opt. Express* **2014**, *22*, 10366–10379. [CrossRef] [PubMed]

62. Sherlock, B.; Fei, Y.; Jim, S.; Sean, W.; Carl, P.; Neil Mark, A.A.; French Paul, M.W.; Jonathan, K.; Chris, D. Tunable fibre-coupled multiphoton microscopy with a negative curvature fibre. *J. Biophotonics* **2016**, *9*, 715–720. [CrossRef] [PubMed]

63. Popenda, M.A.; Stawska, H.I.; Mazur, L.M.; Jakubowski, K.; Kosolapov, A.; Kolyadin, A.; Bereś-Pawlik, E. Application of negative curvature hollow-core fiber in an optical fiber sensor setup for multiphoton spectroscopy. *Sensors* **2017**, *17*, 2278. [CrossRef] [PubMed]

64. Kobayashi, T.; Katagiri, T.; Matsuura, Y. Multi-element hollow-core anti-resonant fiber for infrared thermal imaging. *Opt. Express* **2016**, *24*, 26565–26574. [CrossRef] [PubMed]

65. Arora, N.; Martins, D.; Ruggerio, D.; Tousimis, E.; Swistel, A.J.; Osborne, M.P.; Simmons, R.M. Effectiveness of a noninvasive digital infrared thermal imaging system in the detection of breast cancer. *Am. J. Surg.* **2008**, *196*, 523–526. [CrossRef] [PubMed]

66. Lahiri, B.B.; Bagavathiappan, S.; Jayakumar, T.; Philip, J. Medical applications of infrared thermography: A review. *Infrared Phys. Technol.* **2012**, *55*, 221–235. [CrossRef]

67. Wang, Y.; Jin, W.; Ju, J.; Xuan, H.; Ho, H.L.; Xiao, L.; Wang, D. Long period gratings in air-core photonic bandgap fibers. *Opt. Express* **2008**, *16*, 2784–2790. [CrossRef] [PubMed]

68. Yuan, T.; Zhong, X.; Guan, C.; Fu, J.; Yang, J.; Shi, J.; Yuan, L. Long period fiber grating in two-core hollow eccentric fiber. *Opt. Express* **2015**, *23*, 33378–33385. [CrossRef] [PubMed]

69. Wu, Z.; Wang, Z.; Liu, Y.-G.; Han, T.; Li, S.; Wei, H. Mechanism and characteristics of long period fiber gratings in simplified hollow-core photonic crystal fibers. *Opt. Express* **2011**, *19*, 17344–17349. [CrossRef] [PubMed]

70. Iadicicco, A.; Campopiano, S.; Cusano, A. Long-period gratings in hollow core fibers by pressure-assisted arc discharge technique. *IEEE Photonics Technol. Lett.* **2011**, *23*, 1567–1569. [CrossRef]

71. Terrel, M.; Digonnet, M.J.F.; Fan, S. Polarization controller for hollow-core fiber. *Opt. Lett.* **2007**, *32*, 1524–1526. [CrossRef] [PubMed]

72. Pang, M.; Jin, W. A hollow-core photonic bandgap fiber polarization controller. *Opt. Lett.* **2011**, *36*, 16–18. [CrossRef] [PubMed]

73. Xuan, H.F.; Jin, W.; Ju, J.; Wang, Y.P.; Zhang, M.; Liao, Y.B.; Chen, M.H. Hollow-core photonic bandgap fiber polarizer. *Opt. Lett.* **2008**, *33*, 845–847. [CrossRef] [PubMed]

74. Qian, W.; Zhao, C.-L.; Kang, J.; Dong, X.; Zhang, Z.; Jin, S. A proposal of a novel polarizer based on a partial liquid-filled hollow-core photonic bandgap fiber. *Opt. Commun.* **2011**, *284*, 4800–4804. [CrossRef]

75. Kong, G.-J.; Kim, J.; Choi, H.-Y.; Im, J.E.; Park, B.-H.; Paek, U.-C.; Lee, B.H. Lensed photonic crystal fiber obtained by use of an arc discharge. *Opt. Lett.* **2006**, *31*, 894–896. [CrossRef] [PubMed]

76. Ryu, S.Y.; Choi, H.Y.; Na, J.; Choi, W.J.; Lee, B.H. Lensed fiber probes designed as an alternative to bulk probes in optical coherence tomography. *Appl. Opt.* **2008**, *47*, 1510–1516. [CrossRef] [PubMed]

77. Choi, H.Y.; Ryu, S.Y.; Na, J.; Lee, B.H.; Sohn, I.-B.; Noh, Y.-C.; Lee, J. Single-body lensed photonic crystal fibers as side-viewing probes for optical imaging systems. *Opt. Lett.* **2008**, *33*, 34–36. [CrossRef] [PubMed]

78. Stawska, H.; Popenda, M.; Langer, Ł.; Bereś-Pawlik, E. Application of the hollow core fiber ended with fiber microlens in the multiphoton excitation setup. In Proceedings of the 20th International Conference on Transparent Optical Networks (ICTON 2018), Bucharest, Romania, 1–5 July 2018.

79. Xiao, L.; Jin, W.; Demokan, M.S.; Ho, H.L.; Hoo, Y.L.; Zhao, C. Fabrication of selective injection microstructured optical fibers with a conventional fusion splicer. *Opt. Express* **2005**, *13*, 9014–9022. [CrossRef] [PubMed]

80. Tachikura, M. Fusion mass-splicing for optical fibers using electric discharges between two pairs of electrodes. *Appl. Opt.* **1984**, *23*, 492–498. [CrossRef] [PubMed]

81. Yablon, A.D. Mechanics of fusion splicing. In *Optical Fiber Fusion Splicing*; Springer Series in Optical Sciences; Springer: Berlin/Heidelberg, Germany, 2005; pp. 49–89.

82. Bennett, P.J.; Monro, T.M.; Richardson, D.J. Toward practical holey fiber technology:?Fabrication, splicing, modeling, and characterization. *Opt. Lett.* **1999**, *24*, 1203–1205. [CrossRef] [PubMed]

83. Bourliaguet, B.; Paré, C.; Émond, F.; Croteau, A.; Proulx, A.; Vallée, R. Microstructured fiber splicing. *Opt. Express* **2003**, *11*, 3412–3417. [PubMed]

84. Yablon, A.D. Fusion splicing of specialty fiber. In *Optical Fiber Fusion Splicing*; Springer Series in Optical Sciences; Springer: Berlin/Heidelberg, Germany, 2005; pp. 229–253.

85. Thapa, R.; Knabe, K.; Corwin, K.L.; Washburn, B.R. Arc fusion splicing of hollow-core photonic bandgap fibers for gas-filled fiber cells. *Opt. Express* **2006**, *14*, 9576–9583. [CrossRef] [PubMed]

86. Xiao, L.; Demokan, M.S.; Jin, W.; Wang, Y.; Zhao, C. Fusion splicing photonic crystal fibers and conventional single-mode fibers: Microhole collapse effect. *J. Light. Technol.* **2007**, *25*, 3563–3574. [CrossRef]

87. Wu, C.; Song, J.; Zhang, Z.; Song, N. High strength fusion splicing of hollow-core photonic bandgap fiber and single-mode fiber. In Proceedings of the Photonics and Fiber Technology 2016 (ACOFT, BGPP, NP), Sydney, Australia, 5–8 September 2016.

88. Ma, H.; Chen, Z.; Jin, Z. Single-polarization coupler based on air-core photonic bandgap fibers and implications for resonant fiber optic gyro. *J. Light. Technol.* **2014**, *32*, 46–54. [CrossRef]

89. Huang, X.; Ma, J.; Tang, D.; Yoo, S. Hollow-core air-gap anti-resonant fiber couplers. *Opt. Express* **2017**, *25*, 29296–29306. [CrossRef]

90. Product spec. Sheet, hc-800-02 Photonic Crystal Fiber (nkt Photonics, Birkerød, Denmark). Available online: https://www.nktphotonics.com/wp-content/uploads/sites/3/2015/01/hc-800-1.pdf?1539002002 (accessed on 3 October 2018).

91. Yablon, A.D. Optics of fusion splicing. In *Optical Fiber Fusion Splicing*; Springer Series in Optical Sciences; Springer: Berlin/Heidelberg, Germany, 2005; pp. 91–135.

92. Xu, C.; Webb, W.W. Measurement of two-photon excitation cross sections of molecular fluorophores with data from 690 to 1050 nm. *J. Opt. Soc. Am. B* **1996**, *13*, 481–491. [CrossRef]

fibers

MDPI

Article

A Method to Process Hollow-Core Anti-Resonant Fibers into Fiber Filters

Xiaosheng Huang, Ken-Tye Yong and Seongwoo Yoo *

School of Electrical and Electronic Engineering, The Photonics Institute, Nanyang Technological University,
50 Nanyang Avenue, Singapore 639798, Singapore; xhuang012@e.ntu.edu.sg (X.H.); ktyong@ntu.edu.sg (K.-T.Y.)
* Correspondence: seon.yoo@ntu.edu.sg; Tel.: +65-6592-7597

Received: 25 July 2018; Accepted: 15 August 2018; Published: 22 November 2018

Abstract: Hollow-Core Anti-Resonant Fiber (HC-ARF) shows promising applications. Nevertheless, there has been a persistent problem when it comes to all-fiber integration due to a lack of HC-ARF-based fiber components. In response to this remaining challenge, we investigate a reliable, versatile and efficient method to convert an HC-ARF into a fiber filter. By locally heating an HC-ARF with a CO_2 laser, the fiber structure becomes deformed, and cladding capillaries shrink to produce a thicker wall. This process is analogous to "writing" a new fiber with a thicker wall on the original fiber, resulting in creating new high loss regions in the original transmission bands. Thus, the construction of a fiber filter is realized by "writing" a new fiber on the original fiber. The feasibility of this method is confirmed through experiments, adopting both one- and two-layer HC-ARF. The HC-ARF-based fiber filters are found to have transmission spectra consistent with simulation prediction. Both band pass and band reject fiber filters with more than a 20-dB extinction ratio are obtainable without extra loss. Thus, an in-fiber HC-ARF filter is demonstrated by the CO_2 writing process. Its versatile approach promises controlled band selection and would find interesting applications to be discussed.

Keywords: fiber filters; hollow core fibers; anti-resonant; photonic crystal fibers; fabrication

1. Introduction

Since the first theoretical demonstration in 1995 [1], Hollow-Core Photonic Crystal Fibers (HC-PCFs), as a remarkable breakthrough in fiber optics, have made it possible to guide light in the air core. This unique guiding property promises the potentials of achieving a higher damage threshold, lower Rayleigh scattering, lower material absorption and lower nonlinearity as compared to conventional fibers [2]. Hence, HC-PCFs have promising applications in areas of high power/ultrafast beam delivery [3], pulse compression [4] and communication systems [5], to name a few. One type of HC-PCFs is Hollow-Core Photonic Bandgap Fibers (HC-PBGFs), the record loss of which is 1.2 dB/km at 1.62 μm [6]. The HC-PBGF typically has a relatively narrow transmission band. The other type of HC-PCF is the so-called Hollow-Core Anti-Resonant Fibers (HC-ARFs), the guiding property relies of which on the combination of anti-resonance and inhibited coupling to low density of states cladding modes [7,8]. The HC-ARF has received ever-increasing interest thanks to its multiple broad transmission bands [7,9], simple and flexible cladding structures [10–13] and relatively low transmission loss [8,14–16]. The HC-ARFs show promising prospects in applications such as delivering light with a wide spectral range from ultra violet to mid-infrared [17,18], an optofluidic system [19,20] and light gas interaction [21]. Despite the unique properties of hollow core fibers, their connectivity to conventional fiber components is inefficient due to the mismatch of numerical aperture, as well as core size. As a result, most of the HC-ARF-based optical systems rely on free space optical components that hinder the wide uptake of hollow core fibers at the system level.

An alternative to the attempt of connecting hollow core fibers to solid fiber-based components is to develop HC-ARF-based fiber components. Such hollow core-based components definitely facilitate the simplification of the HC-ARF-based optical system. To date, the reported works on hollow-core fiber-based components are limited, with most of them focusing on fiber couplers [22–24]. Fiber filters, on the one hand, comprise one of the most important fiber-based components that allows the transmission of certain wavelengths [25,26]. Nonetheless, the realization of hollow core-based filters has not yet been demonstrated. In fact, HC-ARFs are of great potential to be fiber filters on account of the core wall thickness-dependent transmission wavelengths and transmission bandwidth. Combining HC-ARFs with different core wall thicknesses will enable customization of the actual transmission bands.

In this work, we report a reliable, flexible and efficient method to process HC-ARFs into fiber filters. By locally heating an HC-ARF with a CO_2 laser, the fiber structure becomes deformed, and cladding capillaries shrink to have thicker walls. This process is analogous to "writing" a new fiber with a thicker wall on the original fiber, resulting in creating new high loss regions (resonant wavelengths) in the original transmission bands. Thus, the control of the transmission wavelengths of the fiber filter is realized by controlling the wall thickness "written" on the original fiber. Furthermore, this method is able to integrate an in-line fiber filter into an HC-ARF without extra loss and is promising in many HC-ARF-based applications, which are also discussed.

2. Methods

As depicted in Figure 1a, the HC-ARF is composed of one layer of capillaries that surrounds the hollow region to form an anti-resonant guidance. The nodeless cladding and negative curvature of the core-cladding boundary are two critical features of HC-ARF that significantly reduce the fiber loss [13,27,28]. The HC-ARF fabricated by the stack and draw technique [9,29] has a good structure, as presented in Figure 1b. Among all the geometric parameters, the most important parameters are core diameter D, capillary size p and capillary wall thickness t. While both D and p relate to the fiber loss, t is the only geometric parameter that determines the transmission wavelengths. Transmission bands of HC-ARFs are determined by the resonant wavelengths. Resonant wavelengths are the central wavelength of high loss regions, and a low loss transmission band exists between every adjacent high loss region. The m-th order resonant wavelength, λ_m, can be calculated from the following equation [30]:

$$\lambda_m = \frac{2t \times \sqrt{n_2{}^2 - n_1^2}}{m}, m = 1, 2, 3...$$ (1)

where t is the wall thickness, n_2 is the refractive index of cladding material and n_1 is the refractive index of core material. In the case of silica-based air core fiber, we set $n_2 = 1.45$ and $n_1 = 1.00$. The low loss region between the m-th and $(m + 1)$-th resonant wavelengths is called the m-th transmission band.

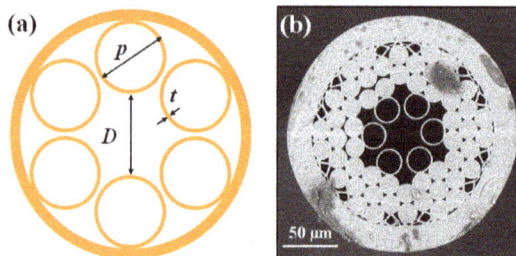

Figure 1. (**a**) Schematic diagram of the cross-sectional view of a negative curvature HC-ARF (Hollow-Core Anti-Resonant Fiber). t is the capillary wall thickness; p is the capillary outer diameter; and D is the core diameter. (**b**) HC-ARF fabricated by the stack and draw method that produces a good structure.

The second and third transmission bands of HC-ARFs with different wall thicknesses are studied by simulation. The results are calculated by a vector wave expansion method using the open source software Polymode [31]. As shown in Figure 2a, combing a fiber ($t = 1.0\,\mu m$) to another fiber possessing a thicker wall ($t = 1.1\,\mu m$) leads to the narrowing of transmission bands. We define the transmission band when its CL is below 3 dB/m. If the wall thickness difference between the fibers becomes larger (e.g., $t = 1.0\,\mu m$ and $t = 1.2\,\mu m$), the ($m+1$)-th transmission band of the thick-wall fiber can overlap with the m-th transmission band of the thin-wall fiber, resulting in multiple and narrow transmission bands (see Figure 2b).

Although combing HC-ARFs with different t can narrow the transmission bands, it is inefficient and troublesome to fabricate multiple HC-ARFs with different wall thicknesses and to splice them together. Instead, it makes more sense to process a piece of uniform fiber to have varied wall thicknesses along its axis. To achieve this, a CO_2 laser-assisted glass processing stage (LZM-100 from Fujikura Ltd., Tokyo, Japan) is used to process the fiber. As illustrated in Figure 3a, a piece of HC-ARF is loaded onto two fiber holders. During the fiber processing, both holders rotate at the same speed to ensure symmetric heating. In parallel, the holders longitudinally move at the same traveling speed toward the same direction in order to avoid any twist or stretch. A section of the HC-ARF is locally heated by the CO_2 laser with tunable power P. Under the CO_2 laser treatment, the exposed section undergoes shrinkage due to surface tension, resulting in increasing wall thickness. The wall thickness of the processed fiber is controllable by adjusting the laser power P. Hence, uniform modification of wall thickness is achievable by moving the laser exposure along the fiber axis, as illustrated in Figure 3b.

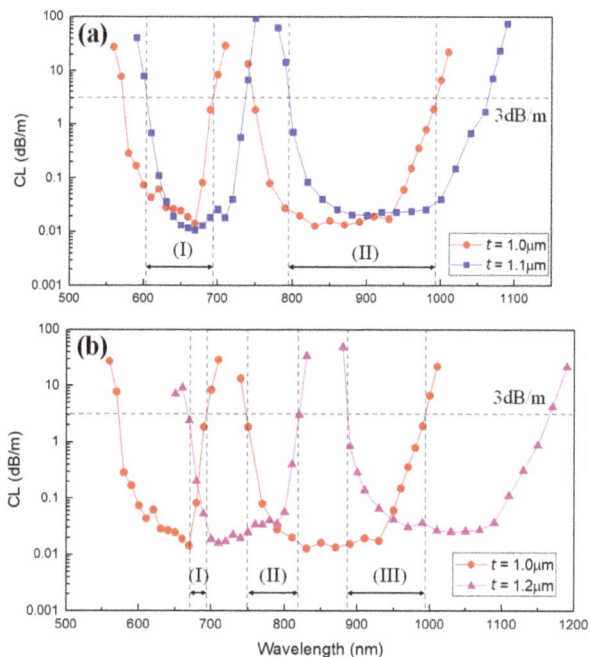

Figure 2. Simulated second and third transmission bands of HC-ARFs with different t. Roman numerals mark the hybrid transmission bands of HC-ARFs with: (**a**) $t = 1.0\,\mu m$ and $t = 1.1\,\mu m$; (**b**) $t = 1.0\,\mu m$ and $t = 1.2\,\mu m$. In all cases, $D = 30.0\,\mu m$, $p = 24.0\,\mu m$.

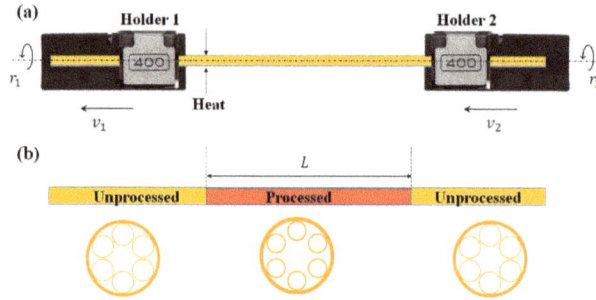

Figure 3. (**a**) Schematic diagram of the HC-ARF processing, $r_1 = r_2 = 50$ rpm, $v_1 = v_2 = 0.5$ μm/ms; heating is realized with a CO_2 laser. (**b**) During the process, the HC-ARF undergoes structural deformation, yielding a thicker wall. L is the length of the processed fiber.

3. Results and Discussion

A 17-cm HC-ARF, Fiber #1 (please, see Figure 4), was processed with the aforementioned method under different heating powers. The original fiber (Fiber #1) had a wall thickness $t = 1.40$ μm and a core diameter $D = 31.2$ μm, while the wall thickness of the processed fiber increased to be 1.49 μm (Fiber #2, processed with 19.2 W of heating power) and 1.63 μm (Fiber #3, processed with 20.1 W of heating power), respectively. As predicted, under high temperature, capillaries shrunk to induce a thicker wall due to surface tension. Transmission spectra of different fiber combinations were also measured, as shown in Figure 4. Resonant wavelengths of Fiber #2 and Fiber #3 were calculated and marked with blue and red dashed lines, respectively. As demonstrated by the measured transmission spectra, writing a new fiber on the original fiber by the CO_2 laser introduced a new high loss region (extra resonant band), realizing a selective transmission/rejection in-fiber filter. We noticed that the writing process did not introduce any significant extra loss. Besides the controllability, the reproducibility of the proposed method was also verified as the fibers processed under the same CO_2 laser power showed similar transmission spectra.

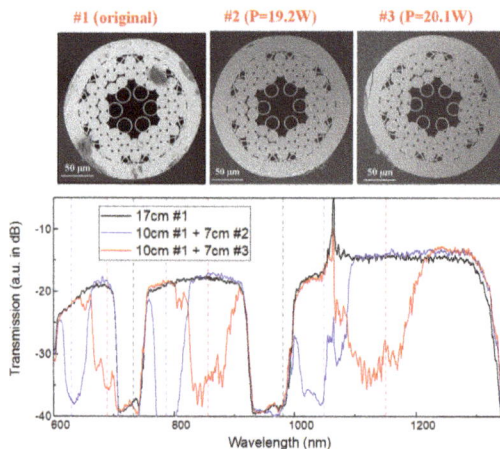

Figure 4. Transmission spectra of different fiber combinations. Fiber #1: unprocessed fiber, $t = 1.40$ μm, $D = 31.2$ μm; Fiber #2: processed under $P = 19.2$ W, $t = 1.49$ μm, $D = 33.2$ μm; Fiber #3: processed under $P = 20.1$ W, $t = 1.63$ μm, $D = 36.5$ μm. Resonant wavelengths of Fiber #1, Fiber #2 and Fiber #3 are calculated from Equation (1) and marked with dashed black lines, dashed blue lines and dashed red lines respectively.

More interestingly, as both t and D were changed, the dispersion curve of the fiber significantly shifted. As shown in Figure 5a, the Effective Refractive Index (ERI) curves were obtained by fitting the simulation values (solid dots), then the Group Velocity Dispersion (GVD) curves were calculated from the ERI curves, as shown in Figure 5b. As the fiber structure was changed from #1–#3, the zero dispersion wavelength shifted from around 860 nm to around 1000 nm. The change of the core size was responsible for this dispersion curve transformation [4]. Therefore, the proposed method also has promising prospects in applications relying on dispersion control, especially in pulse compression.

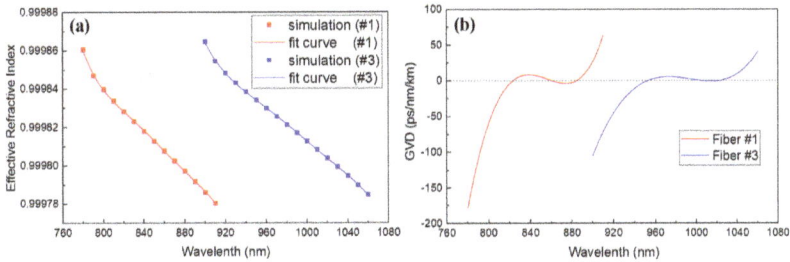

Figure 5. (**a**) Effective refractive index and (**b**) GVD (Group Velocity Dispersion) curve of both Fiber #1 and #3. In (**a**), the dots plot the simulation values, while the solid curves are quintic functions, which fit the simulation values.

The method is also applicable to a multiple layer HC-ARF, as evidenced by the results in Figure 6. The two-layer structure responded to the CO_2 laser writing process to introduce an additional rejection band. As indicated in both Figures 4 and 6, implementing a new single different wall thickness for an existing HC-ARF can made band reject filters. Here, we also demonstrate a band pass filter by introducing multiple different wall thicknesses into an original HC-ARF. The procedure is described in Figure 7. We used a two-layer HC-ARF as a pristine fiber. A section of 7 cm in the 25 cm-long pristine fiber (Fiber #4) was written under exposure power $P = 19.4$ W. Its corresponding transmission is present in the blue curve in Figure 7. Subsequently, another section of the same length was written by the lower power of $P = 18.7$ W to decrease the number of transmission bands (red line in Figure 7). The final fiber had limited transmission bands and worked more like a band pass filter with low excess loss, but 20-dB high extinction ratio.

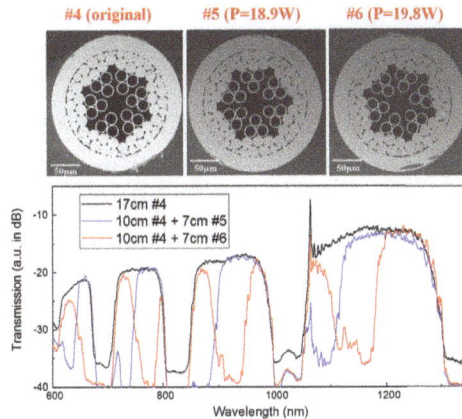

Figure 6. Transmission spectra of different fiber combinations. Fiber #4: unprocessed fiber, $t = 1.97$ μm; Fiber #5: processed under $P = 18.9$ W, $t = 2.09$ μm; Fiber #6: processed under $P = 19.8$ W, $t = 2.21$ μm.

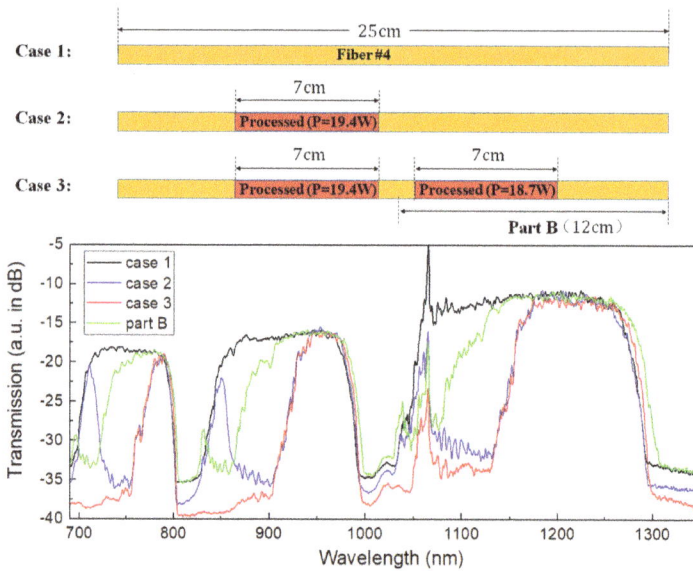

Figure 7. Schematic diagram illustrating the steps to make an HC-ARF-based band pass filter. As multiple different wall thicknesses are "written" into the original HC-ARF, the transmission bands are greatly narrowed.

The demonstrated HC-ARF-based filter could find interesting applications such as pump and signal wavelength separation in a gas-/liquid-filled HC-ARF system. HC-ARFs have been adopted to demonstrate excellent cavities for gas Raman generation and optofluidic systems [19,20] in which for both cases, free space optical filters were selected to filter out the excitation beam. Alternatively, by simply processing the HC-ARF with the method proposed in this work, an HC-ARF-based optical filter can be seamlessly written into the system without extra loss, as illustrated in Figure 8. This would be one step closer to an all-fiberized hollow core fiber system.

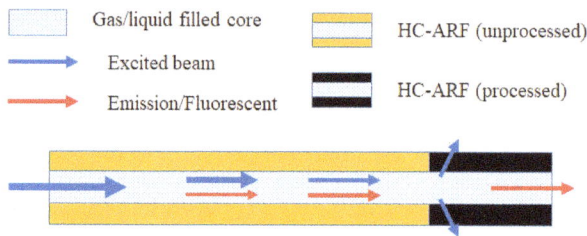

Figure 8. This schematic diagram shows the working principle of the HC-ARF-based filter in gas-/liquid-filled optical systems.

4. Conclusions

HC-ARFs with uniform wall thickness thave multiple and broad transmission bands. We have shown that writing a different wall thickness HC-ARF is feasible by using a CO_2 laser-based glass process stage. Consequently, the written HC-ARF exhibits the AND operation of two transmission characteristics defined by the written and the pristine HC-ARFs' wall thicknesses. In addition, we demonstrated multiple chained AND operations by writing various wall thickness HC-ARFs in series along the fiber axis. On the basis of this principle, a novel method has also been proposed and demonstrated to convert HC-ARFs into in-fiber filters. The HC-ARF filter fabricated by this method benefits from low excess loss, easy integration with the HC-ARF-based system and controllable transmission/rejection wavelengths. We have also suggested a potential application of such fiber filters toward an all-fiberized HC-ARF system.

Author Contributions: X.H. conceived of the idea, performed the experiments, analyzed the data and wrote the manuscript. S.Y. supervised the design of the study and revised the manuscript. K.-T.Y supervised the design and participated in the experimental setup.

Funding: This research received no external funding.

Acknowledgments: S.Y. acknowledges support from KEIP through the Global Research Programme.

Conflicts of Interest: The authors declare no conflict of interest.

Abbreviations

The following abbreviations are used in this manuscript:

HC-PCF	Hollow-Core Photonic Crystal Fiber
HC-PBGF	Hollow-Core Photonic Bandgap Fiber
HC-ARF	Hollow-Core Anti-Resonant Fiber
ERI	Effective Refractive Index
GVD	Group Velocity Dispersion

References

1. Birks, T.A.; Roberts, P.J.; Russell, P.S.J.; Atkin, D.M.; Shepherd, T.J. Full 2-D photonic bandgaps in silica/air structures. *Electron. Lett.* **1995**, *31*, 1941–1943.:19951306. [CrossRef]
2. Smith, C.M.; Venkataraman, N.; Gallagher, M.T.; Müller, D.; West, J.A.; Borrelli, N.F.; Allan, D.C.; Koch, K.W. Low-loss hollow-core silica/air photonic bandgap fibre. *Nature* **2003**, *424*, 657–659. [CrossRef] [PubMed]
3. Jaworski, P.; Yu, F.; Carter, R.M.; Knight, J.C.; Shephard, J.D.; Hand, D.P. High energy green nanosecond and picosecond pulse delivery through a negative curvature fiber for precision micro-machining. *Opt. Express* **2015**, *23*, 8498–8506. [CrossRef] [PubMed]
4. Gérôme, F.; Cook, K.; George, A.K.; Wadsworth, W.J.; Knight, J.C. Delivery of sub-100fs pulses through 8 m of hollow-core fiber using soliton compression. *Opt. Express* **2007**, *15*, 7126–7131. [CrossRef] [PubMed]
5. Poletti, F.; Wheeler, N.V.; Petrovich, M.N.; Baddela, N.; Fokoua, E.N.; Hayes, J.R.; Gray, D.R.; Li, Z.; Slavík, R.; Richardson, D.J. Towards high-capacity fibre-optic communications at the speed of light in vacuum. *Nat. Photonics* **2013**, *7*, 279–284. [CrossRef]
6. Roberts, P.J.; Couny, F.; Sabert, H.; Mangan, B.J.; Williams, D.P.; Farr, L.; Mason, M.W.; Tomlinson, A.; Birks, T.A.; Knight, J.C.; et al. Ultimate low loss of hollow-core photonic crystal fibres. *Opt. Express* **2005**, *13*, 236–244. [CrossRef] [PubMed]
7. Couny, F.; Benabid, F.; Roberts, P.J.; Light, P.S.; Raymer, M.G. Generation and photonic guidance of multi-octave opticalfrequency combs. *Science* **2007**, *318*, 1118–1121. [CrossRef] [PubMed]
8. Gao, S.F.; Wang, Y.Y.; Ding, W.; Jiang, D.L.; Gu, S.; Zhang, X.; Wang, P. Hollow-core conjoined-tube negative-curvature fibre with ultralow loss. *Nat. Common.* **2018**, *9*, 2828–2828. [CrossRef] [PubMed]
9. Huang, X.; Yoo, S.; Yong, K. Function of second cladding layer in hollow core tube lattice fibers. *Sci. Rep.* **2017**, *7*, 1618. [CrossRef] [PubMed]
10. Poletti, F. Nested anti-resonant nodeless hollow core fiber. *Opt. Express* **2014**, *22*, 23807–23828. [CrossRef] [PubMed]
11. Yu, F.; Knight, J.C. Negative curvature hollow core optical fiber. *IEEE J. Sel. Top. Quantum Electron.* **2016**, *22*, 4400610. [CrossRef]
12. Huang, X.; Qi, W.; Ho, D.; Yong, K.T.; Luan, F.; Yoo, S. Hollow core anti-resonant fiber with split cladding. *Opt. Express* **2016**, *24*, 7670–7678. [CrossRef] [PubMed]
13. Belardi, W.; Knight, J.C. Effect of core boundary curvature on the confinement losses of hollow anti-resonant fibers. *Opt. Express* **2013**, *21*, 21912–21917. [CrossRef] [PubMed]

14. Hayes, J.R.; Sandoghchi, S.R.; Bradley, T.D.; Liu, Z.; Slavík, R.; Gouveia, M.A.; Wheeler, N.V.; Jasion, G.; Chen, Y.; Fokoua, E.N.; et al. Antiresonant hollow core fiber with an octave spanning bandwidth for short haul data communications. *J. Lightwave Technol.* **2017**, *35*, 437–442. [CrossRef]

15. Debord, B.; Amsanpally, A.; Chafer, M.; Baz, A.; Maurel, M.; Blondy, J.; Hugonnot, E.; Scol, F.; Vincetti, L.; Gérôme, F.; et al. Ultralow transmission loss in inhibited-coupling guiding hollow fibers. *Optica* **2017**, *4*, 209–217. [CrossRef]

16. Belardi, W.; Knight, J.C. Hollow anti-resonant fibers with reduced attenuation. *Opt. Lett.* **2014**, *39*, 1853–1856. [CrossRef] [PubMed]

17. Gao, S.F.; Wang, Y.Y.; Ding, W.; Wang, P. Hollow-core negative-curvature fiber for UV guidance. *Opt. Lett.* **2018**, *43*, 1347–1350. [CrossRef] [PubMed]

18. Yu, F.; Wadsworth, W.J.; Knight, J.C. Low loss silica hollow core fibers for 3–4 µm spectral region. *Opt. Express* **2012**, *20*, 11153–11158. [CrossRef] [PubMed]

19. Liu, X.L.; Ding, W.; Wang, Y.Y.; Gao, S.F.; Cao, L.; Feng, X.; Wang, P. Characterization of a liquid-filled nodeless anti-resonant fiber for biochemical sensing. *Opt. Lett.* **2012**, *42*, 863–866. [CrossRef] [PubMed]

20. Williams, G.O.; Euser, T.G.; Arlt, J.; Russell, P.S.J.; Jones, A.C. Hollow anti-resonant fibers with reduced attenuation. *ACS Photonics* **2014**, *1*, 790–793. [CrossRef]

21. Russell, P.S.J.; Hölzer, P.; Chang, W.; Abdolvand, A.; Travers, J.C. Hollow-core photonic crystal fibres for gas-based nonlinear optics. *Nat. Photonics* **2014**, *8*, 278–286. [CrossRef]

22. Huang, X.; Ma, J.; Tang, D.; Yoo, S. Hollow-core air-gap anti-resonant fiber couplers. *Opt. Express* **2017**, *25*, 29296–29306. [CrossRef]

23. Liu, X.; Fan, Z.; Shi, Z.; Ma, Y.; Yu, J.; Zhang, J. Dual-core anti-resonant hollow core fibers. *Opt. Express* **2016**, *24*, 17453–17458. [CrossRef] [PubMed]

24. Argyros, A.; Leon-Saval, S.G.; van Eijkelenborg, M.A. Twin-hollow-core optical fibres. *Opt. Commun.* **2009**, *282*, 1785–1788. [CrossRef]

25. Ouellette, F. All-fiber filter for efficient dispersion compensation. *Opt. Lett.* **1991**, *16*, 303–305. [CrossRef] [PubMed]

26. Antonio-Lopez, J.E.; Castillo-Guzman, A.; May-Arrioja, D.A.; Selvas-Aguilar, R.; LiKamWa, P. Tunable multimode-interference bandpass fiber filter. *Opt. Lett.* **2010**, *35*, 324–326. [CrossRef] [PubMed]

27. Pryamikov, A.D.; Biriukov, A.S.; Kosolapov, A.F.; Plotnichenko, V.G.; Semjonov, S.L.; Dianov, E.M. Demonstration of a waveguide regime for a silica hollow-core microstructured optical fiber with a negative curvature of the core boundary in the spectral region >3.5 µm. *Opt. Express* **2011**, *19*, 1441–1448. [CrossRef] [PubMed]

28. Kolyadin, A.N.; Kosolapov, A.F.; Pryamikov, A.D.; Biriukov, A.S.; Plotnichenko, V.G.; Dianov, E.M. Light transmission in negative curvature hollow core fiber in extremely high material loss region. *Opt. Express* **2013**, *21*, 9514–9519. [CrossRef] [PubMed]

29. KolBrilland, L.; Smektala, F.; Renversez, G.; Chartier, T.; Troles, J.; Nguyen, T.N.; Traynor, N.; Monteville, A. Fabrication of complex structures of Holey Fibers in Chalcogenide glass. *Opt. Express* **2006**, *14*, 1280–1285. [CrossRef]

30. Litchinitser, N.M.; Dunn, S.C.; Usner, B.; Eggleton, B.J.; White, T.P.; McPhedran, R.C.; de Sterke, C.M. Resonances in microstructured optical waveguides. *Opt. Express* **2003**, *11*, 1243–1251. [CrossRef] [PubMed]

31. Issa, N.A.; Poladian, L. Vector wave expansion method for leaky modes of microstructured optical fibers. *J. Lightwave Technol.* **2003**, *21*, 1005–1012. [CrossRef]

fibers

MDPI

Article

Hollow Core Optical Fibers for Industrial Ultra Short Pulse Laser Beam Delivery Applications

Sebastian Eilzer * and Björn Wedel

PT Photonic Tools GmbH, Johann-Hittorf-Straße 8, 12489 Berlin, Germany; b.wedel@photonic-tools.de
* Correspondence: s.eilzer@photonic-tools.de; Tel.: +49-30-6392-78000

Received: 14 September 2018; Accepted: 10 October 2018; Published: 16 October 2018

Abstract: Hollow core fibers were introduced many years ago but are now starting to be used regularly in more demanding applications. While first experiments mainly focused on the characterization and analysis of the fibers themselves, they are now implemented as a tool in the laser beam delivery. Owing to their different designs and implementations, different tasks can be achieved, such as flexible beam delivery, wide spectral broadening up to supercontinuum generation or intense gas-laser interaction over long distances. To achieve a constant result in these applications under varying conditions, many parameters of these fibers have to be controlled precisely during fabrication and implementation. A wide variety of hollow core fiber designs have been analyzed and implemented into a high-power industrial beam delivery and their performance has been measured.

Keywords: hollow core fiber; beam delivery; ultrafast lasers

---

1. Introduction

Laser sources with ultrashort pulses have taken a rapid development and broader industrial use over the past couple of years. Of particular interest are material processing technologies, which enable manufacturing of highly precise structures and delicate materials [1]. Thanks to the short light–matter interaction of ultrafast laser pulses, thermal damage of material substrates can be greatly reduced and also new interactions, such as filamentation, can be used.

While the development of laser sources and material processing technologies have steadily evolved (both in terms of available optical parameters as well as industrial performance), system technology has often been identified to be the limiting factor in today's industrial application. This includes beam transport and beam steering and shaping.

The usage of continuous-wave high-power lasers in the industry has made a huge leap forward by incorporating flexible fiber-based beam delivery. There are many advantages of using a fiber beam transport, compared to free space, especially in an industrial environment. The most obvious advantage is the flexible use of the laser at different positions while maintaining the beam properties. While gantry units can be realized with stabilized free-space setups, the option for different angles and flexibility with a fiber are far greater, especially in robot arm applications. These advantages are made accessible for the use of high-power ultrafast lasers by the introduction of hollow core fiber. Their laser beams cannot be transported with traditional step-index fibers as their peak intensity lies above the damage threshold of silica and would therefore destroy the fiber itself. Thanks to the air- or gas-filled core in a hollow core fiber, it is possible to efficiently guide short laser pulses with little dispersion and high peak power over relevant distances.

The first microstructured hollow core fibers were reported by Cregan et al. in 1999 [2]. To contain the laser energy inside a hollow core, distinctly different physical phenomena, compared to those of step-index fibers, have to be applied, since the material surrounding the core has a higher refractive index. The ordered and repetitive structure of a bandgap fiber cladding allows for wavelength areas where the light does not couple into the cladding and therefore remains inside the core. This basic concept, explained in different ways and with varying models, still holds true for any hollow core fiber design.

The integration of hollow core optical fibers into laser light cables has been presented by Photonic Tools in 2015. The ultrafast beam delivery comprises a coupling unit, the laser light cable itself and a modular processing head (see Section 2.3) and can be used for applications ranging from sensing to material processing or nonlinear pulse shaping.

The aim of this work is to summarize the integration of hollow core fiber in an industrial environment and present some representative applications.

2. Materials and Methods

2.1. Types of Fibers

There exists a large variety of hollow core fibers which could in principle be used for transmission of ultrafast laser pulses. However, when considering factors such as required lengths, typical laser application parameters, mechanical tolerances, optical layouts and operation environments, quite a few new boundary conditions play a role in designing the optimal fiber.

A key point, when working with hollow core fibers, concerns the existence of more than one mode inside the core. The amount of dampening of each of these modes determines beam quality at the output when several modes are excited, either during coupling into the fiber or when disturbances influence the transmission. If higher-order modes are highly suppressed, a lower overall transmission can be expected. To transmit higher pulse energies and couple them reliably into a fiber, a larger core is necessary; however, a larger core also allows for less separation between the higher-order modes inside the fiber, leading to possible instability.

To still maintain a high beam quality, special care has to be taken when designing the coupling optics to excite mostly the fundamental mode of the fiber. To contain the light inside the core and reduce coupling into the cladding, core geometries with a negative curvature have been found to be very effective in keeping the high field strength away from the surrounding glass structure [3]. These fibers offer a low loss over a wide wavelength range while still maintaining larger core sizes. There are two main types which are called Kagome fibers, due to the lattice structure, and antiresonant fibers. More well-known are single-ring or revolver fibers which build up the core region by surrounding it with one layer of non-touching rings, or cone-shaped structures (see Figure 1).

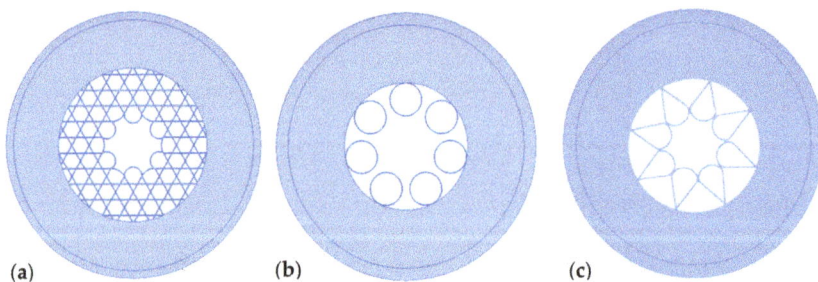

(a) (b) (c)

Figure 1. Depictions of common large-area hollow core fiber designs: (a) Kagome-type fiber; (b) single ring- or revolver-type fiber; and (c) cone-type fiber.

In order to determine the spectral properties and dynamic behavior of these fibers, complex studies of their guiding mechanism are required. Many models of varying complexity have been proposed and compared to analytical calculations as well as experiments. Influences on confinement by the wall thickness [4,5] and by the cladding structure [6] have been investigated, as well as dispersion of these fibers [7]. When taking industrial use into consideration, additional influences on the beam properties, e.g., by bending the fiber, have to be included to achieve an optimized design [8,9]. Therefore, by tuning fiber structure paramters, like core size, web thickness and structure elements, the fibers can be tailored for specific purposes and applications such as spectral broadening up to supercontinuum generation [10] or intense gas-laser interaction over long distances [11].

2.2. Limitations of the Technology

With increasing beam power, one has to look at the limits a hollow core beam transport possesses, determined by the physics involved. These include the maximum energy of lasers pulses and the shortest possible pulse length.

The maximum energy transmission in a hollow core fiber is still limited by the destruction of the silica [12]. The silica structure is not in the center of the beam where the energy density is highest, but rather in the outer areas. It is important to note that while the fundamental mode of the fiber has only very little overlap with the structure, the critical point is coupling into the fiber. The fundamental free-space mode has to excite the fundamental fiber mode, and the matched free-space Gaussian mode has a slightly higher overlap with the unmodified fiber structure than the fundamental fiber mode. A small portion of light will overlap directly with the glass struts of the fiber and interact with it. Using this assumption, one can determine a peak power dependent on the core size and the matched beam that is coupled into it (see Figure 2). In real-life applications, a reduced coupling accuracy and slight movement of the beam have to be taken into account.

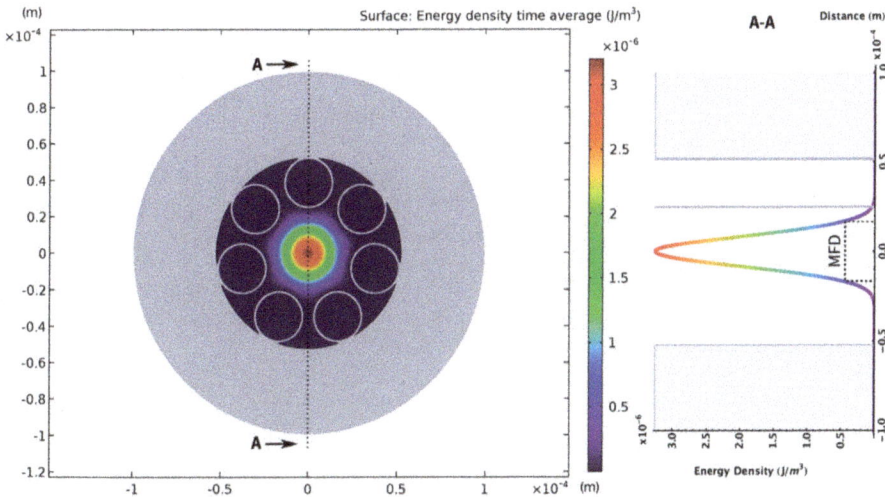

Figure 2. Schematic light field intensity overlapping with the glass structure in an antiresonant fiber design.

The coupling efficiency can be calculated theoretically by calculating the overlap of coupled free-space mode and the fiber eigenmodes. Variations can also be introduced and the results can be compared to experimental results using low-power beams (see Figure 3).

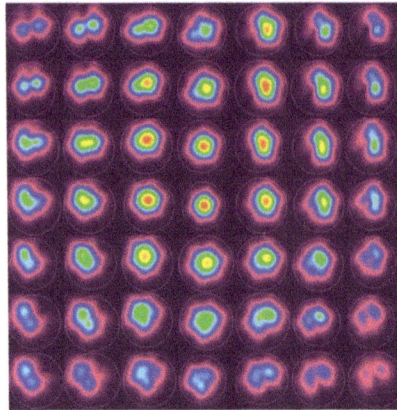

Figure 3. Transmitted mode content under varying coupling conditions of a hollow core fiber.

Typically, high-power pulses travelling along the fiber will often have enough energy to interact with the gas inside the hollow core, which will alter the pulse characteristiocs [13–15]. In order to reduce these effects, the fiber material can be filled with other gases or kept at a vacuum to significantly reduce any gas interaction. However, the fiber itself will still have a small waveguide dispersion, which will lead to pulse broadening and chirping. Therefore, if very short pulses below 100 fs are demanded, the dispersive interaction inside the fiber has to be taken into account and can be precompensated in the compressor of the coupled laser.

Fibers for infrared (IR) light delivery have been produced for a few years, but transmission at shorter wavelengths in the visible or even in the ultraviolet (UV) regime is of high interest. Theoretically, the fundamental guiding mechanism of hollow core fibers allows for structures which transmit in these wavelength range; however, the real-life loss due to scattering is also growing with smaller wavelengths. Due to this, the production tolerances are a key factor when designing fibers in the visible and UV spectral region and currently do not allow for fiber material which can match the attenuation and stability of IR at smaller wavelengths. In general, achieving low-loss and nearly undisturbed beam propagation requires a fiber structure which is optimized for the desired transmission window as well as the production routine.

2.3. Fiber Integration

Bringing the hollow core fiber technology to the industry has been a challenging process. Typically, laboratory experiments hold a fiber freely in a tabletop setup with microadjusters for low power, or in a gas cell for tailored nonlinear processes. In both cases, the fiber is prepared for the investigation which is planned. This, of course, differs vastly from using a fiber as a tool, where it has to perform, mostly without preparation, in a more demanding environment.

A special connector, which houses the fiber tip in a sealed environment, has been designed (see Figure 4). The fiber is protected from environmental influences but can also be filled with gas or evacuated. Differential pumping schemes are also possible by applying different pressures at each side of the laser light cable. Standard safety features, such as a fiber continuity monitoring, are also included, as well as the possibility for active cooling of the connector when used with very high power. The conduit is designed with continuous use in demanding application in mind and can protect the fiber even after years of continuous bending. The fiber itself is prealigned to very few micron precision and can be exchanged in few minutes by the use of a flange connector which is easy to adapt.

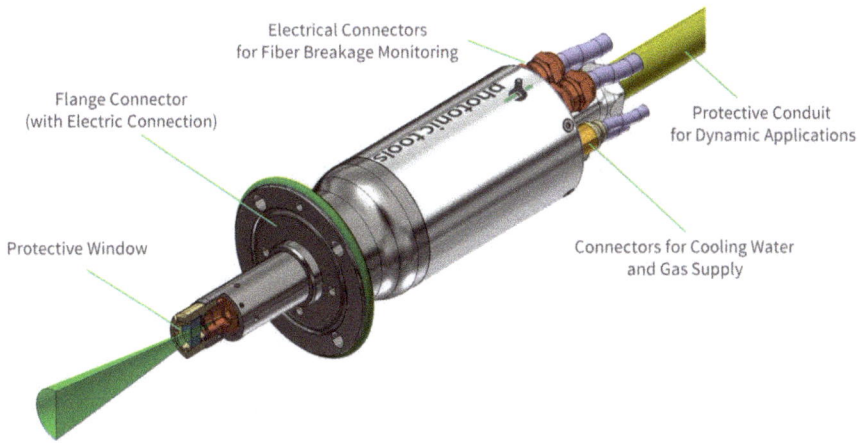

Figure 4. Ultrafast laser light cable connector with connections for water cooling, evacuation and safety circuits.

Together with the laser light cable, a modular system, composed of components specially designed for ultrafast laser beam delivery and applications, has also been introduced (see Figure 5) [16]. These components include a coupling unit, which can accept a wide range of laser beams and adapt them to be optimally coupled into the fiber. The unit uses a flexible modular design and can also be integrated into a laser head. On the output side, different beam shaping tools are available, ranging from collimation units, which prepare a requested beam size, to complex processing heads with diagnostics, observation and variable beam shaping.

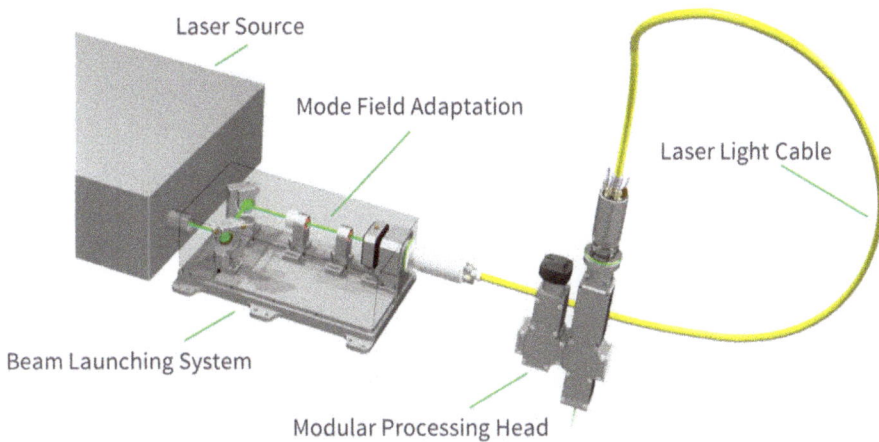

Figure 5. Ultrafast beam delivery system with a beam launching system (BLS) composed of a mode field adaptation, a laser light cable (LLK) and a modular processing head (MPH).

Whilst the ultrafast laser beam can also be delivered to the workpiece by free-space beam delivery components, transporting the laser beam by fiber offers several advantages for the laser user.

Integrating the laser source into the application system or machine will be much simpler, because the flexible beam delivery enables the laser source to be installed separately from the optics, which focuses the beam into the workpiece. For example, this allows for a much simpler mechanical structure and lower moved masses by simply routing the laser light cable, instead of a fixed beam path using multiple mirrors.

Additionally, the service of the system is becoming much simpler because of the reduced number of components in the beam path, and the possibility of installing and exchanging a laser light cable with no or only very little beam alignment. The latter is possible since the fiber is positioned very accurately in the laser light cable connector and therefore provides an excellent optomechanical reference for the laser beam.

In summary, this leads to a lower cost of ownership for the complete laser system not only in capital investment but also running cost. The additional positive side effect is an improved production uptime of the complete laser application system, which is mainly driven by less complexity and simplified serviceability.

The combination of optical design tailored specifically for the use in conjunction with hollow core fibers of different designs as well as rigorous characterization has led to a design, which shows remarkable stability under varying conditions and with most laser sources.

3. Results

3.1. Performance

Numerous integrations of the ultrafast beam delivery system over the years, with varying laser pulse characteristics, have been performed. Peak powers of up to a gigawatt and an average power of 200 W have been transmitted with an efficiency between 85% and 93%, depending on the length of the fiber and the beam quality of the laser coupled into it (see Figure 6 for a selection).

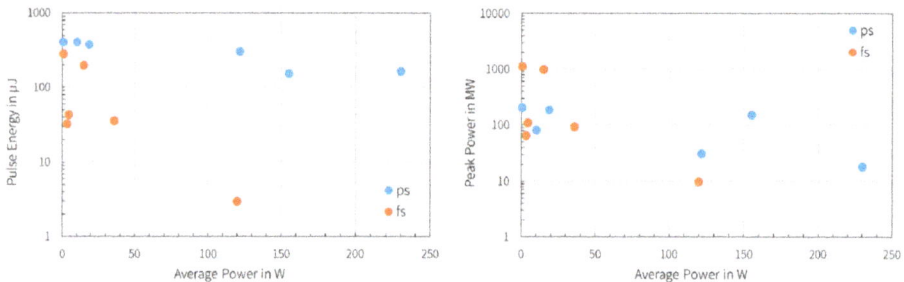

Figure 6. Overview of pulsed laser values which could be coupled and transmitted successfully using the flexible beam delivery. Each dot represents a parameter pair of the laser source. Picosecond and femtosecond laser systems are separately marked.

Higher-modes present in the coupled laser beam are more likely to excite higher-order modes inside the fiber, which reduces transmission and stability. Different fiber designs can influence this behavior in different strengths. An example for a stable result can be seen in Figure 7, where the residual movement of the center of intensity of the laser beam at the workpiece is shown. It was measured while performing strong bending and twist movement of the laser light cable. As can be seen, the residual movement of the beam, even under strong bending of the laser light cable, only amounts to very few percentages of the focus spot size, which hardly has any impact on the typical laser application.

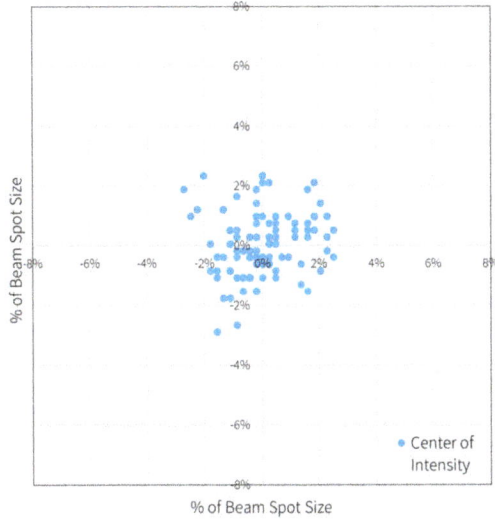

Figure 7. Movement of the focus spot relative to the beam size at the fiber output while moving the fiber in an application.

3.2. Application

Hollow core fibers can transmit ultrafast beams nearly undisturbed. Due to the wide range of pulse paramters possible, different residual influences may affect the transmitted beam slightly. To determine their influence in typical machining applications, many comparative application studies have been conducted. These include sensing, cutting, drilling, ablation and surface structuring and modification.

The key parameters for these applications are the focus intensity and focus shape as well as the pulse shape. All of these can be controlled very precisely when using a hollow core fiber for beam delivery. The output of the fiber is providing a near single-mode beam with minimal fluctuations and fixed positions, which allows for a precise processing.

For this investigation, the fiber beam delivery was integrated in different applications and the results were compared qualitatively with free-space beam paths.

Firstly, cutting results from nitinol stent cutting are shown (see Figure 8). The process parameters are shown in Table 1.

Table 1. Processing parameters of nitinol cutting.

Parameters	Value	Unit
Laser source	Amplitude Satsuma	
Pulse length	350	fs
Power	5	W
Repetition rate	0–2	MHz
Beam delivery	Photonic Tools	
LLK length	3	m
Collimation	100	mm
Air pressue in fiber	5–1000	mbar
Workpiece	Nitinol	
Wavelength	1030	nm
Polarization	circular	
Repetition rate	300	kHz
Pulse energy	~8	µJ
Spot size	~12	µm

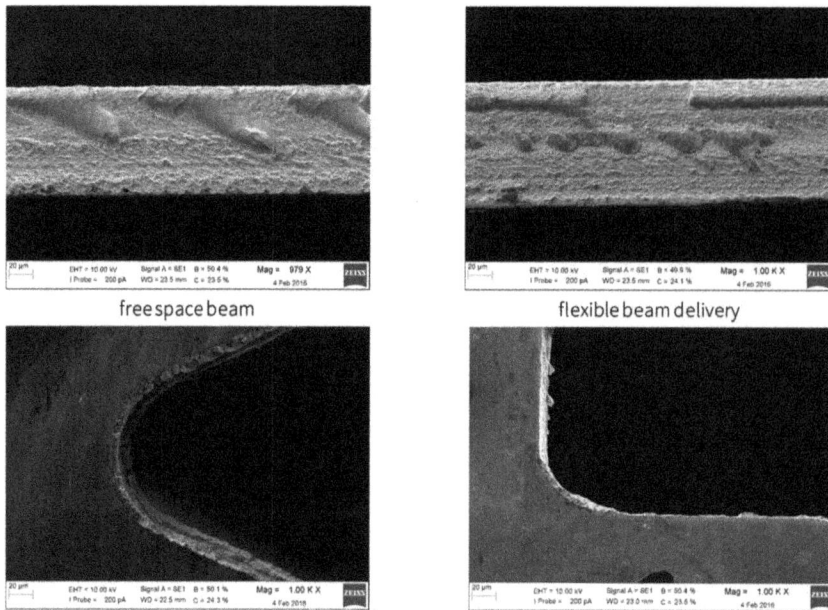

free space beam

flexible beam delivery

Figure 8. Comparison of cutting results. Free-space beam delivery on the left and fiber beam delivery on the right (in co-operation with femtos GmbH).

The comparison shows very similar results with the two different beam delivery methods. The quality of the cutting edge is of the same quality and the final products could be further processed without any alteration.

Next, results from silicon surface modification are presented (see Figure 9). A square was scanned with varying process parameters while incorporating a free-space or fiber beam delivery in the beam path to the galvanometer scanner. See Table 2 for the corresponding process parameters. Additionally, the fiber could be used to widen the spectrum of the pulse due to self-phase modulation, controlled by adjusting the pressure inside the fiber. The spectral width of the pulse was measured with a spectrometer.

Table 2. Processing parameters of silicon surface modification.

Parameters	Value	Unit
Laser source	Amplitude S-Pulse	
Pulse length	750	fs
Power	4	W
Repetition rate	0–300	kHz
Beam delivery	Photonic Tools	
LLK length	2.5	m
Collimation	100	mm
Air pressue in fiber	5–1000	mbar
Workpiece	Polycrystalline silicon	
Wavelength	1028 ± 3 1030 ± 30	nm
Polarization	linear	
Repetition rate	100	kHz
Pulse energy	12.5	μJ
Spot size	16	μm
Scan speed	4	mm/s

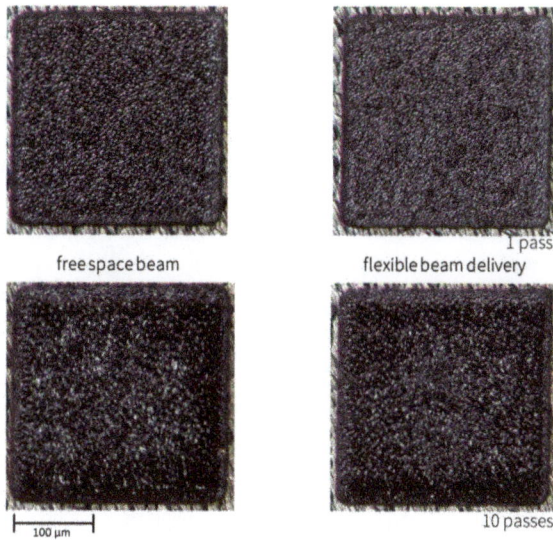

Figure 9. Comparison of surface modification on polychristalline silicon. A single pass of the laser over the area is shown at the top and 10 consecutive passes are shown at the bottom. Free-space beam delivery on the left and fiber beam delivery on the right (in cooperation with Laserzentrum Hannover).

Again, the comparison revealed almost no visible difference between the two beam delivery solutions. The structures, which depended critically on the pulse parameters in size and depth, could be reproduced perfectly with 1 pass over the surface as well as 10 consecutive scans over the same area.

Additionally, a unique feature of the fiber beam delivery was tested and deliberate self-phase modulation could be introduced to the transmitted pulses in varying strengths. Strong increases in ablation depth and structure size were observed when using spectrally broadened pulses (see Figure 10). The pulse bandwidth change from 6 nm to 60 nm revealed a nearly doubled ablation rate in the application.

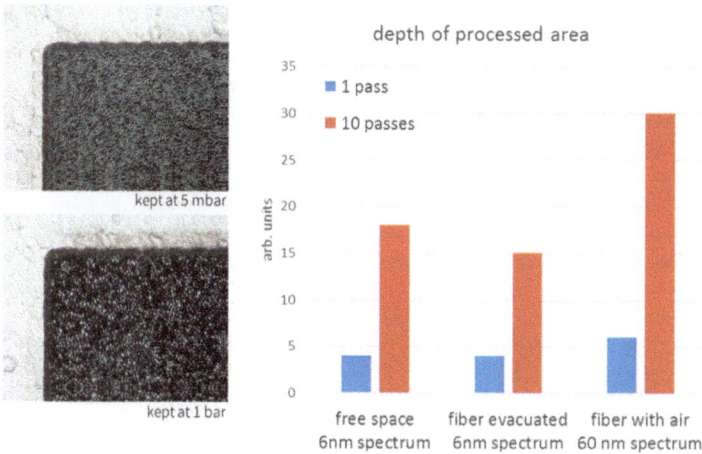

Figure 10. Surface modification on polychristalline silicon with spectrally broadened pulses. Ablation depths and structure sizes with pulse bandwidth change from 6 nm to 60 nm were compared.

4. Conclusions

The use of hollow core fiber for transmission of ultrafast laser pulses in an industrial environment has been shown for a few years in the IR wavelength of YAG laser systems. The completely fiber-based beam delivery system introduced by Photonic Tools already addresses most of the application scenarios, which are actively used. Robot arm applications with hollow core fibers have also been shown. The comparison of free-space beam delivery and fiber-based solutions shows little to no difference in the application results. On the contrary, it can even offer opportunities to increase throughput by using the nonlinear effects inside the hollow core fiber. In a suitable integration, as offered by Photonic Tools, they can be adjusted precisely.

The optimal fiber structure for industrial appplication is still being improved and optimized, so understanding the requirements from applications and characterizing hollow core fibers very rigorously are an essential part of integrating hollow core material into a demanding environment. Recent fiber developments in the green and UV spectral regions show promise for future applications, but cannot reach the same level of performance in high-power applications at the moment.

Taking hollow core fiber out of the scientific environment and tweaking it to perform under diverse conditions is a continuous process, which has made tremendous progress in recent years. Many industrial ultrafast applications can already use the advantages a fiber beam delivery can offer in day-to-day usage with high reliability and unaltered results.

Author Contributions: Conceptualization, S.E. and B.W.; Data curation, S.E.; Funding acquisition, B.W.; Investigation, S.E.; Project administration, B.W.; Resources, B.W.; Supervision, B.W.; Writing—original draft, S.E.; Writing—review & editing, S.E. and B.W.

Funding: This research was funded by the Bundesministerium für Bildung und Forschung, grant numbers [13N13920, 13N13924, 02P14K500].

Acknowledgments: We thank J. Düsing from Laserzentrum Hannover and B. Schöps from femtos GmbH for their cooperation and advise while performing and interpreting the application results.

Conflicts of Interest: The authors declare no conflict of interest. The funders had no role in the design of the study; in the collection, analyses, or interpretation of data; in the writing of the manuscript, and in the decision to publish the results.

References

1. Sugioka, K.; Cheng, Y. Ultrafast lasers—Reliable tools for advanced materials processing. *Light Sci. Appl.* **2014**, *3*, e149. [CrossRef]
2. Cregan, R.F.; Mangan, B.J.; Knight, J.C.; Birks, T.A.; Russell, P.S.J.; Roberts, P.J.; Allan, D.C. Single-mode photonic band gap guidance of light in air. *Science* **1999**, *285*, 1537–1539. [CrossRef] [PubMed]
3. Debord, B.; Alharbi, M.; Bradley, T.; Fourcade-Dutin, C.; Wang, Y.Y.; Vincetti, L.; Gérôme, F.; Benabid, F. Hypocycloid-shaped hollow-core photonic crystal fiber part I: Arc curvature effect on confinement loss. *Opt. Express* **2013**, *21*, 28597–28608. [CrossRef] [PubMed]
4. Litchinitser, N.M.; Abeeluck, A.K.; Headley, C.; Eggleton, B.J. Antiresonant reflecting photonic crystal optical waveguides. *Opt. Lett.* **2002**, *27*, 1592–1594. [CrossRef] [PubMed]
5. Yu, F.; Knight, J.C. Spectral attenuation limits of silica hollow core negative curvature fiber. *Opt. Express* **2013**, *22*, 21466–21471. [CrossRef] [PubMed]
6. Alharbi, M.; Bradley, T.; Debord, B.; Fourcade-Dutin, C.; Ghosh, D.; Vincetti, L.; Gérôme, F.; Benabid, F. Hypocycloid-shaped hollow-core photonic crystal fiber Part II: Cladding effect on confinement and bend loss. *Opt. Express* **2013**, *21*, 28609–28616. [CrossRef] [PubMed]
7. Finger, M.A.; Joly, N.Y.; Weiss, T.; Russell, P.S.J. Accuracy oft he capillary approximation for gas-filled kagomé-style photonic crystal fiber. *Opt. Lett.* **2014**, *39*, 821–824. [CrossRef] [PubMed]
8. Carter, R.M.; Yu, F.; Wadsworth, W.J.; Shephard, J.D.; Birks, T.; Knight, J.C.; Hand, D.P. Measurement of resonant bend loss in anti-resonant hollow core optical fiber. *Opt. Express* **2017**, *25*, 20612–20621. [CrossRef] [PubMed]
9. Michieletto, M.; Lyngsø, J.K.; Jakobsen, C.; Lægsgaard, J.; Bang, O.; Alkeskjold, T.T. Hollow-core fibers for high power pulse delivery. *Opt. Express* **2016**, *24*, 7103–7119. [CrossRef] [PubMed]

10. Ermolov, A.; Mak, K.F.; Frosz, M.H.; Travers, J.C.; Russell, P.S.J. Supercontinuum generation in the vacuum ultraviolet through dispersive-wave and soliton-plasma interatction in a noble-gas-filled hollow-core photonic crystal fiber. *Phys. Rev. A* **2015**, *92*, 033821. [CrossRef]

11. Benabid, F.; Knight, J.C.; Antonopoulos, G.; Russell, P.S.J. Stimulated raman scattering in hydrogen-filled hollow-core photonic crystal fiber. *Science* **2002**, *298*, 399–402. [CrossRef] [PubMed]

12. Smith, A.V.; Do, B.T.; Hadley, G.R.; Farrow, R.L. Optical damage limits to pulse energy from fibers. *IEEE J. Sel. Top. Quantum Electron.* **2009**, *15*, 153–158. [CrossRef]

13. Bhagwat, A.R.; Gaeta, A.L. Nonlinear optics in hollow-core photonic bandgap fibers. *Opt. Express* **2008**, *16*, 5035–5047. [CrossRef] [PubMed]

14. Chang, W.; Nazarkin, A.; Travers, J.C.; Nold, J.; Hölzer, P.; Joly, N.Y.; Russell, P.S.J. Influence of ionizsation on ultrafast gas-based nonlinear fiber optics. *Opt. Express* **2011**, *19*, 21018–21027. [CrossRef] [PubMed]

15. Mousavi, S.A.; Mulvad, H.C.H.; Wheeler, N.V.; Horak, P.; Hayes, J.; Chen, Y.; Bradley, T.D.; Alam, S.; Sandoghchi, S.R.; Richardson, D.J.; et al. Nonlinear dynamic of picosecond pulse propagation in atmospheric air-filled hollow core fibers. *Opt. Express* **2018**, *26*, 8866–8882. [CrossRef] [PubMed]

16. Funck, M.C.; Eilzer, S.; Wedel, B. Ultrafast beam delivery: System technology and industrial application. In Proceedings of the High-Power Laser Materials Processing: Applications, Diagnostics, and Systems VI, San Francisco, CA, USA, 28 January–2 February 2017; p. 100970L.

MDPI

St. Alban-Anlage 66

4052 Basel

Switzerland

Tel. +41 61 683 77 34

Fax +41 61 302 89 18

www.mdpi.com

Fibers Editorial Office

E-mail: fibers@mdpi.com

www.mdpi.com/journal/fibers

www.ingramcontent.com/pod-product-compliance
Lightning Source LLC
Chambersburg PA
CBHW051857210326

41597CB00033B/5926